Green Hydrogen in Energy Systems

Green Hydrogen in Energy Systems: A Computational Fluid Dynamics Approach provides a comprehensive exploration of how CFD modeling techniques are revolutionizing the development, optimization, and safety analysis of green hydrogen technologies. It addresses the critical role hydrogen plays in the global energy transition while offering practical analytical frameworks.

The book systematically examines the entire hydrogen value chain—from innovative production methods like membraneless electrolyzers to storage mechanisms, transportation systems, and end use applications. Through detailed CFD simulations, it demonstrates how computational modeling can solve complex fluid dynamics challenges unique to hydrogen systems, improving efficiency and addressing safety concerns essential for widespread adoption.

This book serves as a reference for graduate students, researchers, and energy practitioners focused on hydrogen technology, fluid mechanics, renewable energy systems, and computational modeling.

Uchenna Godwin Azubuike is a GES researcher in the Department of Mechanical Engineering Sciences, University of Johannesburg, South Africa. He is a registered Engineer with the Council for the Regulation of Engineering in Nigeria (COREN). Dr. Azubuike earned his PhD from the University of Nigeria Nsukka in the Department of Mechanical Engineering, specializing in Energy and Power Technology, and he has published numerous peer-reviewed papers on energy systems analysis in well-regarded research journals. Currently, his focus is on Computational Fluid Dynamics in relation to hydrogen generation and storage in the Atomic Layer Deposition Research Group of the Department of Engineering Science, University of Johannesburg.

Kingsley Ukoba is a lecturer and researcher in the Department of Mechanical Engineering Sciences, University of Johannesburg, South Africa. He obtained a doctoral degree in Mechanical Engineering from the University of KwaZulu-Natal, Durban, South Africa. Dr. Ukoba has won several individual and group grants, awards, and fellowships, including the one million South Africa Rands grant by

SASOL/NRF Collaborative Research grant for water treatment using Atomic Layer deposition in 2021. He is also a 2022 Engineering for Change (E4C) fellow sponsored by the American Society of Mechanical Engineers (ASME) to support an Impact Project around Climate Action. He has authored four books, numerous book chapters, and over 70 journal articles.

Tien-Chien Jen is the South Africa Research Chair (SARCHI) of Green Hydrogen in the whole of South Africa. He received his Ph.D. in Mechanical and Aerospace Engineering from the University of California, Los Angeles (UCLA), USA, specializing in thermal aspects of grinding. He has received several competitive grants for his research, including those from the US National Science Foundation, the US Department of Energy, and the EPA. Professor Jen is the Director of the newly established Atomic Layer Deposition Research Centre of the University of Johannesburg, South Africa, and he is currently leading the drive for the application of Atomic Layer deposition equipment (first of its kind in Africa) for various applications (hydrogen, renewable energy, thin films, etc.). Professor Jen has authored over 360 peer-reviewed articles, 16 book chapters, and 7 books.

Green Hydrogen in Energy Systems

A Computational Fluid Dynamics Approach

Uchenna Godwin Azubuike, Kingsley Ukoba, and Tien-Chien Jen

CRC Press
Taylor & Francis Group
Boca Raton London New York

CRC Press is an imprint of the
Taylor & Francis Group, an **informa** business

Designed cover image: Shutterstock

First edition published 2027
by CRC Press
2385 NW Executive Center Drive, Suite 320, Boca Raton FL 33431

and by CRC Press
4 Park Square, Milton Park, Abingdon, Oxon, OX14 4RN

CRC Press is an imprint of Taylor & Francis Group, LLC

ISBN: 978-1-041-21440-3 (hbk)
ISBN: 978-1-041-21505-9 (pbk)
ISBN: 978-1-003-72719-4 (ebk)

DOI: 10.1201/9781003727194

Typeset in Times
by KnowledgeWorks Global Ltd.

Contents

List of Figures

List of Tables

1 Overview of Global Green Hydrogen Systems

1.1 PREAMBLE

Global climate change, primarily caused by the accumulation of greenhouse gases (GHGs) in the atmosphere, is leading to an unprecedented environmental crisis for the planet. The primary source of carbon dioxide (CO_2) emissions, which account for more than 70% of global GHG emissions, is the burning of fossil fuels for transportation, heating, power generation, and industrial operations. Human health, food security, and economic stability are all seriously threatened by the increased frequency and intensity of extreme weather events, sea level rise, and ecological changes resulting from rising temperatures. The Paris Agreement calls for limiting global warming to less than 1.5°C over pre-industrial levels, and governments and international organizations have responded by underscoring the urgent need for comprehensive decarbonization initiatives. It is increasingly recognized that transitioning to low-carbon energy systems is crucial for achieving long-term economic resilience, national security, and sustainable development [1–3].

Hydrogen has become a viable option for deep decarbonization in this environment. A clean, adaptable, and energy-dense carrier, hydrogen finds application in heavy industries, transportation, and power generation. Hydrogen is a desirable zero-emission energy source because, unlike fossil fuels, it produces only water vapor as a byproduct when burned or used in fuel cells. Renewable-powered electrolysis provides a route to "green hydrogen," or hydrogen produced without carbon emissions. Hydrogen can also be created from a variety of feedstocks and methods. Due to its adaptability, hydrogen can serve as a link between energy sources and end uses, facilitating sector coupling, energy storage, grid balancing, and the integration of fluctuating renewable energy sources. Global investment, research, and policy development are being spurred by the growing recognition of hydrogen's role in the clean energy transition, as countries seek sustainable and scalable alternatives to fossil fuels. Hydrogen can be stored, transported, and converted into multiple energy forms, making it exceptionally adaptable, as shown in Figure 1.1 [4, 5].

1.2 REVIEW OF GLOBAL ENERGY ISSUES

The interaction of increasing demand, environmental restrictions, and geopolitical dynamics has led to a more complex global energy landscape in the modern era. The growing worldwide energy consumption, which is mainly caused by industrialization, population growth, and rapid urbanization, particularly in emerging economies, is one of the most urgent issues. If current trends continue, the International Energy Agency (IEA) projects that global energy demand will increase by more

DOI: 10.1201/9781003727194-1

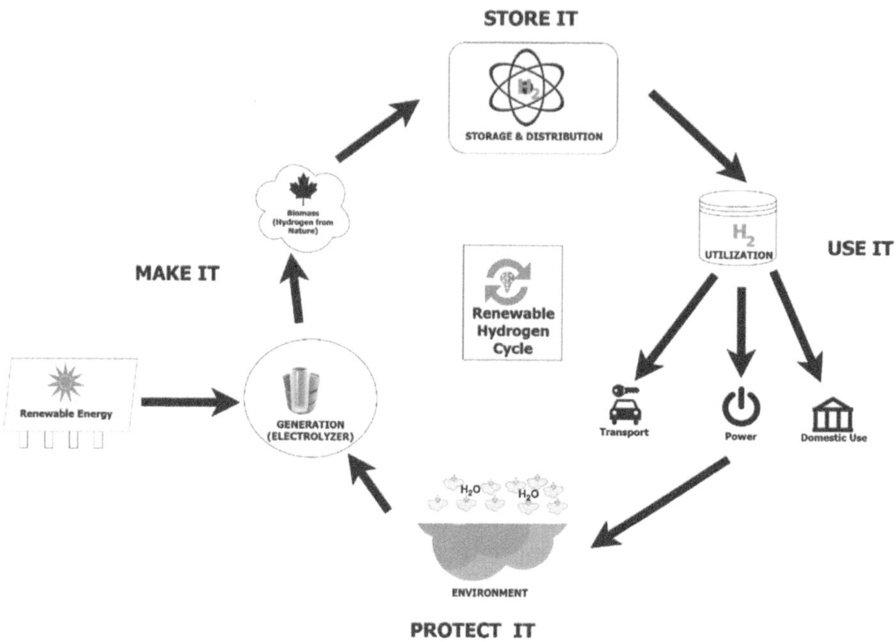

FIGURE 1.1 Schematic of green hydrogen process cycle.

than 25% by 2040. The current energy infrastructure and resource availability are under tremendous strain due to this rise, which calls for a significant shift in the way energy is generated, distributed, and utilized [6–8].

The globe is still largely dependent on fossil fuels, especially coal, oil, and natural gas, which together constitute more than 80% of primary energy consumption, as illustrated in Figure 1.2, even with advancements in renewable energy technologies [9, 10].

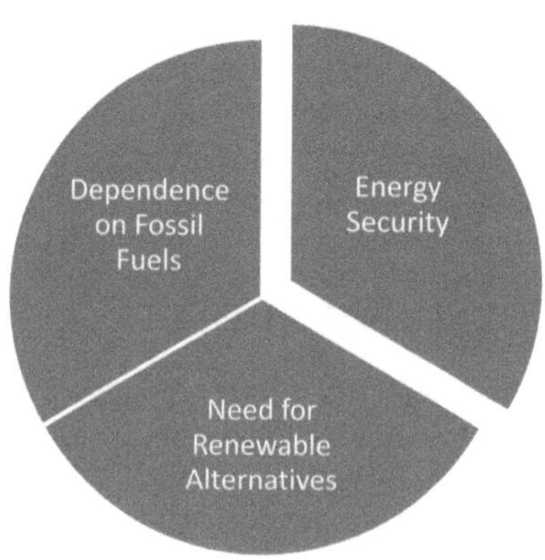

FIGURE 1.2 Review of global energy issues.

Fossil fuels are limited, and their adverse environmental effects—such as air and water pollution, degraded land, and ecosystem disturbance—make this reliance problematic. Fossil fuel mining, processing, and burning are significant contributors to environmental degradation that often disproportionately affect areas and communities already vulnerable to environmental risks. The fundamental chemical equation representing hydrocarbon combustion is shown in Eq. (1.1) [11, 12]:

$$C_xH_\gamma + O_2 \rightarrow CO_2 + H_2O + \text{Heat} \tag{1.1}$$

For instance, the combustion of methane (CH_4), a primary component of natural gas, is shown in Eq. (1.2):

$$CH_4 + 2O_2 \rightarrow CO_2 + 2H_2O \tag{1.2}$$

Using fossil fuels directly contributes to climate change by releasing GHGs, including CO_2, CH_4, and nitrous oxide (N_2O). The energy sector is the most significant contributor to worldwide GHG emissions, accounting for over 73% of the total. Sea level rise, extreme weather, global warming, and extensive biological disruptions have all resulted from the buildup of these gases in the atmosphere. The Earth runs the risk of exceeding critical climate thresholds if mitigating measures are not implemented immediately [13, 14].

Geopolitical tensions and energy security further complicate the energy landscape. Due to their reliance on gas and oil imports from politically unstable areas, many countries are vulnerable to price volatility and supply interruptions. Energy independence is a primary strategic goal for many nations, as control over vital energy resources has historically influenced international relations and conflicts. Diversified and sustainable energy solutions are desperately needed in light of these interrelated issues. Along with advancements in energy efficiency and conservation, the shift to a low-carbon economy must be supported by more investment in renewable energy sources, including solar, wind, hydro, and green hydrogen. Energy portfolio diversification also lessens sensitivity to geopolitical shocks and increases resilience to market swings. To maintain global stability and intergenerational equality, sustainable energy development is not only a technological or financial necessity but also a moral and environmental imperative. Technology development, innovative policy, and concerted international action are needed to address global energy concerns. The world can fulfil growing energy demands while preserving environmental integrity and international security by moving away from reliance on fossil fuels and toward a diversified, low-carbon energy future [15, 16].

1.3 HYDROGEN IN CLEAN ENERGY EVOLUTION

Although hydrogen has long been utilized in scientific research and commercial applications, its role is evolving rapidly in response to international efforts to transition to a low-carbon, sustainable energy system. Hydrogen has been extensively employed in industries such as methanol synthesis, ammonia production, and petroleum refining. These applications primarily utilize steam methane reforming (SMR), which produces hydrogen from natural gas and is a significant contributor to GHG emissions. Hydrogen has played a key role in scientific investigations into materials science, fuel cell development, and combustion chemistry. The features of hydrogen have long indicated a greater potential, which is now being fulfilled in the pursuit of clean energy solutions, despite its historical reliance on production based on fossil fuels. In the Haber-Bosch process, hydrogen reacts with nitrogen under high temperature and pressure to form ammonia as represented in Eq. (1.3) [17, 18]:

$$N_2 + 3H_2 \rightarrow 2NH_3 \tag{1.3}$$

This process, developed in the early 20th century, revolutionized agriculture and remains one of the largest-scale uses of hydrogen today. Hydrogen has also been used in space exploration as a rocket fuel, thanks to its high energy content per unit mass [19, 20].

The energy strategy has undergone a paradigm shift with the advent of hydrogen as a clean energy source. As concerns about air pollution, climate change, and energy security grow, hydrogen is being reconsidered as a key element of the global energy transition, rather than just an industrial fuel. Carbon dioxide is not released during the production or usage of green hydrogen, which is created by electrolyzing water with power from renewable sources. As a result, hydrogen is now viewed as a crucial facilitator of deep decarbonization, particularly in industries such as heavy industry, shipping, and long-distance transportation, where direct electrification is either technically or financially impractical [21, 22].

Hydrogen possesses both special and complementary qualities when compared to other renewable energy technologies, particularly solar, wind, and biomass. Though their intermittent nature presents difficulties for grid dependability and load balancing. Solar and wind energy sources are becoming more and more cost-competitive and essential for decarbonizing electricity generation. Although biomass provides prospects for carbon sequestration and dispatchable electricity, its usage is constrained by feedstock availability and competition from other land uses. On the other hand, excess renewable energy can be efficiently transformed into a transportable and storable energy carrier by synthesizing hydrogen using excess electricity from unreliable sources. The limitations of time and place can be overcome with renewable energy through this "power-to-gas" strategy [23, 24].

Hydrogen's exceptional capacity for sector coupling and long-term energy storage is one of its most alluring features. Hydrogen offers a solution to long-term energy supply variations, as it can store energy over days, weeks, or even seasons, unlike batteries, which are best suited for short-term storage. Due to this, hydrogen is significant for achieving energy system flexibility in regions with fluctuating profiles of renewable power. Hydrogen can also help with sector coupling, which is the integration of industries such as energy, mobility, heating, and power [25, 26].

The idea of integrated energy systems, or "hydrogen economies," in which renewable electricity is utilized not only as a source of electricity but also as a raw material for chemicals and synthetic fuels, is further supported by hydrogen's adaptability. Due to its cross-sectoral utility, hydrogen is a key component of strategies aimed at achieving net-zero emissions. To promote the adoption of green hydrogen technologies, several nations, including South Korea, Japan, and Germany, have already implemented national hydrogen strategies and made substantial investments in infrastructure, research, and demonstration projects [27, 28].

The transformation of hydrogen from a specialized industrial gas to a key component of clean energy is a reflection of both technological advancement and the pressing need for worldwide decarbonization. It is uniquely positioned to overcome many of the drawbacks of existing renewable technologies due to its compatibility with renewable energy systems, potential for large-scale energy storage, and ability to integrate diverse industries. Hydrogen will be crucial in transforming how we produce, distribute, and utilize energy as the world transitions to more robust, sustainable, and inclusive energy systems [29, 30].

1.4 HYDROGEN CATEGORIZATION

As a flexible energy carrier, hydrogen can be produced in various ways, each with distinct energy and environmental characteristics. A color-based categorization system is frequently employed to distinguish between the environmental impact and hydrogen generation methods. As shown in Figure 1.3, these categories, gray, blue, turquoise, and green, reflect not only the technology and source of hydrogen production but also their energy intensities and lifecycle emissions [31, 32].

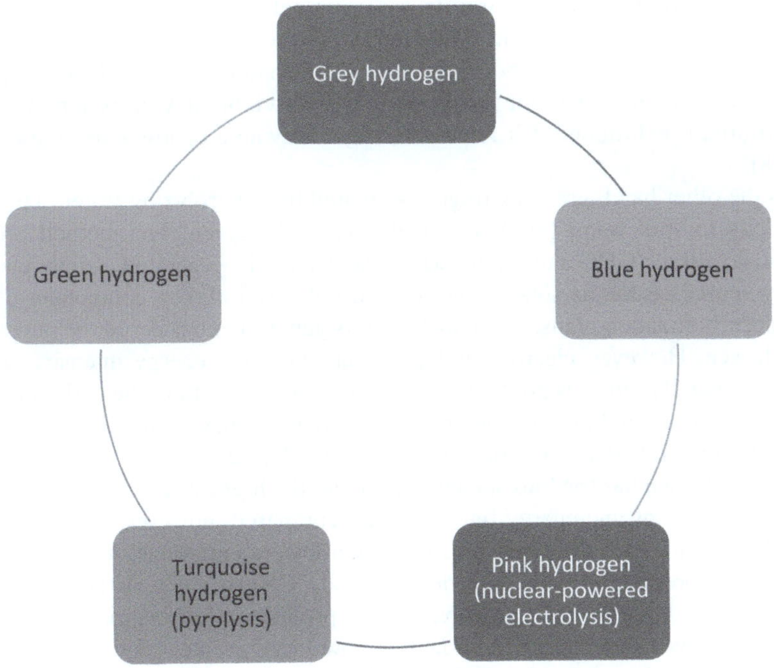

FIGURE 1.3 Hydrogen categorization.

The most common type of hydrogen generated today is gray hydrogen, which is primarily produced through SMR from fossil fuels, particularly natural gas. Methane and high-temperature steam react in this method to produce CO_2 and hydrogen. Gray hydrogen's primary disadvantage is that the CO_2 it produces is released directly into the environment, significantly increasing GHG emissions. About 9 to 12 tons of CO_2 are released for every ton of hydrogen produced by SMR. Despite its current cost-effectiveness, gray hydrogen is incompatible with long-term decarbonization ambitions because of its significant carbon footprint [33, 34].

Many people consider blue hydrogen to be a solution that lies between gray and green hydrogen. Auto-thermal reforming (ATR) or SMR of natural gas is also used to make it; however, carbon capture and storage (CCS) technologies are used to absorb and store the CO_2 emissions that are generated. Generally speaking, CCS may absorb 60–90% of the emissions, depending on the specific installation and technology used. The effectiveness and durability of the carbon storage determine whether blue hydrogen is more environmentally friendly than gray hydrogen, even though it offers lower lifecycle emissions. Furthermore, the climate benefits of CCS can be considerably outweighed by fugitive emissions, which are methane leaks that occur during the extraction and transportation of natural gas [35, 36].

Turquoise hydrogen is a relatively recent and less established manufacturing method that utilizes methane pyrolysis. This thermal breakdown process splits methane into hydrogen gas and solid carbon at high temperatures in the absence of oxygen. Instead of directly emitting CO_2, this process produces elemental carbon, which can be utilized in various industrial processes (e.g., carbon black for rubber and batteries). Regarding its environmental impact, turquoise hydrogen falls somewhere in the middle. The energy source utilized for pyrolysis and the management of the solid carbon byproducts' lifespan have a significant impact on its carbon footprint. Methane pyrolysis may be a viable low-emission alternative to traditional techniques if it is powered by renewable or low-carbon electricity [37, 38].

On the other hand, green hydrogen is created by electrolyzing water, which uses electricity to split water (H_2O) into hydrogen and oxygen. The method produces almost no direct GHG emissions when electricity is generated using renewable energy sources (such as solar, wind, or hydroelectric). A key component of long-term decarbonization plans, green hydrogen is generally considered the purest form of hydrogen. However, electrolysis has a relatively high energy intensity; 1 kg of hydrogen usually requires 50–55 kWh of electricity. Therefore, the carbon intensity of the energy utilized and the electrolyzers' efficiency have a significant impact on the environmental advantages of green hydrogen [39, 40].

Gray hydrogen has the lowest energy cost and the highest carbon emissions when lifecycle emissions and energy intensities are compared; however, it is not environmentally sustainable. Although blue hydrogen lowers emissions, it still uses fossil fuels and is constrained by methane leakage and CCS efficiency. Although turquoise hydrogen is still in its infancy as a commercial product, it delivers reduced emissions when fueled by clean energy. Although it now faces greater production costs and energy requirements, green hydrogen stands out for its potential to have zero emissions [41, 42].

The color-coded classification of hydrogen offers essential information about the environmental costs and benefits of various production methods. The emphasis must shift from gray to green and other low-carbon hydrogen technologies as international efforts to achieve net-zero emissions intensify. To scale up pure hydrogen and guarantee their integration into future energy systems, strategic investment, governmental assistance, and technological innovation are necessary [43, 44].

1.5 GREEN HYDROGEN BENEFITS

A key component of the global shift to clean and sustainable energy systems is green hydrogen, which is created by electrolyzing water with electricity from renewable sources like solar, wind, or hydropower. Green hydrogen is a crucial technology for achieving net-zero climate targets, as it produces and utilizes zero carbon emissions, unlike conventional hydrogen derived from fossil fuels. Since green hydrogen produces only oxygen as a byproduct when generated using renewable electricity, it can be utilized in various industries without contributing to local air pollution or GHG emissions (Figure 1.4) [45, 46].

Green hydrogen's versatility in terms of storage and transportation is one of its most significant benefits. Hydrogen can be transported via pipelines, trucks, or ships, or transformed into synthetic fuels such as methanol and ammonia. It can also be stored as a gas, liquid, or bound in chemical carriers. Due to its adaptability, excess electricity generated during peak hours can be stored as hydrogen and utilized when demand increases or generation decreases, providing a vital solution to the intermittent nature of renewable energy sources. Hydrogen has long-duration and seasonal storage capacities, which enhance the stability and

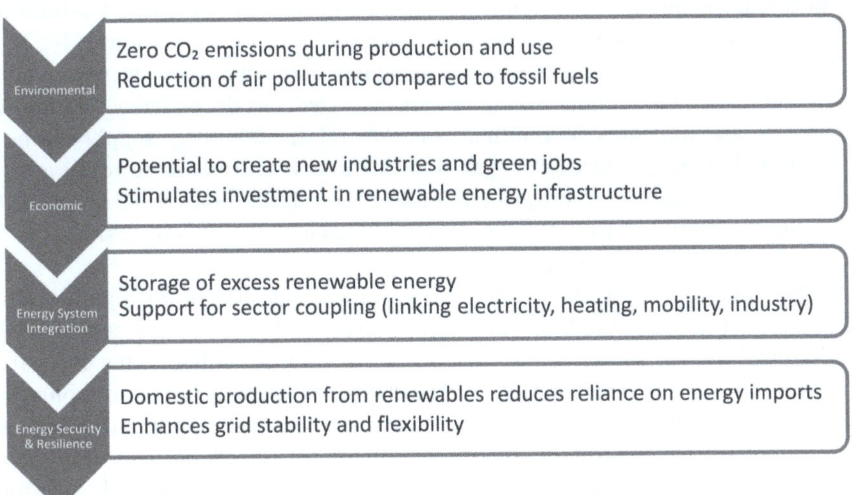

FIGURE 1.4 Green hydrogen benefits.

dependability of the energy system, in contrast to batteries, which are best suited for short-term storage [47].

With some adjustments, green hydrogen also shows a high degree of compatibility with current infrastructure. For example, hydrogen or hydrogen-natural gas mixes can be transported using pipelines that already carry natural gas. Similarly, existing industrial boilers and gas turbines can be converted to operate on hydrogen. Recent developments in materials and safety technologies have enhanced the viability of infrastructure adaptation, despite the challenges posed by hydrogen's small molecular size and flammability, which present storage and transportation issues. This compatibility speeds up implementation and lowers costs by reducing the requirement for completely new infrastructure [48].

Green hydrogen stands out among end-use applications because it can decarbonize a variety of industries, including industrial, transportation, heating, and power generation. In the power industry, hydrogen can be utilized in fuel cells for distributed generation or in gas turbines for the production of electricity. For heavy-duty vehicles, such as buses and trains, and for shipping, where battery electric solutions may have limitations in terms of range, recharging time, and cargo capacity, hydrogen fuel cells are particularly beneficial. Green hydrogen provides a high-temperature, low-emission alternative to coal and natural gas in industrial processes such as cement manufacturing, ammonia synthesis, and steelmaking. Additionally, hydrogen can serve as a zero-emission fuel for district heating systems or residential heating, especially in colder climates where electrification may not be economically feasible [49, 50].

The ability of green hydrogen to decarbonize hard-to-abate sectors, which contribute significantly to global emissions and are challenging to directly electrify, is one of its main advantages. These industries include aviation, long-distance freight transportation, high-temperature industrial processes, and chemical manufacturing. In situations where electrification is impractical due to energy density, technological, or financial limitations, green hydrogen and its derivatives (such as e-fuels and ammonia) provide scalable and sustainable ways to minimize emissions [51–53].

Additionally, green hydrogen makes a substantial contribution to energy diversification and resilience. It enhances energy security, reduces dependency on imported fossil fuels, and mitigates geopolitical concerns related to energy supply chains by enabling nations to produce hydrogen domestically from renewable resources. By expanding the variety of energy vectors and encouraging innovation in renewable technologies, electrolyzer production, and hydrogen logistics, this diversification fortifies national energy systems [54].

In the global quest for sustainable energy, green hydrogen offers a game-changing prospect. It is a key facilitator of deep decarbonization due to its zero-emission profile, flexibility in storage and transport, compatibility with existing infrastructure, and multi-sector applicability. Green hydrogen helps achieve environmental objectives while also enhancing energy security, economic competitiveness, and long-term resilience by targeting the most challenging aspects of the energy system [34, 54]. Green hydrogen has the potential to be a key component of a cleaner, more integrated, and more resilient global energy future, provided that sustained technological advancements and governmental support are in place [55].

1.6 ENVIRONMENTAL AND ECONOMIC ASPECTS

Hydrogen energy is increasingly seen as a pivotal component of a clean energy future due to its potential to decarbonize sectors where direct electrification is challenging. Evaluating both environmental and economic aspects is essential to understanding its viability and scalability. This chapter explores key environmental benefits, cost trends, and the role of policy support in the development of a hydrogen-based energy system [23, 56].

A primary environmental advantage of hydrogen, particularly green hydrogen produced via electrolysis using renewable electricity, is zero emissions at the point of use. When hydrogen is used in fuel cells, the only byproduct is water vapor, making it ideal for reducing air pollution and GHG emissions in transport and industry [34, 54].

However, a comprehensive lifecycle emissions analysis reveals that hydrogen's environmental impact depends heavily on its production method. Gray hydrogen, derived from natural gas via SMR, has high carbon emissions unless combined with CCS to produce blue hydrogen. Green hydrogen, in contrast, can achieve near-zero emissions across its lifecycle, provided the electricity used in electrolysis comes from low-carbon sources. Therefore, the environmental benefit hinges not only on the fuel's usage but also on its entire supply chain [5, 20, 57].

The current cost of hydrogen production varies significantly depending on the method. As of 2024, green hydrogen typically costs between $4 and $6 per kilogram, while gray hydrogen remains cheaper at around $1.50 to $2.50/kg. Blue hydrogen falls in the intermediate range. However, costs are projected to decline sharply by 2030 due to technological advancements, economies of scale, and increased deployment. The International Renewable Energy Agency (IRENA) projects green hydrogen could become cost-competitive with gray hydrogen by 2030 in regions with abundant renewable resources [24, 29, 56].

The levelized cost of hydrogen (LCOH), which accounts for capital, operational, and fuel costs over the asset's lifetime, is a crucial economic metric. The current LCOH for green hydrogen ranges from $3 to $8/kg, but this is expected to drop below $2/kg by 2030 under optimal conditions. Improvements in electrolyzer efficiency, reduced electricity prices, and improved infrastructure will contribute to this trend, enhancing hydrogen's economic competitiveness across sectors such as transportation, industry, and power generation [31, 34].

Government support and regulation play a central role in scaling up hydrogen technologies. Policies such as carbon pricing, clean hydrogen tax credits, subsidies for renewable energy, and mandates for hydrogen use in transport or industry can significantly alter the economic equation. Globally, international strategies are accelerating the adoption of hydrogen. The European Union's Hydrogen Strategy aims to install at least 40 GW of electrolyzer capacity by 2030, supported by a €470 billion investment roadmap. The United States has launched the Hydrogen Energy Earthshot initiative, aiming to achieve a green hydrogen cost of $1/kg by 2031. In Asia, Japan and South Korea are spearheading hydrogen development with national roadmaps that include transportation, industrial use, and power generation. China is also rapidly expanding its hydrogen economy, with a focus on fuel cell vehicles and renewable hydrogen production outlined in its 14th Five-Year Plan [12, 15, 24, 40].

Hydrogen energy presents substantial environmental and economic opportunities, especially in the context of decarbonization. Its zero emissions at the point of use and potential for lifecycle sustainability make it an attractive alternative to fossil fuels. Although current costs are high, projected declines in LCOH, combined with strong policy support and international collaboration, suggest that hydrogen could become a cornerstone of a low-carbon global economy. Continued investment, innovation, and regulatory clarity will be essential to unlocking its full potential [39, 42, 54].

1.7 CONCLUSION

As a key component of the global energy transition, green hydrogen has the potential to be a revolutionary force. Green hydrogen is produced through the electrolysis of water using renewable energy sources, such as solar and wind, in contrast to gray or blue hydrogen, which relies on fossil fuels and often involves carbon emissions. This technique is a key facilitator of deep decarbonization since it produces hydrogen with no direct GHG emissions. In the future energy landscape, green hydrogen can serve as a clean fuel for transportation, a means to store renewable electricity, a tool to balance energy grids, and a replacement for fossil fuels in industrial processes. Its adaptability enables sectoral integration, tackling emissions in industries like steel manufacturing, aviation, and maritime shipping that are challenging to decarbonize using electricity alone [2, 34, 53].

However, achieving the full potential of green hydrogen requires more than just scientific promise; it also necessitates international cooperation, consistent innovation, and a well-coordinated and robust policy framework. To lower production costs, boost demand, and establish favorable market conditions, strategic government measures, including subsidies, carbon pricing, infrastructure investments, and research incentives, are crucial. Enhancing electrolyzer efficiency, reducing capital costs, and creating dependable storage and transportation systems must be the main goals of innovation. International collaboration is equally vital to creating cross-border hydrogen value chains, harmonizing legislation, and setting global standards. The world can accelerate its transition away from fossil fuels and toward a more resilient, sustainable, and carbon-neutral energy future by fostering an environment conducive to green hydrogen. Scaling green hydrogen is one of the critical actions required to reach global climate targets, and the window for effective climate action is closing [12, 55, 56].

REFERENCES

1. M. D. Allendorf, V. Stavila, J. L. Snider, M. Witman, M. E. Bowden, K. Brooks, B. L. Tran, and T. Autrey, 'Challenges to developing materials for the transport and storage of hydrogen', *Nature Chemistry*, vol. 14, no. 11, pp. 1214–1223, 2022.
2. O. C. Anika, S. G. Nnabuife, A. Bello, E. R. Okoroafor, B. Kuang, and R. Villa, 'Prospects of low and zero-carbon renewable fuels in 1.5-degree net zero emission actualisation by 2050: A critical review', *Carbon Capture Science & Technology*, vol. 5, p. 100072, 2022.

3. S. A. Asongu, M. O. Agboola, A. A. Alola, and F. V. Bekun, 'The criticality of growth, urbanization, electricity and fossil fuel consumption to environment sustainability in Africa', *Science of the Total Environment*, vol. 712, p. 136376, 2020.
4. O. O. Babayomi, D. A. Dahoro, and Z. Zhang, 'Affordable clean energy transition in developing countries: Pathways and technologies', *Iscience*, vol. 25, no. 5, 2022. https://www.cell.com/iscience/fulltext/S2589-0042(22)00448-5.
5. C. Bauer, K. Treyer, C. Antonini, J. Bergerson, M. Gazzani, E. Gencer, J. Gibbins, M. Mazzotti, S. T. McCoy, R. McKenna, and R. Pietzcker, 'On the climate impacts of blue hydrogen production', *Sustainable Energy & Fuels*, vol. 6, no. 1, pp. 66–75, 2022.
6. A. Bhaskar, R. Abhishek, M. Assadi, and H. N. Somehesaraei, 'Decarbonizing primary steel production: Techno-economic assessment of a hydrogen based green steel production plant in Norway', *Journal of Cleaner Production*, vol. 350, p. 131339, 2022.
7. R. Bhattacharyya, K. K. Singh, R. B. Grover, and K. Bhanja, 'Nuclear hydrogen production for industrial decarbonization: Creating the business case for the near term', *International Journal of Energy Research*, vol. 46, no. 5, pp. 6929–6943, 2022.
8. D. Bogdanov, M. Ram, A. Aghahosseini, A. Gulagi, A. S. Oyewo, M. Child, U. Caldera, K. Sadovskaia, J. Farfan, L. D. S. N. S. Barbosa, and M. Fasihi, 'Low-cost renewable electricity as the key driver of the global energy transition towards sustainability', *Energy*, vol. 227, p. 120467, 2021.
9. S. Bolwig, G. Bazbauers, A. Klitkou, P. D. Lund, A. Blumberga, A. Gravelsins, and D. Blumberga, 'Review of modelling energy transitions pathways with application to energy system flexibility', *Renewable and Sustainable Energy Reviews*, vol. 101, pp. 440–452, 2019.
10. L. Bouchene, Z. Cassim, H. Engel, K. Jayaram, and A. Kendall, 'Green Africa: A growth and resilience agenda for the continent', *McKinsey Sustainability*, vol. 28, p. 28, 2021.
11. G. Buffo, P. Marocco, D. Ferrero, A. Lanzini, and M. Santarelli, 'Power-to-X and power-to-power routes', in *Solar Hydrogen Production*, Academic Press, 2019, pp. 529–557.
12. W. Cheng, and S. Lee, 'How green are the national hydrogen strategies?', *Sustainability*, vol. 14, no. 3, p. 1930, 2022.
13. S. Cloete, O. Ruhnau, J. H. Cloete, and L. Hirth, 'Blue hydrogen and industrial base products: The future of fossil fuel exporters in a net-zero world', *Journal of Cleaner Production*, vol. 363, p. 132347, 2022.
14. J. Davies, F. Dolci, D. Klassek-Bajorek, R. Ortiz Cebolla, and E. Weidner, Current Status of Chemical Energy Storage Technologies, Publications Office of the European Union: Luxembourg, 2020. https://publications.jrc.ec.europa.eu/repository/handle/JRC118776.
15. R. R. Esily, Y. Chi, D. M. Ibrahiem, and Y. Chen, 'Hydrogen strategy in decarbonization era: Egypt as a case study', *International Journal of Hydrogen Energy*, vol. 47, no. 43, pp. 18629–18647, 2022.
16. O. Faye, J. Szpunar, and U. Eduok, 'A critical review on the current technologies for the generation, storage, and transportation of hydrogen', *International journal of hydrogen energy*, vol. 47, no. 29, pp. 13771–13802, 2022.
17. I. N. Gahlawat, and P. Lakra, 'Global Climate change and its effects', *Integrated Journal of Social Sciences*, vol. 7, no. 1, pp. 14–23, 2020.
18. T. B. Hadj, 'Nonlinear impact of biomass energy consumption on ecological footprint in a fossil fuel–dependent economy', *Environmental Science and Pollution Research*, vol. 28, no. 48, pp. 69329–69342, 2021.
19. E. Holden, K. Linnerud, and B. J. Rygg, 'A review of dominant sustainable energy narratives', *Renewable and Sustainable Energy Reviews*, vol. 144, p. 110955, 2021.

20. R. W. Howarth, and M. Z. Jacobson, 'How green is blue hydrogen?', *Energy Science & Engineering*, vol. 9, no. 10, pp. 1676–1687, 2021.
21. O. K. Jogdand, 'Study on the effect of global warming and greenhouse gases on environmental system', in *Green Chemistry and Sustainable Technology*, Apple Academic Press, 2020, pp. 275–306.
22. M. J. B. Kabeyi, and O. A. Olanrewaju, 'Sustainable energy transition for renewable and low carbon grid electricity generation and supply', *Frontiers in Energy Research*, vol. 9, p. 743114, 2022.
23. A. Kovač, M. Paranos, and D. Marciuš, 'Hydrogen in energy transition: A review', *International Journal of Hydrogen Energy*, vol. 46, no. 16, pp. 10016–10035, 2021.
24. A. Krupnick, and A. Bergman, 2022. Incentives for clean hydrogen production in the inflation reduction act. *Resources for the Future*. https://media.rff.org/documents/Report_22-13_UNfLJLS.pdf
25. S. Krupnik, A. Wagner, O. Vincent, T. J. Rudek, R. Wade, M. Mišík, S. Akerboom, C. Foulds, K. S. Stegen, Ç Adem, and S. Batel, 'Beyond technology: A research agenda for social sciences and humanities research on renewable energy in Europe', *Energy Research & Social Science*, vol. 89, p. 102536, 2022.
26. L. Lescrauwaet, H. Wagner, C. Yoon, and S. Shukla, 'Adaptive legal frameworks and economic dynamics in emerging tech-nologies: Navigating the intersection for responsible innovation', *Law and Economics*, vol. 16, no. 3, pp. 202–220, 2022. https://cir.nii.ac.jp/crid/1360023717936124544
27. H. Li, X. Cao, Y. Liu, Y. Shao, Z. Nan, L. Teng, W. Peng, and J. Bian, 'Safety of hydrogen storage and transportation: An overview on mechanisms, techniques, and challenges', *Energy Reports*, vol. 8, pp. 6258–6269, 2022.
28. R. T. Madsen, L. E. Klebanoff, S. A. M. Caughlan, J. W. Pratt, T. S. Leach, T. B. Appelgate Jr, S. Z. Kelety, H. C. Wintervoll, G. P. Haugom, A. T. Y. Teo, and S. Ghosh, 'Feasibility of the Zero-V: A zero-emissions hydrogen fuel-cell coastal research vessel', *International Journal of Hydrogen Energy*, vol. 45, no. 46, pp. 25328–25343, 2020.
29. G. Maggio, G. Squadrito, and A. Nicita, 'Hydrogen and medical oxygen by renewable energy based electrolysis: A green and economically viable route', *Applied Energy*, vol. 306, p. 117993, 2022.
30. C. Mapelli, G. Dall'Osto, D. Mombelli, S. Barella, and A. Gruttadauria, 'Future scenarios for reducing emissions and consumption in the Italian steelmaking industry', *Steel Research International*, vol. 93, no. 5, p. 2100631, 2022.
31. J. W. McNamara, V. DeAngelis, R. H. Byrne, A. Benson, B. R. Chalamala, and R. Masiello, 'Long-duration energy storage in a decarbonized future: Policy gaps, needs, and opportunities', *MRS Energy & Sustainability*, vol. 9, no. 2, pp. 142–170, 2022.
32. A. Monk, D. Rook, and R. Sharma, 2021. Transparency and Innovation in Institutional Investing: How to Unlock Beneficial Investment Organizations' Capital for Society's Critical Projects. https://papers.ssrn.com/sol3/papers.cfm?abstract_id=3796535
33. H. Nazir, N. Muthuswamy, C. Louis, S. Jose, J. Prakash, M. E. Buan, C. Flox, S. Chavan, X. Shi, P. Kauranen, and T. Kallio, 'Is the H_2 economy realizable in the foreseeable future? Part II: H_2 storage, transportation, and distribution', *International Journal of Hydrogen Energy*, vol. 45, no. 41, pp. 20693–20708, 2020.
34. A. M. Oliveira, R. R. Beswick, and Y. Yan, 'A green hydrogen economy for a renewable energy society', *Current Opinion in Chemical Engineering*, vol. 33, p. 100701, 2021.
35. A. O. Oni, K. Anaya, T. Giwa, G. Di Lullo, and A. Kumar, 'Comparative assessment of blue hydrogen from steam methane reforming, autothermal reforming, and natural gas decomposition technologies for natural gas-producing regions', *Energy Conversion and Management*, vol. 254, p. 115245, 2022.

36. G. Palmer and J. Floyd, 2020. Energy Storage and Civilization: A Systems Approach. Springer.
37. I. Panić, A. Cuculić, and J. Ćelić, 'Color-coded hydrogen: Production and storage in maritime sector', *Journal of Marine Science and Engineering*, vol. 10, no. 12, p. 1995, 2022.
38. R. Priyadarshi, P. Ezati, and J. W. Rhim, 'Recent advances in intelligent food packaging applications using natural food colorants', *ACS Food Science & Technology*, vol. 1, no. 2, pp. 124–138, 2021.
39. M. G. Rasul, M. A. Hazrat, M. A. Sattar, M. I. Jahirul, and M. J. Shearer, 'The future of hydrogen: Challenges on production, storage and applications', *Energy Conversion and Management*, vol. 272, p. 116326, 2022.
40. F. Razi, and I. Dincer, 'Renewable energy development and hydrogen economy in MENA region: A review', *Renewable and Sustainable Energy Reviews*, vol. 168, p. 112763, 2022.
41. J. Rogelj, O. Geden, A. Cowie, and A. Reisinger, 'Three ways to improve net-zero emissions targets', *Nature*, vol. 591, no. 7850, pp. 365–368, 2021.
42. G. S. Seck, E. Hache, J. Sabathier, F. Guedes, G. A. Reigstad, J. Straus, O. Wolfgang, J. A. Ouassou, M. Askeland, I. Hjorth, and H. I. Skjelbred, 'Hydrogen and the decarbonization of the energy system in Europe in 2050: A detailed model-based analysis', *Renewable and Sustainable Energy Reviews*, vol. 167, p. 112779, 2022.
43. M. Shen, W. Huang, M. Chen, B. Song, G. Zeng, and Y. Zhang, '(Micro) plastic crisis: Un-ignorable contribution to global greenhouse gas emissions and climate change', *Journal of Cleaner Production*, vol. 254, p. 120138, 2020.
44. Singh, S. (2021). Energy crisis and climate change. In *Energy* (eds P. Singh, S. Singh, G. Kumar and P. Baweja). https://doi.org/10.1002/9781119741503.ch1
45. B. K. Sovacool, and S. Griffiths, 'Culture and low-carbon energy transitions', *Nature Sustainability*, vol. 3, no. 9, pp. 685–693, 2020.
46. K. Svobodova, J. R. Owen, J. Harris, and S. Worden, 'Complexities and contradictions in the global energy transition: A re-evaluation of country-level factors and dependencies', *Applied Energy*, vol. 265, p. 114778, 2020.
47. S. Timmerberg, M. Kaltschmitt, and M. Finkbeiner, 'Hydrogen and hydrogen-derived fuels through methane decomposition of natural gas–GHG emissions and costs', *Energy Conversion and Management: X*, vol. 7, p. 100043, 2020.
48. R. K. Upadhyay, 'Markers for global climate change and its impact on social, biological and ecological systems: A review', *American Journal of Climate Change*, vol. 9, no. 03, p. 159, 2020.
49. Y. S. Wang, and J. D. Gu, 'Ecological responses, adaptation and mechanisms of mangrove wetland ecosystem to global climate change and anthropogenic activities', *International Biodeterioration & Biodegradation*, vol. 162, p. 105248, 2021.
50. M. Wei, C. A. McMillan, and S. de la Rue Du Can, 'Electrification of industry: Potential, challenges and outlook', *Current Sustainable/Renewable Energy Reports*, vol. 6, pp. 140–148, 2019.
51. M. P. Weijnen, Z. Lukszo, and S. Farahani, *Shaping an Inclusive Energy Transition*, Springer Nature, 2021, p. 258.
52. X. Yang, C. Nielsen, S. Song, M. McElroy, and X. Wang, 2021. Breaking the "hard-to-abate" bottleneck in China's path to carbon neutrality. *Nature Energy*, vol. 7, no. 10, pp.955–965.
53. X. Yang, C. P. Nielsen, S. Song, and M. B. McElroy, 'Breaking the hard-to-abate bottleneck in China's path to carbon neutrality with clean hydrogen', *Nature Energy*, vol. 7, no. 10, pp. 955–965, 2022.

54. M. Yue, H. Lambert, E. Pahon, R. Roche, S. Jemei, and D. Hissel, 'Hydrogen energy systems: A critical review of technologies, applications, trends and challenges', *Renewable and Sustainable Energy Reviews*, vol. 146, p. 111180, 2021.
55. B. Zakeri, K. Paulavets, L. Barreto-Gomez, L. Gomez Echeverri, S. Pachauri, J. Rogelj, F. Creutzig, D. Urge-Vorsatz, D. Victor, B. Boza-Kiss, and C. Zimm, 2021. Transformations within reach: Pathways to a sustainable and resilient world-Rethinking energy solutions. IIASA Report. IIASA-ISC. https://covid19.iiasa.ac.at/isc/
56. A. Acharya, 'Scaling-up green hydrogen development with effective policy interventions', *Journal of Sustainable Development*, vol. 15, no. 5, pp. 135–149, 2022.
57. A. Ajanovic, M. Sayer, and R. Haas, 'The economics and the environmental benignity of different colors of hydrogen', *International Journal of Hydrogen Energy*, vol. 47, no. 57, pp. 24136–24154, 2022.

2 Green Hydrogen Generation Technologies

2.1 HYDROGEN GENERATION APPROACHES

Environmental decline, increasing oil costs, limited fuel reserves, and impending climate shifts are orchestrating a targeted effort for sustainable energy sources, such as renewable energy sources [1]. In sustainable energy systems, establishment and development pose the most critical challenges that today's society has to address [2]. These energy systems must be able to perform in a comprehensive, reliable manner, thereby minimizing environmental impacts.

Hydrogen is an energy medium with an energy content of about 120 MJ/kg (i.e., LHV). It is a gas that is colorless, non-toxic, and tasteless, as well as odorless, which generates water when burned in the presence of oxygen. The combustion of hydrogen (H_2) and oxygen (O_2) produces water as a byproduct.

$$H_{2\,(g)} + \frac{1}{2}O_{2\,(g)} \rightarrow H_2O_{(g)} \quad \Delta H = -242 \text{ kJ / mol} \tag{2.1}$$

Hydrogen is recorded as the most abundant element in nature, described by the number of atoms (92%) and mass (75%). H_2 abundance is due to its lightness and high reactivity in the universe. It can be found in water, hydrocarbons, and organisms in bond forms [3–5]. H_2 generated from renewable sources, referred to as clean fuel, has become a foundational element in addressing environmental emissions, ensuring energy reliability, and promoting sustainability. This section highlights the primary aspects of H_2 generation approaches, focusing on the underlying principles of each approach.

Hydrogen is generated in three broad classifications. These are carbon-based or thermochemical processes, electrolysis-based or electrochemical processes, and sunlight-based or photolytic approaches. Global H_2 generation reached 97 million tons (Mt) as of 2023, based on classifications. Fossil fuels (mainly natural gas and coal) dominate, accounting for 85% of the total, as depicted in Figure 2.1. Low-emission H_2 generation, which involves mainly electrolysis-based methods, accounts for less than 1%. However, low emission H_2 generation is projected to reach 49 Mt per annum by 2030, an increase from 38 Mt per annum in the 2023 global hydrogen review [6].

2.1.1 Carbon-Based Hydrogen Generation

Carbon-based H_2 generation, which is a thermochemical process of generation, includes Steam Methane Reforming (SMR), coal gasification, and biomass pyrolysis [7]. The SMR is characterized as a high-temperature reaction between steam

DOI: 10.1201/9781003727194-2

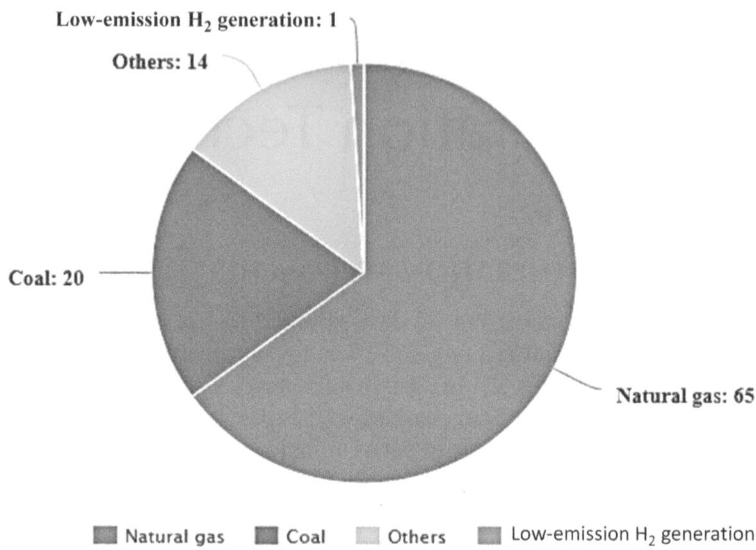

FIGURE 2.1 Global hydrogen generation sources.

and hydrocarbons, which is often undertaken at temperatures of 500°C to 900°C with pressure less than 20 bar, per the hydrocarbon feedstock utilized [8]. Due to SMR's cost-effectiveness, it remains the prevalent industrial approach. However, it is carbon-intensive and environmentally harmful if technologies for carbon capture are not fully utilized.

Carbon dioxide (CO_2) emissions from SMR are approximately 7 kg of CO_2 per H_2 generated, which constitutes roughly 3% of the world's industrial CO_2 emissions. The following are the general steam reforming with hydrocarbon (Eq. (2.2)) and those for SMR reactions (Eqs. (2.3) and (2.4)) [9]:

$$C_xH_{y(g)} + xH_2O_{(g)} \rightarrow xCO_{(g)} + \left(x + \frac{1}{2}y\right)H_{2(g)} \quad \Delta H > 0 \text{ kJ/mol} \tag{2.2}$$

$$CH_{4(g)} + H_2O_{(g)} \rightarrow CO_{(g)} + 3H_{2(g)} \quad \Delta H = 206 \text{ kJ/mol} \tag{2.3}$$

$$CO_{(g)} + H_2O_{(g)} \rightarrow CO_{2(g)} + H_{2(g)} \tag{2.4}$$

The conventional approach of SMR involves four main components: the desulfurizing unit, the shift reactor, the reforming unit, and the separation unit. The purpose of the desulfurizing section is to remove sulfur from the natural gas entering, primarily to inhibit the generation of sulfur oxides, which impacts the environment negatively and can also poison the reformer catalysts, potentially harming the separation unit if a H_2 lead-alloy membrane is employed [8].

Gasification is the thermochemical process of converting carbon into synthetic gas utilizing air, steam, or O_2 [10, 11]. Waste and raw materials, like the wood and plastic sawdust wastes, or sewage, can reliably and conveniently be converted into useful outputs [12, 13]. Despite the abundance of coal in nature, it is not a reliable source of energy. The substantial CO_2 emission it produces makes it unreliable. Coal is converted into gaseous byproducts (primarily CO and H_2) during the coal gasification process [14, 15]. The fundamental chemical reactions of the coal gasification processes are as follows [16]:

$$C + O_{2(g)} \rightarrow CO_{2(g)} \ \Delta H = -394.9 \ KJ/mol \tag{2.5}$$

$$2C + O_{2(g)} \rightarrow 2CO_{(g)} \ \Delta H = -226 \ KJ/mol \tag{2.6}$$

$$2C + O_{2(g)} \rightarrow 2CO_{(g)} \ \Delta H = -564 \ KJ/mol \tag{2.7}$$

The primary byproducts of the reactions between carbon and O_2 in gasification are CO_2 and CO (Eqs. (2.5) to (2.7)). Furthermore, the CO_2 and steam react endothermically, i.e.,

$$C + CO_{2(g)} \rightarrow 2CO_{(g)} \ \Delta H = 168.9 \ KJ/mol \tag{2.8}$$

$$C + H_2O_{(g)} \rightarrow CO_{(g)} + H_{2(g)} \ \Delta H = 136 \ KJ/mol \tag{2.9}$$

An exothermic reaction of coal and H_2 occurs, which generates methane (CH_4):

$$C + 2H_{2(g)} \rightarrow CH_{4(g)} \ \Delta H = -91 \ KJ/mol \tag{2.10}$$

The water shift reaction occurs in the gasification channel:

$$CO_{(g)} + H_2O_{(g)} \rightarrow CO_{2(g)} + H_{2(g)} \ \Delta H = -33.2 \ KJ/mol \tag{2.11}$$

and the dismutation of the CO reaction occurs:

$$2CO_{(g)} \rightarrow CO_{2(g)} + C \ \Delta H = -169 \ KJ/mol \tag{2.12}$$

Other applications of gasification are depicted in Figure 2.2. The thermochemical approaches for converting biomass include pyrolysis, liquefaction, gasification, and combustion. Biomass pyrolysis involves a thermochemical process of degrading biomass feedstock while starving O_2 to generate H_2 [17]. Biomass pyrolysis has emerged as an effective energy conversion method in the past because of its minimal environmental impact [18].

2.1.2 Electrochemical Processes

Electrochemical processes for H_2 generation split water into H_2 and O_2 using electricity. These approaches have regained interest as a substitute for SMR in the H_2

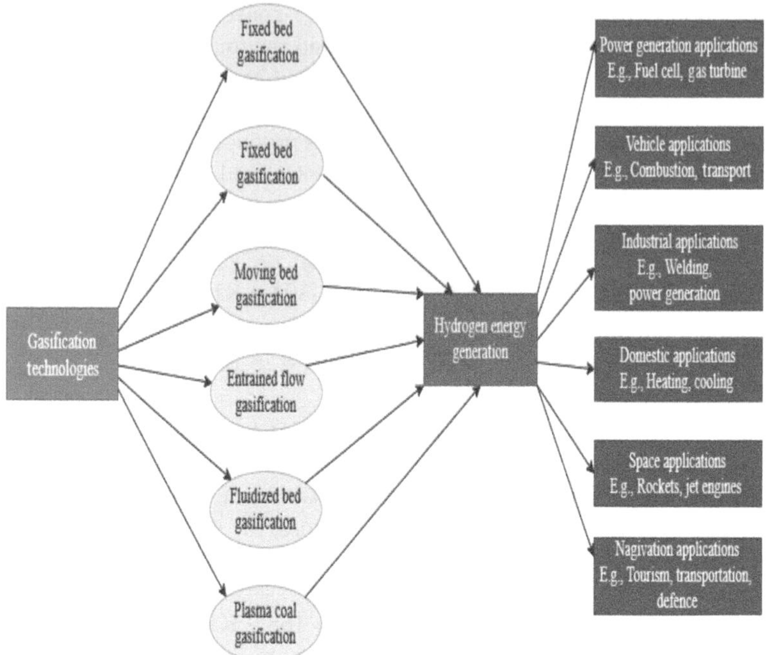

FIGURE 2.2 Gasification technologies.

generation. Renewable energy electrolysis offers a clean alternative for water splitting. The ability to continuously create H_2 as a byproduct is a significant benefit. Hydrogen generation via the electrochemical route is an economically viable option that increases stakeholder confidence. Other benefits include its simplicity in operation, the absence of chemical requirements, a shorter processing time, and the lower amount of sludge produced compared to chemical methods [20].

2.1.3 Photolytic and Biological Approaches

Energy harnessed from the sun has enormous potential to generate clean and efficient H_2. Two technologies exist in solar conversion: solar thermal and solar photon conversion. The solar photon conversion is categorized into three solar cells: solid-state photovoltaic, metal-semiconductor-based Schottky barrier, and semiconductor liquid junction-based photoelectrochemical solar cells [21]. Photoelectrochemical approaches utilize sunlight to split water or drive microbial H_2 production. Some of the microbes (such as the strains and substrates) that can aid in H_2 generation are depicted in Table 2.1.

The biological H_2 generation approach is an encouraging method of generation from resources that are renewable, low-cost, and environmentally friendly under atmospheric pressure and temperature [40]. In this approach, water molecules and organic substrates are converted into H_2 by microorganisms via two main enzymes,

TABLE 2.1
Microbial Strains and Substrates That Generate Hydrogen under Photofermentation Systems

Microbial Strains	Substrates	Hydrogen Generation
Rhodobium marinum A-501 [22]	Glycerol	8.3 mmol/l
Rhodobacter sphaeroides [23]	Hexose	8.35 mol H_2/mol
Rhodopseudomonas palustris [24]	Lactate	9.1 ml H_2/l h
Rhodobacter sphaeroides [25]	Acetate	90 ml H_2/l h
Rhodobacter sphaeroides [25]	Malate	92 ml H_2/l h
Rhodobacter sphaeroides [25]	Succinate	94 ml H_2/l h
Rhodobium marinum A-501 [22]	Acetic acid	0.2 mmol H_2/l
Rhodopseudomonas palustris [26]	Glucose	0.2 mol H_2/mol
Rhodobacter capsulatus [27]	Acetate	0.6 mol H_2/mol
Rhodopseudomonas sp. [24]	Malate	1.1 ml H_2/l h
Rhodopseudomonas sp. [24]	Lactate	10.7 ml H_2/l h
Rhodobacter sphaeroides [25]	Butyrate	110 ml H_2/l h
Rhodobium marinum A-501 [22]	Sucrose	12.3 mmol H_2/l
Rhodobium marinum A-501 [22]	Malate	13.6 mmol H_2/l
Rhodopseudomonas palustris [24]	Acetate	2.2 ml H_2/l h
Rhodobacter sphaeroides [28]	Wastewaters	2.24 l H_2/l medium
Rhodobacter sphaeroides [29]	Succinate	2.3 mol H_2/mol
Rhodobacter sphaeroides [30]	Sodium lactate	2.4 mg/l
Rhodopseudomonas faecalis [31]	Acetate	2.61 mol H_2/mol
Rhodopseudomonas faecalis [32]	Acetate	2.64 mol H_2/mol
Rhodobacter sphaeroides [33]	Acetate	20 ml H_2/l h
Rhodobacter sphaeroides [33]	Butyrate	20 ml H_2/l h
Rhodobacter sphaeroides [33]	Lactate	20 ml H_2/l h
Rhodobium marinum [34]	Soy sauce	200 ml H_2
Rhodopseudomonas palustris [35]	Wastewater	205 ml H_2 l/d
Rhodobium marinum A-501 [22]	Glucose	21.6 mmol H_2/l
Rhodobacter sphaeroides [33]	Propionate	22 ml H_2/l h
Rhodobium marinum A-501 [22]	Malic acid	23.4 mmol H_2/l
Rhodobacter sphaeroides [33]	Malate	24 ml H_2/l h
Rhodopseudomonas sp. [24]	Acetate	25.2 ml H_2/l h
Rhodopseudomonas faecalis [36]	Acetate	3.12 mol H_2/mol
Rhodopseudomonas faecalis [37]	Acetate	3.17 mol H_2/mol
Rhodobacter sphaeroides [38]	Succinate	3.7 mol H_2/mol
Rhodobium marinum A-501 [22]	Lactic acid	37.3 mmol H_2/l
Rhodobacter capsulatus [39]	Acetate	3752.7 ml H_2/l
Rhodopseudomonas palustris [24]	Malate	5.8 ml H_2/l h
Rhodopseudomonas sp. [24]	Butyrate	7.6 ml H_2/l h

FIGURE 2.3 Issues in hydrogen generation through biological processes.

nitrogenase and hydrogenase [41]. Some of the biohydrogen generation techniques are photo-fermentation, photolysis, CO gas-fermentation, and dark-fermentation. These techniques and their combinations have shortcomings that inhibit their sustainability, and these are summarized in Figure 2.3.

2.2 BASICS OF WATER ELECTROLYSIS

Electrolysis is regarded as a distinguished method for generating fuels of the future. Many industries, such as pharmaceutical, food, steel, and chemical, utilize electrolysis to generate H_2 [42]. Water electrolysis generates H_2 safely and in a green manner; however, the cost of electricity consumption for the H_2 generation represents over 75% of the total costs of generation [43].

An electrolysis cell consists of two metallic electrodes (an anode for water oxidation and a cathode for water reduction) arranged in series with a separator (membrane), an electrolyte solution, and an electric power supply (Figure 2.4). In operating the cell, a continuous DC power supply flows through the cell, the charge carriers in the electrodes are the electrons, and the charge carriers in the electrolyte are the mobile ions.

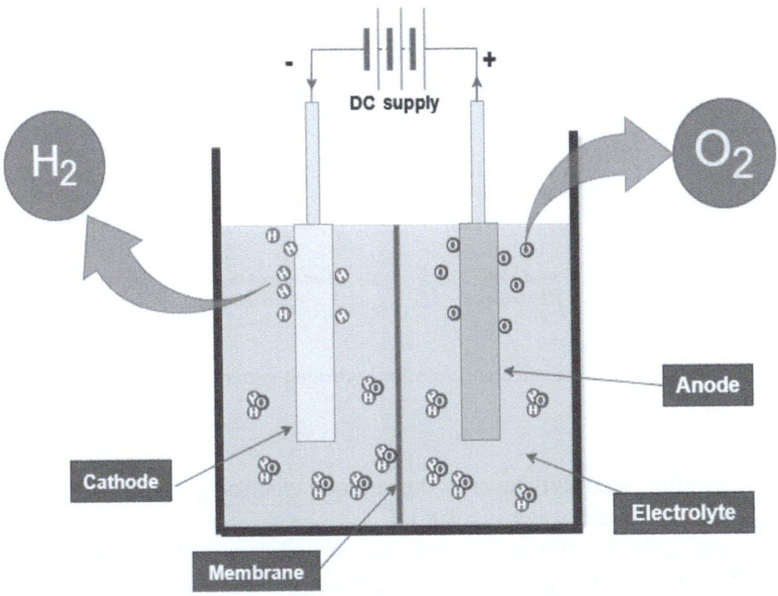

FIGURE 2.4 A simple water electrolytic cell.

An electrolyzer is, therefore, an instrument consisting of a simple electrolysis cell or several cells used to carry out endergonic reactions (i.e., nonspontaneous chemical processes that require the input of electrical work to proceed) [44]. Each electrolysis cell is connected in series in a filter-press configuration commonly used to enhance compactness and reduce unwanted ohmic losses. Water in liquid and vapor forms can be dissociated into H_2 and O_2 elemental components, as given in the following equations:

$$H_2O_{(l)} \rightarrow H_{2(g)} + \frac{1}{2}O_{2(g)} \tag{2.13a}$$

$$H_2O_{(vap)} \rightarrow H_{2(g)} + \frac{1}{2}O_{2(g)} \tag{2.13b}$$

When current is passed through the electrolysis cell, the cathode and the anode reactions are thus given by:

$$\text{Anode (oxidation): } H_2O_{(l)} \rightarrow \frac{1}{2}O_{2(g)} + 2H^+ + 2e^- \tag{2.14}$$

$$\text{Cathode (reduction): } 2H^+ + 2e^- \rightarrow H_{2(g)} \tag{2.15}$$

In the Oxygen Evolution Reaction (OER), protons are formed at the O_2-evolving anode in the aqueous solution, which are dissolved to create hydronium ions (H_3O^+). As a result of the electric current flowing through the electrolysis cell, the H_3O^+ ions

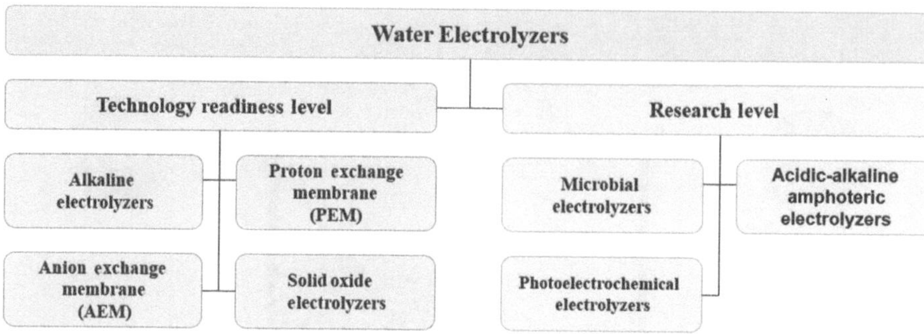

FIGURE 2.5 Types of water electrolyzers for hydrogen generation.

move through the electrolyte layer to the cathode electrode, where they are reduced into H_2, and the solvated water is released.

There are different types of water electrolyzers, which are summarized in Figure 2.5. These are grouped based on technology and research levels.

The common electrolyzers are the alkaline, Proton Exchange Membrane (PEM), Anion Exchange Membrane (AEM), and solid oxide electrolyzers. Those still under research are the microbial, acidic-alkaline amphoteric, and photoelectrochemical electrolyzers [45].

2.2.1 Alkaline Electrolysis

Alkaline electrolysis cells are low-temperature electrolyzers which utilize liquid alkaline electrolytes such as NaCl [45]. These electrolyzers are technologically mature, readily available in the market, and cheaper [46]. They are cheap primarily due to their inexpensive electrolytes, lower-cost catalysts, low-cost electrodes, and separators [47]. The fundamental reactions in an alkaline electrolysis cell are as follows:

$$\text{Anode: } 4OH^-_{aq} \rightarrow O_{2(g)} + 2H_2O_{(l)} + 4e^- \quad (2.16)$$

$$\text{Cathode: } 4H_2O_{(l)} + 4e^- \rightarrow 2H_{2(g)} + 4OH^-_{(aq)} \quad (2.17)$$

The charge carrier or ion conducted is the hydroxide ion (OH^-). However, despite the affordability of alkaline electrolyzers, there are still concerns about their low system efficiency. A simple configuration of the alkaline electrolyzer is shown in Figure 2.6a.

2.2.2 Proton Exchange Membrane Electrolysis

PEM electrolyzers use a solid polymer membrane as their type of electrolyte and operate at higher current densities with faster response times; however, cheap

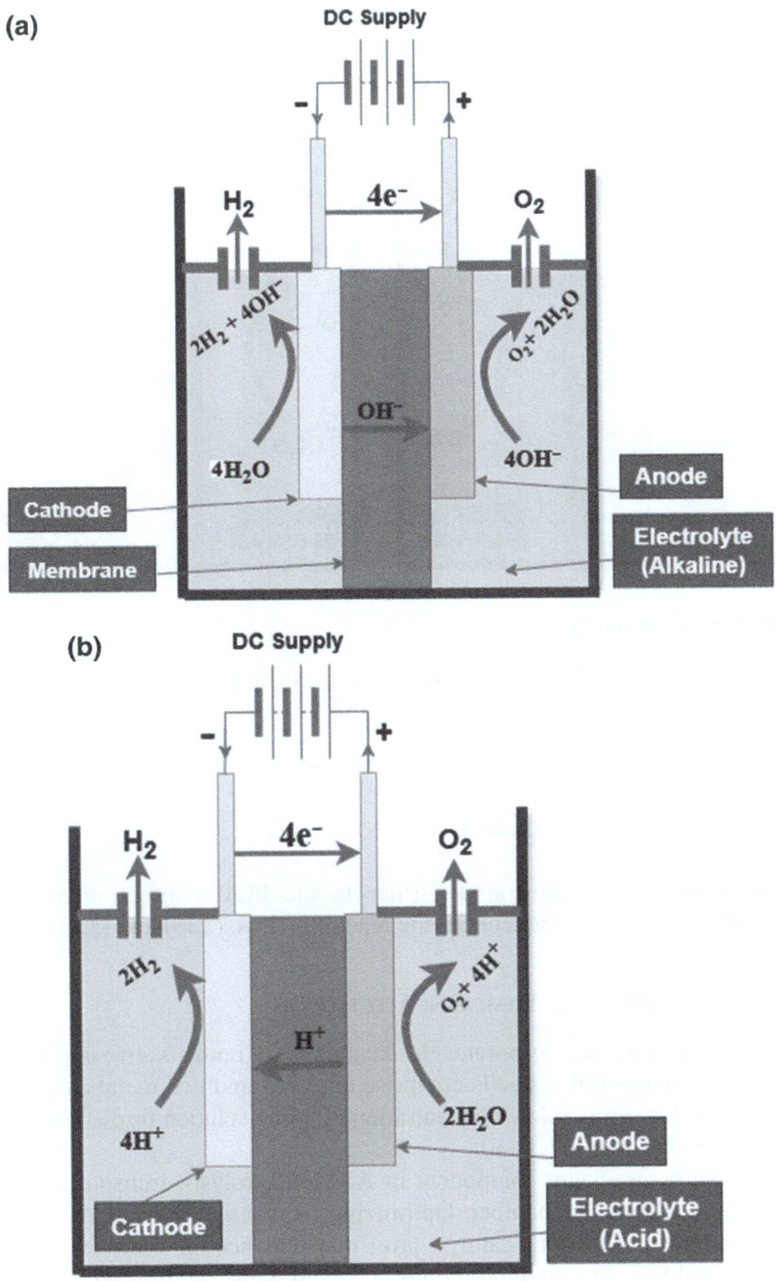

FIGURE 2.6 Reactions in (a) alkaline, (b) proton exchange membrane (PEM), (c) anion exchange membrane (AEM), and (d) solid oxide, electrolyzers. (*Continued*)

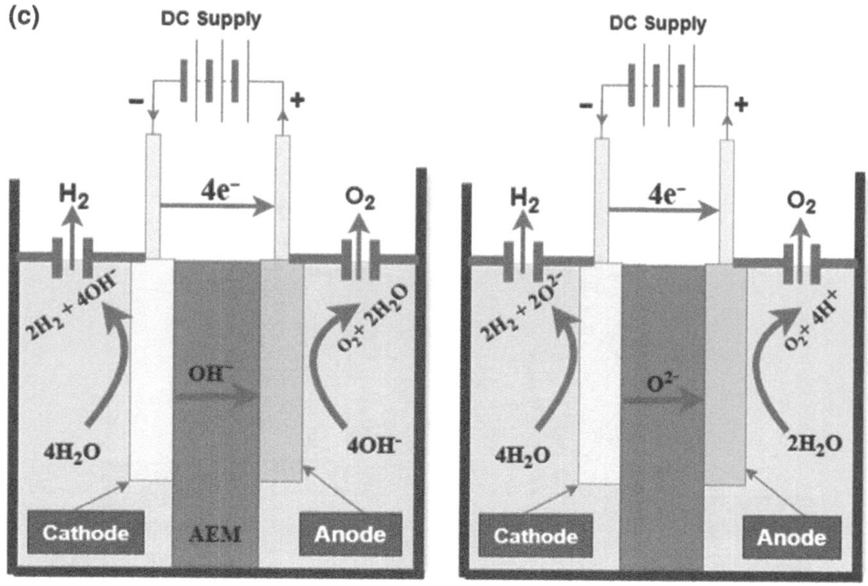

FIGURE 2.6 *(Continued)*

membranes are needed for this electrolyzer type [45, 48]. The reactions occurring in the electrolyzer are given by:

$$\text{Anode: } 2H_2O_{(l)} \rightarrow O_{2(g)} + 4H^+ + 4e^- \tag{2.18}$$

$$\text{Cathode: } 2H^+ + 2e^- \rightarrow H_{2(g)} \tag{2.19}$$

These reactions are illustrated further in the PEM configuration shown in Figure 2.6b. The ion, as conducted in the reactions (Eqs. (2.18) and (2.19)), is H^+.

2.2.3 Anion Exchange Membrane Electrolysis

This type of electrolysis incorporates the advantages of both alkaline and PEM electrolysis [49]. The electrolysis cells comprise an AEM and two metal catalyst-based electrodes. In this setup, a low-concentration alkaline solution or distilled water as an electrolyte may be utilized [50].

The AEM, a significant component in AEM electrolysis, transports OH^- from the cathode to the anode chamber. Optimizing the configuration of the water feed setup and ionomer within the catalyst layer may enhance the lifetime of the AEM electrolysis cell [51]. The configuration of a typical AEM electrolyzer is depicted in Figure 2.6c, and their reactions are given by:

$$\text{Anode: } 4OH^-_{(aq)} \rightarrow O_{2(g)} + 2H_2O_{(l)} + 4e^- \tag{2.20}$$

$$\text{Cathode: } 4H_2O_{(l)} + 4e^- \rightarrow 2H_{2(g)} + 4OH^-_{(aq)} \tag{2.21}$$

2.2.4 Solid Oxide Electrolysis

The Solid Oxide Electrolyzer (SOE) is distinguished from other electrolyzers as a high-temperature electrolyzer, which is usually operated at a temperature variation between 500°C and 850°C [45]. It has high potential for economic and efficient H_2 generation, and it is theoretically more efficient compared to other electrolyzers (PEM and alkaline). The SOEs utilize ceramic electrolyte materials, and their configuration is shown in Figure 2.6d. The system comprises two porous electrodes and a dense, ionic-conducting electrolyte. Unlike other electrolytes, in the SOE, the cathode is fed with steam as given in the following reactions:

$$\text{Anode: } 2H_2O_{(vap)} \rightarrow O_{2(g)} + 4H^+ + 4e^- \tag{2.22}$$

$$\text{Cathode: } 2H_2O_{(vap)} + 4e^- \rightarrow 2H_{2(g)} + 2O^{2-} \tag{2.23}$$

Water at the interface of the cathode and electrolyte dissociates to form oxide ions and H_2 gas. The oxide ions are then diffused to the anode, where they are oxidized to O_2, while the H_2 generated at the cathode is collected. Comparisons of the different electrolyzers discussed, along with their benefits and limitations, are summarized in Table 2.2.

TABLE 2.2
The Benefits and Limitations of the Different Electrolyzers [45, 52–54]

Electrolysis	Benefits	Limitations
Alkaline	a. Highest maturity level and well-established technology b. Less expensive components c. Long-term durability d. Uses non-noble catalysts e. Commercialized at megawatt scale f. Environmentally friendly if powered by renewables	a. Lower efficiency (~70%) b. Low current density and pressure operation c. Requires corrosive liquid electrolytes (KOH, NaOH) d. Gas crossover leads to lower H_2 purity e. Limited partial load operation f. Membranes dissolved under long-term electrolysis, indicating poor durability
PEM	a. High current density and voltage efficiency b. High-purity H_2 due to gas separation by membrane c. Compact design with rapid dynamic response d. Good performance under partial load e. Faraday efficiency is often >99% f. Can operate at high pressure	a. Expensive materials (noble catalysts like Ir, Pt; Nafion membranes) b. Acidic environment causes durability issues c. Durability is lower than that of alkaline electrolyzers d. Limited membrane alternatives (Nafion is costly) e. Commercialization is still developing (below MW scale) f. PEM electrolyzers depend on iridium and platinum; scaling up is limited by global supply

(Continued)

TABLE 2.2 (*Continued*)
The Benefits and Limitations of the Different Electrolyzers [45, 52–54]

Electrolysis	Benefits	Limitations
AEM	a. Uses abundant materials (low-cost materials) b. AEM has good durability c. Has zero gap management d. No addition of liquid electrolyte required e. Improved lifetime f. Charge density is high g. Nickel electrodes have a high surface area	a. Still a laboratory-phase technology b. Current efficiency is low c. Current density is low d. Generation yield is low e. Power rate is low f. Catalyst loading is excessive g. More progress in synthesis is required h. It has both mechanical and chemical stability issues
SOE	a. Very high efficiency (~100% or more with thermal energy input) b. Uses steam—less electrical energy needed due to favorable thermodynamics c. Non-noble catalysts can be used d. Suitable for integration with nuclear, solar, geothermal, or waste heat sources e. High operating pressure reduces energy loss due to gas bubble resistance	a. Still in laboratory or pilot-scale stage b. Requires complex, bulky system design c. Durability issues due to thermal stress and material degradation d. System components (ceramic electrolytes) are sensitive to thermal cycling

2.2.5 Membraneless Electrolysis

Unlike conventional electrolyzers that require membranes, electrolyzers that do not utilize membranes are known as membraneless electrolyzers. These electrolyzers represent a promising technology for effective H_2 generation [62]. The configurations of membraneless electrolyzers permit operation of the systems at any pH, reduce ionic resistance, and decrease the system's exclusion of membranes [63]. An attempt has been made to classify these membraneless electrolyzers as shown in Figure 2.7.

Membraneless electrolyzers have been broadly classified as quasi-membraneless and membraneless electrolyzers. Decoupled reaction water electrolysis is a quasi-membraneless electrolysis that mitigates some limitations of the conventional water electrolysis by splitting the generation of H_2 and O_2 into two separate phases that can occur at varying times, rates, and locations [64].

In the decoupled reaction electrolysis, an appropriate redox mediator is employed to facilitate the transfer of H_2 and OH^- between the anode and the cathode. Buoyancy-driven membraneless electrolyzers utilize buoyancy forces to move O_2 and H_2 bubbles generated in the electrolyte into separate streams in angled cell geometries [65, 66].

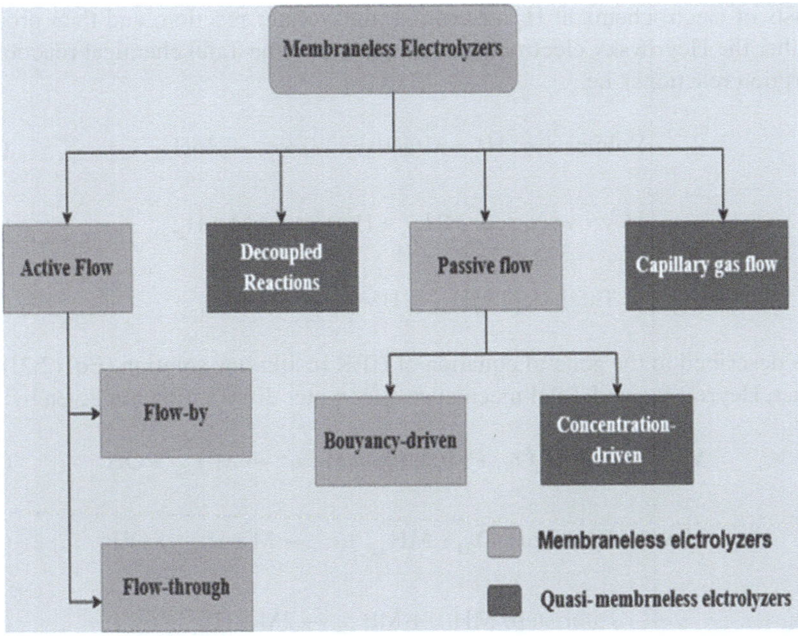

FIGURE 2.7 Categories of quasi-membraneless and membraneless electrolyzers.

The active flow electrolyzers are the flow-by and flow-through types of electrolyzers. The flow-by electrolyzers use fluid dynamics to induce gas separation along parallel electrodes. In this type, laminar H_2 and O_2 bubbles from the cathode and anode, respectively, flow in microfluidic cells [67]. In the flow-through configuration of the active category, the movement of gases for the catholyte and anolyte is directed in opposite directions, utilizing flow-through cells equipped with porous cathode and anode electrodes.

2.3 HYDROGEN EVOLUTION REACTION MECHANISMS

The Hydrogen Evolution Reaction (HER) is central to the H_2 generation process. It is often examined in the processes of electrocatalysis. It is a reaction that occurs at the cathode of a water electrolysis cell. HER involves the adsorption of H_2 on the electrode surface, and since the energy of adsorption is influenced by the type of electrode materials utilized, the kinetics of HER are also affected [55]. HER mechanisms differ based on the electrolyte environment, which is characterized by acidic and alkaline media.

HER can occur in highly concentrated acidic or basic solutions; however, it is occasionally examined at moderate pH levels and in buffered solutions. When using unbuffered solutions, the pH at the electrode fluctuates continuously with both current and time, resulting in the absence of a steady-state condition [55]. From the general equation of HER in acidic solutions described by Eq. (2.19), the mechanism consists of electrochemical H_2 adsorption, the Volmer reaction, and then proceeds by either the Heyrovsky electrochemical reaction or the Tafel chemical reaction (H_2 desorption reactions), i.e.,

$$\text{Volmer step: } H^+ + M(\text{active site}) + e^- \rightarrow MH_{ads} \tag{2.24}$$

$$\text{Heyrovsky step: } MH_{ads} + H^+ + e^- \rightarrow M + H_{2(g)} \tag{2.25}$$

$$\text{Tafel step: } MH_{ads} + HM_{ads} \rightarrow 2M + H_{2(g)} \tag{2.26}$$

As described in the general equation of HER in alkaline solution (Eq. (2.21)), the Volmer, Heyrovsky, and Tafel mechanisms for water dissociation are given by:

$$\text{Volmer step: } H_2O_{(l)} + M(\text{active site}) + e^- \rightarrow MH_{ads} + OH^- \tag{2.27}$$

$$\text{Heyrovsky step: } H_2O_{(l)} + MH_{ads} + e^- \rightarrow M + H_{2(g)} + OH^- \tag{2.28}$$

$$\text{Tafel step: } MH_{ads} + MH_{ads} \rightarrow 2M + H_{2(g)} \tag{2.29}$$

The Tafel steps in acidic and alkaline media are the same as those obtained in Eqs. (2.26) and (2.29), respectively. The HER steps emphasize the need for: a high surface area, active sites that offer favorable access to reactants and that can aid in the collection of products, and effective gas desorption [55, 56]. All these requirements are influenced by electrode design.

Limitations of electrolyzers, such as bubble management, mass transport, and electrode surface accessibility, have prompted the exploration of novel electrode configurations. The development of novel electrode substrates is crucial for enhancing performance for industrial applications [57]. Also, the effect of bubble dynamics on the performance and mechanical stability of electrolyzers needs further investigation [58].

While conventional electrolyzers employ planar or porous electrode structures, their efficiency is often hindered by non-uniform current distribution, limited reactant accessibility, poor gas bubble removal, high ohmic, and high concentration overpotentials [59, 60]. To address all these limitations, advanced electrode configurations such as those of the membraneless electrolyzers offer spatial manipulation of electric field lines, improved flow distribution, and enhanced mass transport. The membraneless approach (especially the flow through) introduces geometrical divergence between electrodes, creating unique fluid flow patterns that can substantially influence electrochemical kinetics and overall system efficiency [61].

2.4 CONCEPT OF FLOW-THROUGH ELECTRODE

The Divergent Electrodes Flow-Through (DEFT) concept describes an electrode architecture in which the flow of electrolyte is directed through a spatially divergent

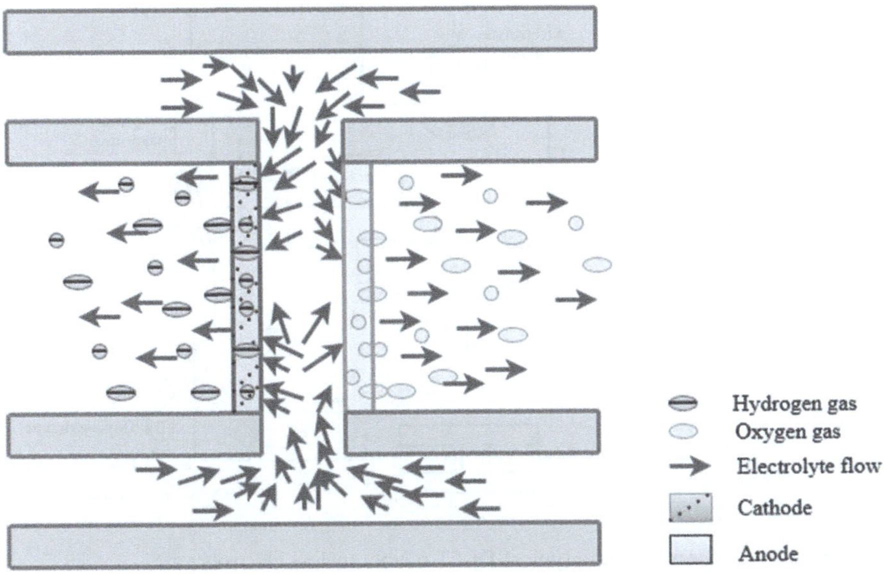

FIGURE 2.8 A typical DEFT electrolyzer configuration.

electrode pair. In 2013, Hydrox Holdings Ltd. patented the membraneless DEFT technology (DEFT™). This technology utilizes a flow-through technique relying on electrolyte flow through opposite and parallel porous electrodes for both H_2 and O_2 gas separation. Figure 2.8 shows the schematic diagram of the DEFT electrolyzer configuration. This configuration aims to address limitations in HER diffusion-limited reactions, where uniform access to reactants and removal of products are critical.

Enhanced electrolytic flows facilitate the diverging of gases to the porous electrodes, initiating a separation process where O_2 gas flows through the anode and H_2 through the cathode electrodes in opposite directions, ensuring negligible gas formation between the electrodes [68]. In this way, it reduces the effects of void fracture and the associated increase in resistance. These divergent flows of H_2 through the cathode and O_2 through the anode are bi-phase or multiphase flows, which require a gas separation technique.

In DEFT, the geometry of divergent flow involves the electrolyte being pushed through electrode structures that are angled or expanding, generating non-uniform velocity fields that enhance convective transport and reduce diffusion limitation. With this, there is good reactant distribution and gas removal, aiding efficient HER [68].

The design of the divergent channel promotes localized turbulence and mixing, especially near the electrode surfaces, leading to lower mass transfer resistance, improved bubble detachment, and reduced boundary layer thickness [69]. Unlike conventional planar cells, where stagnant zones can form, DEFT ensures continuous electrolyte removal across all active areas. This helps maintain consistent reaction

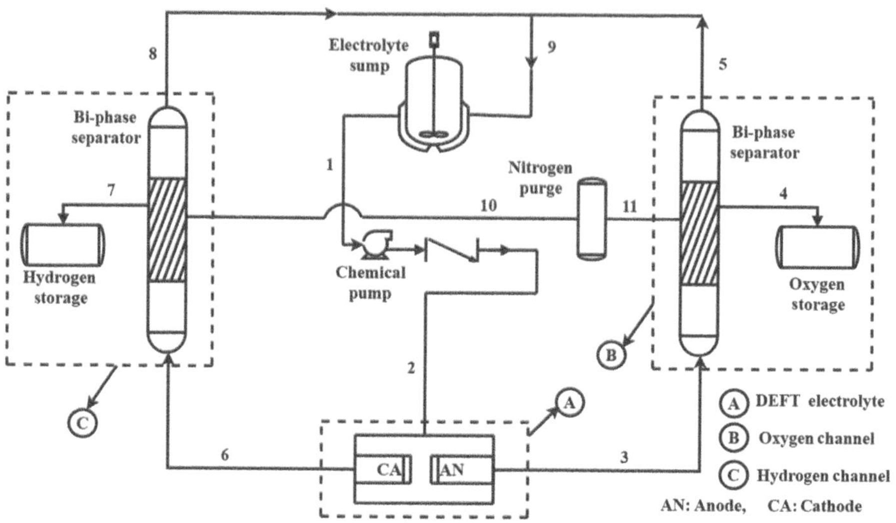

FIGURE 2.9 Process flow of a typical DEFT membraneless electrolyzer.

Note: 1, 2—fresh electrolyte; 3, 6—b-phase collection (liquiq/gas); 4—oxygen gas; 7—hydrogen gas; 5, 8, 9—return electrolyte; 10, 11—nitrogen gas.

kinetics and current density throughout the electrode surface. Moreover, pressure drops are typically lower in DEFT systems because the flow path widens, reducing flow resistance and enabling energy-efficient operation.

The process flow diagram of a typical membranless electrolysis system, such as that of the DEFT, is depicted in Figure 2.9. The system starts by pumping fresh electrolyte to the electrolyzer from the electrolyte sump to undergo the electrolysis process. Inside the electrolyzer, the electrolyte splits into two parts, one part to the anode and the other to the cathode.

Each of the two output channels contains a bi-phase (or multiphase) flow of gas species and liquid electrolyte. A gas separator is used in conjunction with nitrogen gas to separate and collect the various gas species. The electrolyte exiting the two separators is recirculated back to the electrolyte sump for recirculation.

2.4.1 Electrochemical Principles Underlying DEFT

The DEFT configuration and those of other electrolyzers operate on key electrochemical principles such as Ohm's law, Nernst equation, and Butler-Volmer kinetics. The standard-state reversible voltage (E^0) for the electrolytic cell is given by [70]:

$$E^0 = \frac{\Delta g^0_{rxn}}{nF} \tag{2.30}$$

Here Δg^0_{rxn} is the standard Gibbs free energy change of the electrolysis, n is the number of moles of electrons, while F is the Faraday constant. Using the Nernst equation, the reversible voltage in Eq. (2.30) is rewritten as:

$$E = E^0 - \frac{RT}{nF} ln \frac{\prod a_{prod}^{vi}}{\prod a_{react}^{vi}} \tag{2.31}$$

where R is the ideal gas constant and T is the operating temperature. a_{react} and a_{prod} are the activities of the reactant and product species of the reaction, respectively. While νi is the stoichiometric coefficient.

The reversible voltage is inadequate to operate a typical electrolysis cell, primarily due to ohmic, activation, and concentration losses. The actual voltage required to operate the cell, V_{cell}, is given by:

$$V_{cell} = E + \eta_{act} + \eta_{ohm} + \eta_{conc} \tag{2.32}$$

where $\eta_{act},$ $\eta_{ohm},$ and η_{conc} are denoted as the activation, ohmic, and concentration overpotentials, respectively. Since the reaction occurring in the DEFT electrolyzer is a symmetric reaction, from the Butler-Volmer kinetics equation, the activation overpotential can be obtained for both cathode and anode electrodes, given by [70]:

$$\eta_{act}^{CA,AN} = \frac{RT}{2\alpha_{CA,AN}F} ln\left(\frac{j}{j_{o;\,AN,AN}}\right) \tag{2.33}$$

where $\eta_{act}^{CA,AN}$ is the activation overpotential for both cathode and anode electrodes of the DEFT cell, and $\alpha_{CA,AN}$ is the transfer coefficient. The transfer coefficient conveys how variations in the electrical potential at the reaction interface affect the dimensions of the activation barriers for both forward and reverse reactions. j is the cell current density, while $j_{o;\,AN,AN}$ is the exchange current density for both electrodes. The size of the exchange current density may be obtained empirically and experimentally.

The ohmic overpotential, η_{ohm}, in Eq. (2.32) may be obtained as the product of the area-normalized resistance, i.e., the area-specific resistance and the current density:

$$\eta_{ohm} = j\beta_R \tag{2.34}$$

where β_R is the area-specific resistance of the DEFT electrolysis cell. The sum of the resistances in the electrolyzer is given by [71]:

$$\beta_{R,tot} = \beta_{R,Ex}^- + \beta_{R,CA} + \beta_{R,bub,H_2} + \beta_{ions} + \beta_{R,bub,O_2}\ \beta_{R,AN} + \beta_{R,Ex}^+ \tag{2.35}$$

where $\beta_{R,tot}$ is the total electrical resistance in the cell. $\beta_{R,Ex}^-$ and $\beta_{R,Ex}^+$ are the external resistances to the cathode and anode electrodes, respectively. $\beta_{R,CA}$ and $\beta_{R,AN}$ are the resistances of the cathode and anode electrodes, respectively. β_{R,bub,H_2} and

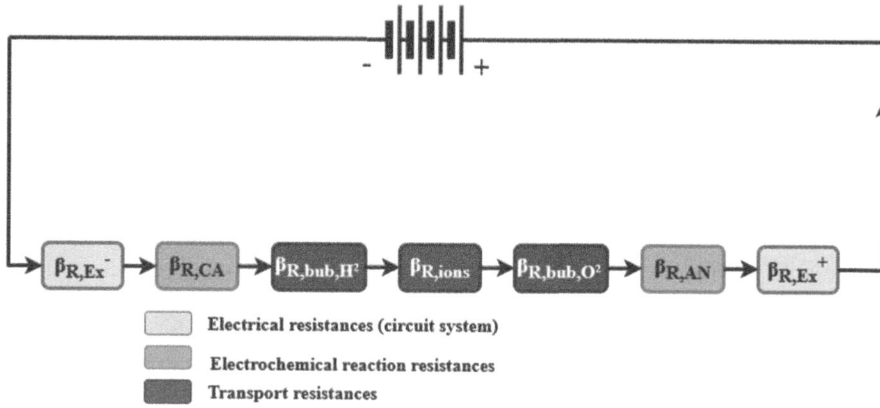

FIGURE 2.10 Resistances in a typical DEFT electrolysis cell.

β_{R,bub,O_2} are the resistances of H_2 and O_2 gas bubbles, respectively. β_{ions} is the resistance of the ions.

The electrical resistances in the cell consist of three sections: the electrical resistances due to the circuit components (such as connectors and wires), the electrochemical reaction resistances of the electrodes, and the transport resistances. The ionic and bubble resistances constitute the transport resistance of the electrolyte (Figure 2.10).

From Eq. (2.34), the ohmic overpotential of the electrolytic cell is then written as:

$$\eta_{ohm} = j\left(\beta_{R,Ex}^{-} + \beta_{R,CA} + \beta_{R,bub,H_2} + \beta_{ions} + \beta_{R,bub,O_2} + \beta_{R,AN} + \beta_{R,Ex}^{+}\right) \quad (2.36)$$

The resistances in the external components of the cell circuit (i.e., $\beta_{R,Ex}^{-}$ and $\beta_{R,Ex}^{+}$) and the resistances in the cathode and anode electrodes are computed as:

$$\sum\left(\beta_{R,Ex}^{-} + \beta_{R,Ex}^{+} + \beta_{R,CA} + \beta_{R,AN}\right) = \sum \frac{L_i}{A_i k_i} \quad (2.37)$$

where L, k, and A are the length, specific conductivity, and cross-sectional area of the ith conductor.

The resistance due to ionic transfer, β_{ions} (Eq. (2.36)), is given by [71]:

$$\beta_{ions} = \varphi_0 K \sqrt{C} \quad (2.38)$$

where φ_0 is the mole conductivity extrapolated to infinite dilution, K is the Kohlrauch coefficient [72], and C is the electrolyte concentration in $m^2/(\Omega\ mol)$. The H_2 and O_2 gas bubble resistances (β_{R,bub,H_2} and β_{R,bub,O_2}) given in Eq. (2.36) are obtained from the relation expressed as [73]:

$$\beta_{R,bub} = \beta_{R,bub_0}\left(1-\varnothing\right)^{-\frac{3}{2}} \quad (2.39)$$

where β_{R,bub_0} is the bubble-free specific resistivity of the electrolyte and $\varnothing$ is the percentage of the bubble on the surface of the electrodes.

The concentration overpotential, η_{conc} (Eq. (2.32)), is expressed as [70]:

$$\eta_{conc} = \left(\frac{RT}{nF}\right)\left(1 + \frac{1}{\alpha}\right)\ln\left(\frac{j_L}{j_L - j}\right) \tag{2.40}$$

where j_L is the limiting current density.

In the purity determination of gas generated, using Faraday's law of electrolysis, the generated molar flow rate of the dissolved gases in the electrolyte, $\dot{n}_i$, of the ith gas species is expressed as:

$$\dot{n}_i = \frac{\eta^F I_{cell}}{\Lambda F} = \frac{p_i \dot{V}}{RT} \tag{2.41}$$

where η^F is the Faraday efficiency, I_{cell} is the electric current of the cell, and Λ is the number of electrons transferred per mole of the gas species. $\dot{V}$ is the total gas volumetric flow rate. With this, the purity of the generated gas species is evaluated as [74]:

$$\text{Purity}_i = \frac{\dot{n}_{i,gas}^{out}}{\sum_{i=1}^{x} \dot{n}_{i,gas}^{out}} = \frac{p_i}{p_{ov}} \tag{2.42}$$

$\dot{n}_{i,gas}^{out}$ is the molar flow rate of the dissolved i^{th} gas generated. p_i and p_{ov} are the partial pressure of the ith gas and the overall pressure of the gases, respectively. The p_i is evaluated by Henry's law, given by [74]:

$$p_i = H_i \Psi_{lte,i} \tag{2.43}$$

where H_i is the Henry coefficient of the ith gas and $\Psi_{lte,\,i}$ is the molar concentration of the ith gas. The latter is computed from the Setchenov relation:

$$\Psi_{lte,\,i} = \frac{\tau_{pw,i}}{10^{\gamma_i \omega}} \tag{2.44}$$

$\tau_{pw,i}$ is the concentration of the ith gas in pure water, γ_i is the Setchenov constant for the ith gas, while ω is the weight percentage of the primary electrolyte, such as salts. Substituting Eq. (2.44) into Eq. (2.43), we have:

$$p_i = H_i \frac{\tau_{pw,i}}{10^{\gamma_i \omega}} \tag{2.45}$$

Applying Eqs. (2.40) to (2.45) for the purity of the gas generated in the H_2 channel using Figure 2.9, the purities of the gases from Eq. (2.42) are estimated as:

$$\text{Purity}_{H_2,H_2-ch} = \frac{\dot{n}_{H_2,g,H_2-ch}^{out}}{\dot{n}_{H_2,g,H_2-ch}^{out} + \dot{n}_{O_2,g,H_2-ch}^{out}} = \frac{p_{H_2,H_2-ch}}{p_{ov}} \tag{2.46a}$$

$$\text{Purity}_{O_2,H_2-ch} = \frac{\dot{n}^{out}_{O_2,g,H_2-ch}}{\dot{n}^{out}_{H_2,g,H_2-ch} + \dot{n}^{out}_{O_2,g,H_2-ch}} = \frac{p_{O_2,H_2-ch}}{p_{ov}} \tag{2.46b}$$

where $\dot{n}^{out}_{H_2,g,H_2-ch}$ and $\dot{n}^{out}_{O_2,g,H_2-ch}$ are the output molar flow rates of the H_2 and O_2, respectively. p_{H_2,H_2-ch} and p_{O_2,H_2-ch} are the partial pressures of hydrogen and oxygen gases, respectively.

From the general molar balance equation:

$$\text{Generation} + \text{Input} = \text{Output} \tag{2.47a}$$

$$\dot{n}_i + \dot{n}_{i,in} = \dot{n}_{i,out} \tag{2.47b}$$

With Eq. (2.47b), the molar balance for the H_2 channel is given by:

$$\text{For } H_2\text{: } \dot{n}_{H_2} + \frac{1}{2}\dot{n}^{in}_{H_2,l} = \dot{n}^{out}_{H_2,g,H-ch} + \dot{n}^{out}_{H_2,l,H-ch} \tag{2.48}$$

$$\text{For } O_2\text{: } \frac{1}{2}\dot{n}^{in}_{O_2,l} = \dot{n}^{out}_{O_2,g,H-ch} + \dot{n}^{out}_{O_2,l,H-ch} \tag{2.49}$$

where $\dot{n}_{H_2}$ is the molar flow rate of H_2 generation in the cell. $\dot{n}^{in}_{H_2,l}$ and $\dot{n}^{in}_{O_2,l}$ are the molar flow rates of the dissolved H_2 and O_2 entering the cell, respectively. A half fraction (1/2) is assigned to $\dot{n}^{in}_{H_2,l}$ and $\dot{n}^{in}_{O_2,l}$ due to the electrolyte being divided into two halves, one to each channel. $\dot{n}^{out}_{H_2,g,H-ch}$ and $\dot{n}^{out}_{O_2,g,H-ch}$ are the molar flow rates of H_2 and O_2 gases leaving the cell, respectively. $\dot{n}^{out}_{H_2,l,H-ch}$ and $\dot{n}^{out}_{O_2,l,H-ch}$ are the molar flow rates of the dissolved H_2 and O_2 leaving the cell, respectively. $\dot{n}^{out}_{O_2,g,H-ch}$ is obtained from Eq. (2.41), i.e.,

$$\dot{n}^{out}_{O_2,g,H-ch} = \frac{p_{O_2,H_2-ch}\dot{V}_{g,H_2-ch}}{RT} \tag{2.50}$$

while $\dot{n}^{out}_{O_2,l,H-ch}$ is given by:

$$\dot{n}^{out}_{O_2,l,H-ch} = \Psi_{lte,O_2,H_2-ch}\frac{1}{2}\dot{n}_{lte} \tag{2.51}$$

$\dot{V}_{g,H_2-ch}$ is the volumetric flow rate of the gas at the H_2 channel, Ψ_{lte,O_2,H_2-ch} is the molar fraction of the dissolved O_2 in the electrolyte at the H_2 channel, and $\dot{n}_{lte}$ is the molar flow rate of the electrolyte. When Eqs. (2.50) and (2.51) are substituted in Eq. (2.49) along with Henry's law equation (Eq. (2.43)), $\Psi_{lte,\,O_2,H_2-ch}$ is formulated as:

$$\Psi_{lte,O_2,H_2-ch} = \left[\frac{\frac{1}{2}\dot{n}^{in}_{O_2,l}}{\frac{1}{2}\dot{n}_{lte} + \text{H}_{O_2}\left(\frac{\dot{V}_{g,H_2-ch}}{RT}\right)}\right]\frac{1}{10^{\gamma_i\omega}} \tag{2.52}$$

Substituting Eq. (2.52) into Eq. (2.43), we will have:

$$p_{O_2,H_2-ch} = H_{O_2}\left[\frac{\frac{1}{2}\dot{n}_{O_2,l}^{in}}{\frac{1}{2}\dot{n}_{lte} + \mathrm{H}_{O_2}\left(\frac{\dot{V}_{g,H_2-ch}}{RT}\right)}\right]\frac{1}{10^{\gamma_i\omega}} \tag{2.53}$$

Since p_{O_2,H_2-ch} is obtained from Eq. (2.53), p_{H_2,H_2-ch} (Eq. (2.46a)) can be obtained from Dalton's law of partial pressure, i.e.,

$$p_{H_2,H_2-ch} = p_{ov} - p_{O_2,H_2-ch} \tag{2.54}$$

The Henry's coefficient, H_{O_2}, in Eqs. (2.52) and (2.53) and for H_{H_2} are given by [75]:

$$\ln\left(\frac{H_i}{p_{vap}}\right) = \frac{A_{1,i}}{T_R} + \frac{B_{1,i}\left(1 - T_R\right)^{0.355}}{T_R} + C_{1,i}T_R^{-0.41}e^{1-T_R} \tag{2.55}$$

where p_{vap} is the vapor pressure of water, obtained as [76]:

$$\ln\left(\frac{P_{vap}}{\mathrm{kPa}}\right) = A_2 - \frac{B_2}{\frac{T - 273.15}{K} + C_2} \tag{2.56}$$

A_1, B_1, C_1, A_2, B_2 and C_2 are constants obtained empirically as given in Table A2.1. T_R is obtained as the ratio of temperature and critical temperature of the species ($T_R = T / T_C$).

The efficiency of the DEFT electrolyzer, η_{DEFT}, is computed as:

$$\eta_{DEFT} = \frac{\dot{n}_{H_2,g}^{out} M_{H_2}\epsilon_{H_2}}{P_{sys} + P_{cp}} \tag{2.57}$$

M_{H_2} is the molar mass of H_2, ϵ_{H_2} is the higher heating value of H_2, P_{sys} is the electrical power required to operate the system, and P_{cp} is the chemical pump power required for circulating the electrolyte in the system. The electrical power requirements for the system and chemical pump are, respectively, computed as:

$$P_{sys} = V_{cell}I_{cell} \tag{2.58}$$

$$P_{cp} = \frac{\dot{m}_{lte}H_{h,ov}g}{\eta_{h-cp}} \tag{2.59}$$

where V_{cell} is the voltage of the cell, $\dot{m}_{lte}$ is the mass flow rate of the electrolyte, g is the acceleration due to gravity, η_{h-cp} is the hydraulic efficiency of the chemical pump. $H_{h,ov}$ is the overall head loss of the hydraulic circuit, computed as [74]:

$$H_{h,ov} = H_{h,m} + H_{h,n} + H_{h,d} \tag{2.60}$$

where $H_{h,m}$, $H_{h,n}$, $H_{h,d}$ are the main, minor, and device losses, respectively. $H_{h,m}$ is computed from the Darcy-Weisbach equation, given by [78]:

$$H_{h,m} = \lambda_{h,m} \frac{L_{pipe}}{D_{pipe}} \frac{s^2}{2g} \tag{2.61}$$

L_{pipe} is the length of the pipes, s is the electrolyte speed in the pipes, D_{pipe} is the diameter of the pipes, and $\lambda_{h,m}$ is the loss factor. The loss factor is computed based on the fluid flow. For laminar flow $\lambda_{h,m} = 64 / Re$ and for turbulent, the Colebrook equation is adopted, given by [79]:

$$\frac{1}{\sqrt{\lambda_{h,m}}} = -2 \log\left(\frac{\varepsilon / D_{pipe}}{3.7} + \frac{2.51}{Re\sqrt{\lambda_{h,m}}}\right) \tag{2.62}$$

where ε is the pipe roughness. $\lambda_{h,m}$ is solved iteratively. The minor head loss, $H_{h,n}$, in Eq. (2.60) is given by:

$$H_{h,n} = Z_n N_n \frac{s^2}{2g} \tag{2.63}$$

where Z_n is the circuit discontinuous, or the curve loss factor, and N_n is the count of the minor losses. The device loss, $H_{h,d}$, in Eq. (2.60) is given by:

$$H_{h,d} = Z_n \frac{s^2}{2g} \tag{2.64}$$

Hydrogen diffusion is a critical process in electrolysis. It involves the movement of H_2 atoms or molecules through metals (such as the porous electrodes) or electrolytes. During HER at the cathode electrode, H_2 gas is generated and passes through the porous electrode. The H_2 produced at the active sites diffuses through the porous electrodes to the H_2 channel. Additionally, the dissolved H_2 diffuses through the electrolyte toward the electrodes.

Fick's laws of diffusion govern the diffusion of H_2. For a steady-state diffusion,

$$J_{H_2} = -D_{H_2} \frac{\Delta C_{H_2}}{E_{gap}} \tag{4.65}$$

where J_{H_2} is the diffusion flux of H_2, D_{H_2} is the diffusion coefficient of H_2 in the electrolyte (for KOH solution, see Ref. [80]), ΔC_{H_2} is the H_2 concentration difference in the cell between the two channels, while E_{gap} is the distance between the two electrodes.

2.5 MATERIAL AND STRUCTURAL CONSIDERATIONS IN DEFT ELECTROLYZERS

The selection of electrode materials and the design of their structure have a profound impact on the efficiency and lifetime of DEFT systems [68]. These elements determine the system's electrical conductivity, catalytic effectiveness, mechanical strength, and its ability to resist corrosion, particularly in challenging alkaline or acidic electrolysis environments.

In selecting materials for DEFT electrodes, they must possess characteristics such as excellent electrical conductivity to facilitate effective charge transfer, resistance to corrosion in various electrolyte environments, sufficient mechanical strength to withstand stresses caused by fluid flow, and compatibility of the surface with electrocatalyst coatings [81]. Common base materials comprise nickel-based alloys, which are preferred because of their benefits, such as abundant presence in the earth, excellent resistance to corrosion, high levels of electrical conductivity, their stability, and moderate catalytic performance in alkaline environments [82, 83].

Stainless steel and titanium offer mechanical durability and resistance to corrosion, making them suitable as substrates [84], while carbon-based materials (such as carbon foam, carbon paper, and graphene-coated felts) provide a lightweight, conductive, and porous support, though they often need catalyst enhancements to improve HER activity [85]. These materials are frequently treated with active catalytic layers, including platinum, ruthenium, molybdenum disulfide, or transition metal phosphides, to enhance the efficiency of the HER.

DEFT electrodes necessitate a 3D porous framework to support electrolyte flow and distribution. This structure must ensure an even flow distribution across the surface of the electrode, provide a significant electrochemical surface area to optimize active sites, and sustain a low pressure drop for effective pumping [86]. Various techniques exist for developing these structures, including sintering metal powders into foam-like formations, utilizing 3D printing or additive manufacturing for custom divergent shapes, and employing electrospinning and templating approaches to develop nanostructured surfaces. These porous frameworks must be resilient against fluid shear stress and retain their shape over prolonged use.

Nanoscale electrocatalysts applied as surface coatings are one of the prevalent methods employed to enhance the performance of the HER. Techniques like electrodeposition, hydrothermal synthesis, or atomic layer deposition (ALD) can be utilized to apply these nanocoatings [87]. To improve catalytic performance and fluid contact, surfaces such as NiCoP, Pt/C, or RuO_2 are commonly covered with nanocatalysts for optimizing the release of gas bubbles [88].

As DEFT systems operate with flow and pressure gradients, the electrodes must possess mechanical strength to withstand deformation, chemical stability for long-term electrolysis, and scalability for modular integration in industrial hydrogen production systems. Recent advancements in polymer composite electrodes, laser-patterned metal foils, and graphene-based frameworks are designed to enhance performance and cost efficiency for commercial DEFT applications.

2.6 LIMITATIONS AND OPPORTUNITIES IN DEFT TECHNOLOGY

DEFT systems constitute a new category of electrolyzer designs that provide considerable performance improvement for hydrogen generation applications, mainly because of their effective mass transport, improved gas separation, and potential for scalability. Despite these benefits, DEFT systems encounter various technical, economic, and material challenges that need to be resolved for broader commercial implementation. At the same time, there are many opportunities to address these challenges through innovations in design, materials, and system integration.

2.6.1 Limitations in DEFT Technology

Porous materials like nickel foam, stainless steel mesh, and carbon-based structures are commonly utilized, but they are susceptible to corrosion, fouling, and mechanical deterioration over time, particularly under dynamic multiphase flow conditions [89]. Furthermore, catalyst coatings can deteriorate due to forces from bubble detachment, fluctuations in pH, or temperature variations [90].

The diverging flow design improves transport but also leads to uneven current distribution, variations in pressure drops, and flow maldistribution on a larger scale. The presence of bubble blockages in micro-porous areas and localized dry-out zones can significantly impact the efficiency and safety of the system [91, 92].

Effective management of flow in divergent geometries is not yet fully optimized, and DEFT electrodes require meticulous manufacturing approaches to control porosity, divergence angles, and catalyst arrangement. This usually results in higher production costs and hinders scalability, mainly when advanced 3D structuring or functional coatings are utilized [93]. DEFT prototypes are still at the laboratory scale and not yet widely tested under industrial conditions, compared to existing PEM or alkaline electrolyzers [94].

Additionally, data on durability, life cycle performance, and economic metrics are scarce, which limits stakeholder confidence in adopting the technology. Long-term field tests are still required, especially under fluctuating renewable energy input scenarios.

2.6.2 Opportunities for Advancement in DEFT Technology

Despite these constraints, DEFT systems present considerable unexplored potential for advanced H_2 generation technologies, particularly given the growing focus on decentralized and renewable energy systems [6]. DEFT setups are especially advantageous for intermittent, low-pressure electrolyzers powered by solar or wind energy [94]. Their low cell resistance and dynamic flow characteristics can enhance the response to starting and stopping, as well as overall energy efficiency. Integrating DEFT systems with photovoltaic arrays or hybrid wind-PV microgrids can facilitate local H_2 generation for transportation or as a backup energy source.

There is potential to utilize 3D printing, laser ablation, or templated synthesis to create multi-scale porous electrodes featuring customized flow paths and catalytic gradients [95]. Such designs can significantly improve electrochemical surface area while optimizing the flow distribution. Functionally graded porous electrodes can facilitate the uniform distribution of gas and liquid, thereby increasing operational lifespan [96].

Future DEFT systems might benefit from coatings that self-repair, layers that resist fouling, or materials that change in response to stimuli, allowing adjustments in wettability or functionality based on operating conditions. These advanced surfaces may help reduce degradation and improve performance stability. Coatings that respond to local pH levels or gas evolution rates can reduce fouling and prevent catalyst deactivation, thereby enhancing catalyst performance.

Advanced simulations, machine learning algorithms, and digital twin technologies can facilitate real-time optimization of DEFT systems. By integrating Computational Fluid Dynamics (CFD) with electrochemical modeling, it is possible to tailor cell design according to particular operating conditions [97].

Given their compact and stackable design, DEFT systems are ideally suited for on-site H_2 production in urban, rural, or off-grid settings. Their modularity allows for adaptable deployment in specific markets such as home hydrogen generators, refueling stations, and backup solutions for remote infrastructures. This supports global initiatives aimed at democratizing access to clean energy and reducing the costs associated with H_2 logistics.

Although DEFT systems encounter both technical and market issues, the foundational ideas of improved flow-through design, modular integration, and advanced electrode designs make a compelling argument for future advancements. It is crucial in tackling issues concerning longevity of material, flow consistency, and large-scale production to shift DEFT from an encouraging concept to a tangible commercial product.

REFERENCES

1. P. C. Hallenbeck, 'Fermentative hydrogen production: Principles, progress, and prognosis', *International Journal of Hydrogen Energy*, vol. 34, no. 17, pp. 7379–7389, 2009. doi: 10.1016/j.ijhydene.2008.12.080.
2. J. A. Turner, 'Sustainable hydrogen production', *Science*, vol. 972, no. 4, pp. 972–974, 2004. doi: 10.1126/science.1103197.
3. M. Balat, 'Potential importance of hydrogen as a future solution to environmental and transportation problems', *International Journal of Hydrogen Energy*, vol. 33, pp. 4013–4029, 2008.
4. P. Nikolaidis, and A. Poullikkas, 'A comparative overview of hydrogen production processes', *Renewable and Sustainable Energy Reviews*, vol. 67, pp. 597–611, 2017.
5. N. Armaroli, and V. Balzani, 'The hydrogen issue', *ChemSusChem*, vol. 4, pp. 21–36, 2011.
6. IEA, 'Global Hydrogen Review 2024', 2024. [Online]. Available: https://www.iea.org/reports/global-hydrogen-review-2024
7. A. Midilli, H. Kucuk, M. E. Topal, U. Akbulut, and I. Dincer, 'A comprehensive review on hydrogen production from coal gasification: Challenges and Opportunities', *International Journal of Hydrogen Energy*, vol. 46, no. 50, pp. 25385–25412, 2021. doi: 10.1016/j.ijhydene.2021.05.088.

8. G. Voitic, B. Pichler, A. Basile, A. Iulianelli, K. Malli, S. Bock, V. Hacke, 'Hydrogen production', in *Fuel Cells and Hydrogen*, Elsevier, 2018, pp. 215–241. doi: 10.1016/B978-0-12-811459-9.00010-4.
9. I. Dincer, M. A. Rosen, and M. Al-Zareer, 'Chemical energy production', in *Comprehensive Energy Systems*, vol. 3, 2018. [Online]. Available: 10.1016/B978-0-12-809597-3.00326-6
10. M. M. Sarafraz, and F. C. Christo, 'Thermodynamic assessment and techno-economic analysis of a liquid indium-based chemical looping system for biomass gasification', *Energy Conversion and Management*, vol. 225, p. 113428, 2020. doi: 10.1016/j.enconman.2020.113428.
11. E. Delikonstantis *et al.*, 'Biomass gasification in microwave plasma: An experimental feasibility study with a side stream from a fermentation reactor', *Chemical Engineering and Processing – Process Intensification*, vol. 141, p. 107538, 2019. doi: 10.1016/j.cep.2019.107538.
12. F. Z. Mansur, C. K. M. Faizal, M. U. Monir, N. A. Samad, S. M. Atnaw, and S. A. Sulaiman, 'Co-gasification between coal/sawdust and coal/wood pellet: A parametric study using response surface methodology', *International Journal of Hydrogen Energy*, vol. 45, no. 32, pp. 15963–76, 2020.
13. V. E. Messerle, A. B. Ustimenko, and O. A. Lavrichshev, 'Plasma coal conversion including mineral mass utilization', *Fuel*, vol. 203, pp. 877–83, 2017.
14. T. Sutardi, L. Wang, N. Karimi, and M. C. Paul, 'Utilization of H_2O and CO_2 in coal particle gasification with an impact of temperature and particle size', *Energy Fuels*, vol. 34, no. 10, pp. 12841–12852, 2020. doi: 10.1021/acs.energyfuels.0c02280.
15. L. Emami Taba, M. F. Irfan, W. A. M. Wan Daud, and M. H. Chakrabarti, 'The effect of temperature on various parameters in coal, biomass and CO-gasification: A review', *Renewable and Sustainable Energy Reviews*, vol. 16, no. 8, pp. 5584–5596, 2012. doi: 10.1016/j.rser.2012.06.015.
16. D. Strmcnik, P. P. Lopes, B. Genorio, V. R. Stamenkovic, and N. M. Markovic, 'Design principles for hydrogen evolution reaction catalyst materials', *Nano Energy*, vol. 29, pp. 29–36, 2016. doi: 10.1016/j.nanoen.2016.04.017.
17. B. V. Babu, 'Biomass pyrolysis: A state-of-the-art review', *Biofuels, Bioproducts and Biorefining*, vol. 2, no. 5, pp. 393–414, 2008. doi: 10.1002/bbb.92.
18. O. Azeta, A. O. Ayeni, O. Agboola, and F. B. Elehinafe, 'A review on the sustainable energy generation from the pyrolysis of coconut biomass', *Scientific African*, vol. 13, p. e00909, 2021. doi: 10.1016/j.sciaf.2021.e00909.
19. A. Keçebaş, M. Kayfeci, and M. Bayat, 'Electrochemical hydrogen generation' [19], in *Solar Hydrogen Production*, Elsevier, 2019, pp. 299–317. doi: 10.1016/B978-0-12-814853-2.00009-6.
20. İ. A. Şengil, and M. Özacar, 'The decolorization of C.I. Reactive Black 5 in aqueous solution by electrocoagulation using sacrificial iron electrodes', *Journal of Hazardous Materials*, vol. 161, no. 2–3, pp. 1369–1376, 2009. doi: 10.1016/j.jhazmat.2008.04.100.
21. C. Acar, and I. Dincer, 'A review and evaluation of photoelectrode coating materials and methods for photoelectrochemical hydrogen production', *International Journal of Hydrogen Energy*, vol. 41, no. 19, pp. 7950–7959, 2016. doi: 10.1016/j.ijhydene.2015.11.160.
22. A. Ike, T. Murakawa, H. Kawaguchi, K. Hirata, and K. Miyamoto, 'Photoproduction of hydrogen from raw starch using a halophilic bacterial community', *Journal of Bioscience and Bioengineering*, vol. 88, no. 1, pp. 72–77, 1999. doi: 10.1016/S1389-1723(99)80179-0.
23. D.-H. Kim, and M.-S. Kim, 'Development of a novel three-stage fermentation system converting food waste to hydrogen and methane', *Bioresource Technology*, vol. 127, pp. 267–274, 2013. doi: 10.1016/j.biortech.2012.09.088.

24. M. J. Barbosa, J. M. S. Rocha, J. Tramper, and R. H. Wijffels, 'Acetate as a carbon source for hydrogen production by photosynthetic bacteria', *Journal of Biotechnology*, vol. 85, no. 1, pp. 25–33, 2001. doi: 10.1016/S0168-1656(00)00368-0.
25. Y. Tao, 'Characteristics of a new photosynthetic bacterial strain for hydrogen production and its application in wastewater treatment', *International Journal of Hydrogen Energy*, vol. 33, no. 3, pp. 963–973, 2008. doi: 10.1016/j.ijhydene.2007.11.021.
26. X. Tian *et al.*, 'Characteristics of a biofilm photobioreactor as applied to photo-hydrogen production', *Bioresource Technology*, vol. 101, no. 3, pp. 977–983, 2010. doi: 10.1016/j.biortech.2009.09.007.
27. E. Boran, E. Özgür, J. Van Der Burg, M. Yücel, U. Gündüz, and I. Eroglu, 'Biological hydrogen production by Rhodobacter capsulatus in solar tubular photo bioreactor', *Journal of Cleaner Production*, vol. 18, pp. S29–S35, 2010. doi: 10.1016/j.jclepro.2010.03.018.
28. K. Seifert, M. Waligorska, and M. Laniecki, 'Brewery wastewaters in photobiological hydrogen generation in presence of Rhodobacter sphaeroides O.U. 001', *International Journal of Hydrogen Energy*, vol. 35, no. 9, pp. 4085–4091, 2010. doi: 10.1016/j.ijhydene.2010.01.126.
29. D.-H. Kim, J. Cha, S. Kang, and M.-S. Kim, 'Continuous photo-fermentative hydrogen production from lactate and lactate-rich acidified food waste', *International Journal of Hydrogen Energy*, vol. 38, no. 14, pp. 6161–6166, 2013. doi: 10.1016/j.ijhydene.2012.12.072.
30. H. Zhu, H. Fang, T. Zhang, and L. Beaudette, 'Effect of ferrous ion on photo heterotrophic hydrogen production by Rhodobacter sphaeroides', *International Journal of Hydrogen Energy*, vol. 32, no. 17, pp. 4112–4118, 2007. doi: 10.1016/j.ijhydene.2007.06.010.
31. B.-F. Liu, N.-Q. Ren, J. Ding, G.-J. Xie, and G.-L. Cao, 'Enhanced photo-H_2 production of R. faecalis RLD-53 by separation of CO_2 from reaction system', *Bioresource Technology*, vol. 100, no. 3, pp. 1501–1504, 2009. doi: 10.1016/j.biortech.2008.08.027.
32. G-J. Xie, B-F. Liu, H-Q. Wen, Q. Li, C-Y. Yang, W-L. Han, J. Nan, N-Q. Ren 'Bioflocculation of photo-fermentative bacteria induced by calcium ion for enhancing hydrogen production' *International Journal of Hydrogen Energy*, vol. 38, no. 19, pp. 7780–7788, 2013. doi. 10.1016/j.ijhydene.2013.04.099.
33. B. Uyar, I. Eroglu, M. Yücel, and U. Gündüz, 'Photofermentative hydrogen production from volatile fatty acids present in dark fermentation effluents', *International Journal of Hydrogen Energy*, vol. 34, no. 10, pp. 4517–4523, 2009. doi: 10.1016/j.ijhydene.2008.07.057.
34. K. Anam, M. S. Habibi, T. U. Harwati, and D. Susilaningsih, 'Photofermentative hydrogen production using Rhodobium marinum from bagasse and soy sauce wastewater', *International Journal of Hydrogen Energy*, vol. 37, no. 20, pp. 15436–15442, 2012. doi: 10.1016/j.ijhydene.2012.06.076.
35. C.-M. Lee, G.-J. Hung, and C.-F. Yang, 'Hydrogen production by Rhodopseudomonas palustris WP 3-5 in a serial photobioreactor fed with hydrogen fermentation effluent', *Bioresource Technology*, vol. 102, no. 18, pp. 8350–8356, 2011. doi: 10.1016/j.biortech.2011.04.072.
36. G.-J. Xie *et al.*, 'Feasibility studies on continuous hydrogen production using photo-fermentative sequencing batch reactor', *International Journal of Hydrogen Energy*, vol. 37, no. 18, pp. 13689–13695, 2012. doi: 10.1016/j.ijhydene.2012.02.107.
37. N.-Q. Ren *et al.*, 'Strategy for enhancing photo-hydrogen production yield by repeated fed-batch cultures', *International Journal of Hydrogen Energy*, vol. 34, no. 18, pp. 7579–7584, 2009. doi: 10.1016/j.ijhydene.2009.07.030.

38. J. K. Kim, B.-H. Um, and T. H. Kim, 'Bioethanol production from micro-algae, Schizocytrium sp., using hydrothermal treatment and biological conversion', *Korean Journal of Chemical Engineering*, vol. 29, no. 2, pp. 209–214, 2012. doi: 10.1007/s11814-011-0169-3.
39. C. Ma, X. Wang, L. Guo, X. Wu, and H. Yang, 'Enhanced photo-fermentative hydrogen production by Rhodobacter capsulatus with pigment content manipulation', *Bioresource Technology*, vol. 118, pp. 490–495, 2012. doi: 10.1016/j.biortech.2012.04.105.
40. N. Srivastava *et al.*, 'Nanoengineered cellulosic biohydrogen production via dark fermentation: A novel approach', *Biotechnology Advances*, vol. 37, no. 6, p. 107384, 2019. doi: 10.1016/j.biotechadv.2019.04.006.
41. J. S. Silva, J. S. Mendes, J. A. C. Correia, M. V. P. Rocha, and L. Micoli, 'Cashew apple bagasse as new feedstock for the hydrogen production using dark fermentation process', *Journal of Biotechnology*, vol. 286, pp. 71–78, 2018. doi: 10.1016/j.jbiotec.2018.09.004.
42. P. Cavaliere, *Water Electrolysis for Hydrogen Production*. Springer International Publishing, 2023. doi: 10.1007/978-3-031-37780-8.
43. M.-J. Zhao *et al.*, 'Automatic operation of decoupled water electrolysis based on bipolar electrode', *Renewable Energy*, vol. 203, pp. 583–591, 2023. doi: 10.1016/j.renene.2022.12.083.
44. P. Millet, 'Fundamentals of water electrolysis', in *Hydrogen Production*, Ed. A. Godula-Jopek, Wiley, 2015, pp. 33–62. doi: 10.1002/9783527676507.ch2.
45. M. E. Şahin, 'An overview of different water electrolyzer types for hydrogen production', *Energies*, vol. 17, no. 19, p. 4944, 2024. doi: 10.3390/en17194944.
46. M. Bodner, A. Hofer, and V. Hacker, 'H_2 generation from alkaline electrolyzer', *WIREs Energy and Environment*, vol. 4, no. 4, pp. 365–381, 2015. doi: 10.1002/wene.150.
47. A. Badgett, M. Ruth, and B. Pivovar, 'Economic considerations for hydrogen production with a focus on polymer electrolyte membrane electrolysis', in *Electrochemical Power Sources: Fundamentals, Systems, and Applications*, Elsevier, 2022, pp. 327–364. doi: 10.1016/B978-0-12-819424-9.00005-7.
48. H. Gorgun, 'Dynamic modelling of a proton exchange membrane (PEM) electrolyzer', *International Journal of Hydrogen Energy*, vol. 31, no. 1, pp. 29–38, 2006. doi: 10.1016/j.ijhydene.2005.04.001.
49. M. K. Cho *et al.*, 'Alkaline anion exchange membrane water electrolysis: Effects of electrolyte feed method and electrode binder content', *Journal of Power Sources*, vol. 382, pp. 22–29, 2018. doi: 10.1016/j.jpowsour.2018.02.025.
50. C. Li, and J.-B. Baek, 'The promise of hydrogen production from alkaline anion exchange membrane electrolyzers', *Nano Energy*, vol. 87, p. 106162, 2021. doi: 10.1016/j.nanoen.2021.106162.
51. Y. Leng, G. Chen, A. J. Mendoza, T. B. Tighe, M. A. Hickner, and C.-Y. Wang, 'Solid-state water electrolysis with an alkaline membrane', *Journal of American Chemical Society*, vol. 134, no. 22, pp. 9054–9057, 2012. doi: 10.1021/ja302439z.
52. M. Rashid, M. K. A. Mesfer, H. Naseem, and M. Danish, 'Hydrogen production by water electrolysis: A review of alkaline water electrolysis', *PEM Water Electrolysis and High Temperature Water Electrolysis*, vol. 4, no. 3, 2015.
53. International Renewable Energy Agency (IRENA). *Green hydrogen cost reduction: Scaling up electrolysers to meet the 1.5° C climate goal*. Irena, 2020.
54. J. D. Holladay, J. Hu, D. L. King, and Y. Wang, 'An overview of hydrogen production technologies', *Catalysis Today*, vol. 139, no. 4, pp. 244–260, 2009. doi: 10.1016/j.cattod.2008.08.039.
55. A. Lasia, 'Mechanism and kinetics of the hydrogen evolution reaction', *International Journal of Hydrogen Energy*, vol. 44, no. 36, pp. 19484–19518, 2019. doi: 10.1016/j.ijhydene.2019.05.183.

56. M. A. Khan *et al.*, 'Recent progresses in electrocatalysts for water electrolysis', *Electrochemical Energy Reviews*, vol. 1, no. 4, pp. 483–530, 2018. doi: 10.1007/s41918-018-0014-z.
57. H. G. Xu *et al.*, 'Rational design of hydrogen evolution reaction electrocatalysts for commercial alkaline water electrolysis', *Small Structures*, vol. 4, no. 8, p. 2200404, 2023. doi: 10.1002/sstr.202200404.
58. D. Jang, J. Kim, D. Kim, W.-B. Han, and S. Kang, 'Techno-economic analysis and Monte Carlo simulation of green hydrogen production technology through various water electrolysis technologies', *Energy Conversion and Management*, vol. 258, p. 115499, 2022. doi: 10.1016/j.enconman.2022.115499.
59. N. M. Palakkool *et al.*, 'Conformally-coated Nickel-Carbon Nitride on Structured Pyrolytic Carbon Electrodes for Enhanced Hydrogen Evolution', May 10, 2025. doi: 10.48550/arXiv.2505.06627.
60. T. Chu, S. Park, and K. K. Fu, '3D printing-enabled advanced electrode architecture design', *Carbon Energy*, vol. 3, no. 3, pp. 424–439, 2021. doi: 10.1002/cey2.114.
61. P. Hadikhani, S. M. H. Hashemi, S. A. Schenk, and D. Psaltis, 'A membrane-less electrolyzer with porous walls for high throughput and pure hydrogen production', *Sustainable Energy & Fuels*, vol. 5, no. 9, pp. 2419–2432, 2021. doi: 10.1039/D1SE00255D.
62. A. Manzotti, M. J. Robson, and F. Ciucci, 'Recent developments in membraneless electrolysis', *Current Opinion in Green and Sustainable Chemistry*, vol. 40, p. 100765, 2023. doi: 10.1016/j.cogsc.2023.100765.
63. J. Newman, 'Scaling with Ohm's Law; wired vs. wireless photoelectrochemical cells', *Journal of the Electrochemical Society*, vol. 160, no. 3, pp. F309–F311, 2013. doi: 10.1149/2.020304jes.
64. M. D. Symes, and L. Cronin, 'Decoupling hydrogen and oxygen evolution during electrolytic water splitting using an electron-coupled-proton buffer', *Nature Chemistry*, vol. 5, no. 5, pp. 403–409, 2013. doi: 10.1038/nchem.1621.
65. J. T. Davis, J. Qi, X. Fan, J. C. Bui, and D. V. Esposito, 'Floating membraneless PV-electrolyzer based on buoyancy-driven product separation', *International Journal of Hydrogen Energy*, vol. 43, no. 3, pp. 1224–1238, 2018. doi: 10.1016/j.ijhydene.2017.11.086.
66. M. A. Modestino, D. Fernandez Rivas, S. M. H. Hashemi, J. G. E. Gardeniers, and D. Psaltis, 'The potential for microfluidics in electrochemical energy systems', *Energy & Environmental Science*, vol. 9, no. 11, pp. 3381–3391, 2016. doi: 10.1039/C6EE01884J.
67. S. M. H. Hashemi, M. A. Modestino, and D. Psaltis, 'A membrane-less electrolyzer for hydrogen production across the pH scale', *Energy & Environmental Science*, vol. 8, no. 7, pp. 2003–2009, 2015. doi: 10.1039/C5EE00083A.
68. M. I. Gillespie, F. Van Der Merwe, and R. J. Kriek, 'Performance evaluation of a membraneless divergent electrode-flow-through (DEFT) alkaline electrolyser based on optimisation of electrolytic flow and electrode gap', *Journal of Power Sources*, vol. 293, pp. 228–235, 2015. doi: 10.1016/j.jpowsour.2015.05.077.
69. Y. Chen, J. Chen, K. Bai, Z. Xiao, and F. Senqing, 'A flow-through electrode for hydrogen production from water splitting by mitigating bubble induced overpotential', *Journal of Power Sources*, vol. 561, p. 232733, 2023.
70. R. O'Hayre, S. Cha, W. Colella, and F. B. Prinz, *Fuel Cell Fundamentals*, 1st ed., Wiley, 2016. doi: 10.1002/9781119191766.
71. K. Zeng, and D. Zhang, 'Recent progress in alkaline water electrolysis for hydrogen production and applications', *Progress in Energy and Combustion Science*, vol. 36, no. 3, pp. 307–326, 2010. doi: 10.1016/j.pecs.2009.11.002.
72. P. H. Rieger, *Electrochemistry*, 1st ed., Prentice-Hall, 1987.

73. F. Hine, and K. Murakami, 'Bubble effects on the solution IR drop in a vertical electrolyzer under free and forced convection', *Journal of the Electrochemical Society*, vol. 127, no. 2, pp. 292–297, 1980. doi: 10.1149/1.2129658.
74. A. Manzotti, E. Quattrocchi, A. Curcio, S. C. T. Kwok, M. Santarelli, and F. Ciucci, 'Membraneless electrolyzers for the production of low-cost, high-purity green hydrogen: A techno-economic analysis', *Energy Conversion and Management*, vol. 254, p. 115156, 2022. doi: 10.1016/j.enconman.2021.115156.
75. R. Fernández-Prini, J. L. Alvarez, and A. H. Harvey, 'Henry's constants and vapor–liquid distribution constants for gaseous solutes in H_2O and D_2O at high temperatures', *Journal of Physical and Chemical Reference Data*, vol. 32, no. 2, 2013.
76. A. Roosta, and J. Hekayati, 'A simple generic model for estimating saturated vapor pressure', *Chemical Engineering Communications*, vol. 203, no. 8, pp. 1020–1028, 2016. doi: 10.1080/00986445.2015.1135428.
77. P. Haug, *Experimental and theoretical investigation of gas purity in alkaline water electrolysis systems*, PhD thesis, Clausthal University of Technology, Clausthal-Zellerfeld, Germany, 2019.
78. B. R. Munson, D. F. Young, T. H. Okiishi, and W. W. Huebsch, *Fundamentals of Fluid Mechanics*, 7th ed., Wiley, 2013.
79. C. F. Colebrook, 'Turbulent flow in pipes, with particular reference to the transition region between the smooth and rough pipe laws', *Journal of the Institution of Civil Engineers*, vol. 11, no. 4, pp. 133–156, 1939. doi: 10.1680/ijoti.1939.13150.
80. M. J. Tham, R. D. Walker, and K. E. Gubbins, 'Diffusion of oxygen and hydrogen in aqueous potassium hydroxide solutions', *Journal of Physical Chemistry*, vol. 74, no. 8, pp. 1747–1751, 1970. doi: 10.1021/j100703a015.
81. M. S. Ratsoma *et al.*, 'Application of nickel foam in electrochemical systems: A review', *Journal of Electronic Materials*, vol. 52, no. 4, pp. 2264–2291, 2023. doi: 10.1007/s11664-023-10244-w.
82. X. Liu *et al.*, 'Recent advances in the electrochemical applications of Ni-based metal organic frameworks (Ni-MOFs) and their derivatives', *Chemosphere*, vol. 307, p. 135729, 2022. doi: 10.1016/j.chemosphere.2022.135729.
83. C. Hu *et al.*, 'Recent advances in Ni-based electrocatalysts for hydrogen evolution reaction', *Energy Technology*, vol. 11, no. 1, p. 2201048, 2023. doi: 10.1002/ente.202201048.
84. F. O. Al-Qwairi, S. Shaheen Shah, A. H. Shabi, A. Khan, and M. A. Aziz, 'Stainless steel mesh in electrochemistry: Comprehensive applications and future prospects', *Chemistry: An Asian Journal*, vol. 19, no. 20, p. e202400314, 2024. doi: 10.1002/asia.202400314.
85. K. Sakaushi, and M. Antonietti, 'Carbon- and nitrogen-based organic frameworks', *Accounts of Chemical Research*, vol. 48, no. 6, pp. 1591–1600, 2015. doi: 10.1021/acs.accounts.5b00010.
86. R. Kumari, A. Sammi, Shubhangi, A. Srivastava, U. P. Azad, and P. Chandra, 'Emerging 3D nanomaterials as electrocatalysts for water splitting reactions', *International Journal of Hydrogen Energy*, vol. 74, pp. 214–231, 2024. doi: 10.1016/j.ijhydene.2024.06.014.
87. P. O. Oviroh, S. T. Oyinbo, S. Karimzadeh, P. E. Imoisili, T.-C. Jen, *Emerging Atomic Layer Deposition for Hydrogen Energy*, Springer, 2025.
88. D. Chen, Y. Zou, and S. Wang, 'Surface chemical-functionalization of ultrathin two-dimensional nanomaterials for electrocatalysis', *Materials Today Energy*, vol. 12, pp. 250–268, 2019. doi: 10.1016/j.mtener.2019.01.006.
89. X. Zou, and Y. Zhang, 'Noble metal-free hydrogen evolution catalysts for water splitting', *Chemical Society Reviews*, vol. 44, no. 15, pp. 5148–5180, 2015. doi: 10.1039/C4CS00448E.

90. K. Chen and S. P. Jiang, 'Review—materials degradation of solid oxide electrolysis cells', *Journal of the Electrochemical Society*, 163 (11), F3070-F3083 (2016). doi: 10.1149/2.0101611jes
91. B. T. Sangtam, and H. Park, 'Review on bubble dynamics in proton exchange membrane water electrolysis: Towards optimal green hydrogen yield', *Micromachines*, vol. 14, no. 12, p. 2234, 2023. doi: 10.3390/mi14122234.
92. X. Zhao, H. Ren, and L. Luo, 'Gas bubbles in electrochemical gas evolution reactions', *Langmuir*, vol. 35, no. 16, pp. 5392–5408, 2019. doi: 10.1021/acs.langmuir.9b00119.
93. A. M. Boyce *et al.*, 'Design of scalable, next-generation thick electrodes: Opportunities and challenges', *ACS Nano*, vol. 15, no. 12, pp. 18624–18632, 2021. doi: 10.1021/acsnano.1c09687.
94. Hydrox Holdings, 'DEFT™ (Divergent Electrode Flow Through) Electrolysis', 2023. [Online]. Available: https://hydroxholdings.co.za/wp-content/uploads/2020/04/DEFT-Technology-Showcase-Presentation_Website_RevB-1.pdf
95. C. Zhu, H. B. Gemeda, E. B. Duoss, and C. M. Spadaccini, 'Toward multiscale, multimaterial 3D printing', *Advanced Materials*, vol. 36, no. 34, p. 2314204, 2024. doi: 10.1002/adma.202314204.
96. M. J. Kermani *et al.*, 'Application of a foam-based functionally graded porous material flow-distributor to PEM fuel cells', *Energy*, vol. 254, p. 124230, 2022. doi: 10.1016/j.energy.2022.124230.
97. Z. Feng, Y. Luo, D. Li, J. Pan, R. Tan, and Y. Chen, 'Integrating digital twins and machine learning for advanced control in green hydrogen production', *CHAIN*, vol. 2, no. 1, pp. 1–14, 2025. doi: 10.23919/CHAIN.2025.000003.

APPENDIX

TABLE A2.1
Constants Obtained Empirically for the Computations

Species	Parameter	Value	Reference
Hydrogen	A_{1,H_2}	–4.73284	[75]
	B_{1,H_2}	6.08954	
	C_{1,H_2}	6.06066	
Oxygen	A_{1,O_2}	–9.44833	
	B_{1,O_2}	4.43822	
	C_{1,O_2}	11.42005	
	A_2	16.3872	[76]
	B_2	3885.7	
	C_2	230.17	
	γ_{O_2}	3.66	[77]

3 Hydrogen Storage Mechanisms, Transportation, and Safety

3.1 STORAGE SPECIFICATIONS AND ISSUES

An economy based on H_2 is frequently suggested as a solution to energy security challenges of costs, infrastructure, availability, environmental concerns, and efficiency [1]. This is due to the exceptional qualities exhibited by H_2, such as adaptability, gravimetric energy density, and potential to decarbonize production and transportation industries, as well as several other sectors [2]. Hydrogen does not exist naturally in the free molecular form; unlike conventional fossil fuels, it must be generated, stored, and transported before it can be used. One of the most crucial and technically demanding aspects of the H_2 supply chain is storage [3–5].

Hydrogen exhibits several unique characteristics that make its storage particularly challenging. Among all known elements, it has the lowest molecular weight (~2 g/mol) and the lowest boiling point (~253°C), making it difficult to store at ambient conditions [6]. Moreover, compared to liquid hydrocarbons, its volumetric energy density is significantly lower (e.g., gaseous H_2 at 700 bar has a volumetric energy density of ~6 MJ/l, but gasoline has about 32 MJ/l). These huge discrepancies in volumetric energy density require compression, liquefaction, or hydrogen binding to materials.

For the hydrogen economy to be feasible, efficient, economical, and safe, H_2 storage technologies must be developed. As shown in Figure 3.1, production, storage, safety, and use are the four interrelated and interdependent stages of the contemporary H_2 economy, which is proposed by Dawood *et al.* [7]. This section outlines the essential specifications required for H_2 storage systems and discusses the significant challenges that must be addressed.

Hydrogen storage systems are numerous for various types of applications. They are primarily classified into two categories: surface and subsurface systems, as these are illustrated in Figure 3.2. Based on H_2 occurrence, subsurface storage materials can be further categorized into naturally occurring porous structures and artificial caverns, and further groups. The surface storage solutions can be classified according to their technology for H_2 storage [2]. Several performance specifications are adopted for evaluating H_2 storage technologies. These measures assess a storage method's viability for a variety of applications, including bulk transportation, stationary power generation, fuel cell vehicles, and portable electronics [8].

DOI: 10.1201/9781003727194-3

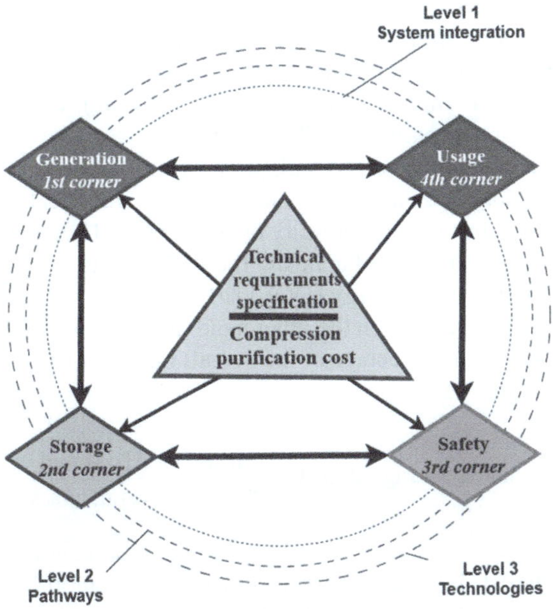

FIGURE 3.1 Interrelated and interdependent stages of the contemporary hydrogen economy [7].

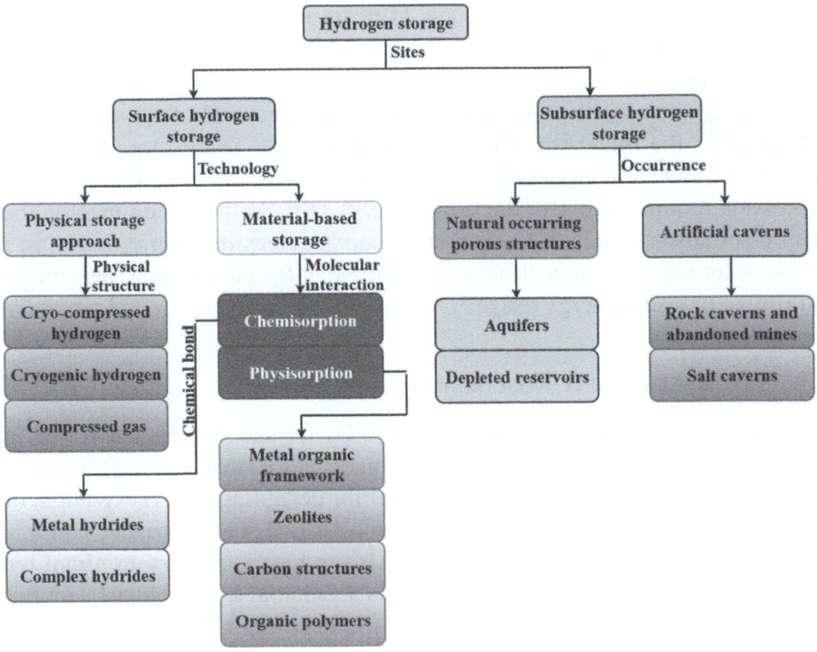

FIGURE 3.2 Hydrogen storage systems for various applications [2].

Volumetric storage density measures the amount of H_2 stored per unit volume of the storage system. For aircraft applications, for instance, where space is a crucial consideration, this criterion is essential. H_2 gas has a very low volumetric density (0.0899 g/l at standard ambient temperature and pressure), whereas gasoline has a volumetric density of approximately 737 g/l [9].

Hydrogen must be compressed to high pressures, liquified at cryogenic temperatures, or stored in solid or chemical forms to achieve practical volumetric storage densities. The volumetric storage capacity of compressed H_2 at ambient temperature is 2990 ± 40 kJ/l, which is still far below the 2025 target of 4800 kJ/l set by the United States Department of Energy (DOE) [10].

Gravimetric storage density, on the other hand, is a measure of the H_2 mass stored in the storage system as a percentage of the overall mass. It may as well be expressed in a weight percentage, i.e.,

$$\text{Gravimetric storage density} = \frac{\text{Mass of } H_2 \text{ stored}}{\text{System overall mass}} \times 100 \tag{3.1}$$

A higher gravimetric storage density or weight percentage shows a lighter storage system, which is essential for aircraft and automobile applications. For onboard H_2 storage systems such as light-duty vehicles, DOE set a target of 5.5 wt% gravimetric capacity and also an ultimatum target of 6.5 wt% for 2025 [8, 11]. H_2 has a high gravimetric energy density of 120 to 142 MJ/kg for both liquid and compressed forms compared to most fossil fuels, such as methanol (22.7 MJ/kg), methane (50–55 MJ/kg), gasoline (44–46 MJ/kg), natural gas (42–55 MJ/kg), diesel (42–46 MJ/kg), and other fuels [12]. Its high gravimetric energy density is a significant advantage; however, its low volumetric storage density remains a major challenge.

The energy needed to compress, liquefy, or release H_2 from storage materials determines the round-trip efficiency of H_2 storage. Due to the intense cooling required for liquefaction, usually ~30% of the energy stored in the H_2 is needed; compression requires ~10% of the stored energy [13]. At the same time, the thermodynamics of adsorption and desorption reactions determine energy losses in material-based storage. To keep hydrogen competitive with conventional fuels, the goal is to minimize parasitic energy usage.

3.1.1 Hydrogen Safety Requirements

Strict safety requirements must be adhered to H_2 storage systems due to the special physicochemical properties and potential dangers associated with high-pressure and cryogenic storage. H_2 flammability is crucial for designing safe H_2 systems, including the generation, storage, and transportation [14]. H_2 has a broader flammability range, from 4% to 75% by volume in air. Also, it has a very low minimal ignition energy (~0.02 mJ) compared to other fuels; an accidental leakage may be more likely to result in a fire [15–17]. Additionally, due to the colorless flame of H_2, it is nearly invisible, which can be challenging to detect and pose a risk to workers in the event of a leak or accident.

The primary safety design considerations for the central H_2 systems are the systems' structural integrity and mechanical resistance, leak prevention and detection, embrittlement resistance, thermal management and pressure relief components, and compliance with international safety codes. H_2 systems must be designed to withstand high internal pressures and temperatures, as well as cycling related to filling and emptying activities (which can cause material wear), and mechanical impacts from collisions or falls in mobile storage applications, such as transport containers and automobiles. In accordance with ISO and SAE regulations, cylinders and tanks undergo extensive testing, including penetration, drop, and burst tests [18, 19].

Given H_2's propensity for rapid ignition and the risks associated with its combustion, detecting even tiny H_2 leaks is essential for fulfilling health and safety regulations [20]. Even microscopic defects can lead to leakage due to their small molecular size and high diffusivity. Quantitative and gas-selective detectors are the relevant methods for efficiently detecting tiny leaks in the gas grid [21]. Advanced sealing technologies and multilayered composites are used to minimize permeation. Continuous leak detection systems employing sensors based on catalytic, electrochemical, or thermal conductivity principles are required for high-pressure installations.

Various leak detectors exist which measure H_2 leaks based on the concentration in the air. These detectors could be stationary or portable. The preferred tools for detecting leaks are portable detectors [20]. Their readings are contingent upon the leak geometry, the angle between the probe and the flow direction, and the probe's position with respect to the leak source.

H_2 leak rate ($\dot{L}_{H_2}$) is formulated in terms of concentration (c_{H_2}), i.e.,

$$\dot{L}_{H_2} = c_{H_2} \times p \times S \tag{3.2}$$

where S is the speed of the detector's pump, and p is the atmospheric pressure. $\dot{L}_{H_2}$ may also be obtained if the leakage point is appropriately covered with a bag or hood and the volume (V) of the hood is known,

$$\dot{L}_{H_2} = \frac{c_{H_2} V}{t} \tag{3.3}$$

where t is the time accumulated. Ventilation in storage spaces is essential to prevent the accumulation of H_2 above the lower flammability limit (LFL).

H_2 embrittlement, which occurs when H_2 diffuses into metals, can cause structural components to lose their ductility, leading to cracking and potentially catastrophic failure [22]. The diffusion properties of H_2 within the material under specific temperature and stress circumstances, as well as the impact of H_2 on the cohesive forces of the material's crystal lattice, are significant variables. Moreover, H_2 embrittlement is greatly influenced by the interaction of H_2 with oxide layers or metal atoms at the interface between the steel and the gaseous or liquid media.

Austenitic stainless steels, certain aluminum alloys, and composites are common materials selected for embrittlement risks minimization due to their resistance to H_2-assisted degradation. The interactions between absorbed H_2 atoms and microstructural trap sites are the main factors influencing the embrittlement behavior of steels exposed to H_2. These traps include surfaces, fracture tips, grain boundaries, dislocations, substitutional atoms, point defects, and vacancies. The primary embrittlement mechanism is influenced by the nature, density, and distribution of these trapping sites, which also govern the mobility and concentration of H_2 inside the material [23, 24]. Standards (ISO 11114-4 [18] and ASTM G142 [25]) guide the selection and qualification of materials exposed to H_2 environments.

Boil-off gas must be efficiently managed in cryogenic H_2 storage tanks to prevent hazardous pressure buildup brought either by heat ingress or the typical evaporation of liquid H_2 [26]. Under unusual circumstances, such as fire exposure, system failure, or overfilling, pressure relief devices are integrated to release excess pressure and maintain safety.

International standard guidelines for H_2 technologies, such as requirements for tank design, safety, and operation, are provided by ISO/TC 197 [27]. The NFPA 2 specifies best practices for the production, handling, and storage of H_2 in commercial, industrial, and public settings [28]. To ensure safe operation in the event of a collision or fire, SAE J2579 specifies stringent verification tests for SAE J2579 storage systems used in road vehicles [18]. These standards collectively ensure that cryogenic H_2 tanks are designed and tested to mitigate the risks of fire, excessive pressure, and H_2 boil-off.

3.1.2 Storage Challenges and Issues

For H_2 storage to be widely adopted, several operational, economic, and technical challenges must be overcome. As was previously mentioned, H_2's intrinsically low volumetric energy density makes it challenging to transport and store on board [29]. This results in large tanks that restrict passenger room and vehicle design.

The cost of H_2 storage is a significant barrier to the widespread adoption of H_2. Numerous H_2 storage approaches have been compared based on their costs of storage, as shown in Table 3.1 [30, 31].

By 2035, compressed H_2 storage systems are predicted to make up 67% of all H_2 storage technologies, and by 2050, that percentage will increase to 56% [32] and the

TABLE 3.1
Cost of Different Hydrogen Storage Approaches [30, 31]

Hydrogen Storage Technologies	Operating Temperature (K)	Operating Pressure (bar)	Cost (US$/kg H_2)
Cryogenic	20	1	200 to 700
Cryo-compressed	20	350	400
Compressed gas	Ambient	350 to 700	400 to 700
Metal hydrides	Ambient to 553	1 to 30	500 and above
Complex hydrides	423 to 573	1 to 50	300 to 450
Chemical hydrides	353 to 473	1	160 to 270

cost of storing a kg of H_2 using this technology is projected to rise to $700 from $400 [33], as presented in Table 3.1.

As was previously mentioned, H_2 embrittlement and permeation loss are caused by the small size and strong diffusivity of H_2 atoms. Specialized alloys and protective coatings that are more resistant to embrittlement and H_2 diffusion are frequently used in H_2 storage systems to lessen these effects.

Furthermore, many H_2 storage methods exhibit notable temperature impacts during use. The temperature in the storage tank quickly rises as a result of gas compression during continuous filling, especially at high pressures [34]. Thermal control is much more critical for material-based storage systems, such as those used in metal hydrides (as a result of H_2 absorption and desorption) [35]. These approaches may result in decreased system efficiency and safety risks in the absence of proper heat control.

Overcoming these barriers, advanced metal hydride storage systems incorporate thermal management features, including active heat transfer fluids, fins, and high-conductivity materials, which improve heat dissipation during absorption and offer regulated heating during desorption [35]. This guarantees reliable, effective, and safe hydrogen storage.

Boil-off gas is a recurring operational issue for storing liquid H_2 [36]. Heat ingress is unavoidable even in well-insulated cryogenic tanks, which causes the stored H_2 to partially evaporate. To avoid safety issues, the pressure buildup caused by this vaporization in the tank must be controlled by either reliquefying the gas or safety venting it. Continuous boil-off loss over extended storage can significantly increase operating costs and reduce system efficiency.

Beyond H_2 storage, there are other logistical problems due to interoperability between systems and geographical areas, standards for H_2 purity (which vary by region), and a small number of global networks of refueling stations, which limit the widespread adoption and large-scale deployment of H_2 vehicles [37]. Addressing these storage issues needs investment in H_2 distribution networks, harmonized international standards, and improved storage technologies which minimize losses on boil-off gas.

3.2 PHYSICAL STORAGE MECHANISMS

Hydrogen is a special kind of energy carrier, primarily due to its high gravimetric energy density and low molecular weight. However, storing it poses a significant challenge due to its extremely low volumetric energy density at ambient conditions. The majority of pilot-scale and commercial storage systems now rely on physical storage systems, which comprise liquid H_2 and compressed gaseous H_2 storage. The thermodynamic concepts, design considerations, benefits, drawbacks, and practical uses of these two storage pathways are covered in detail next.

3.2.1 Liquid Hydrogen

The process of H_2 liquefaction forms a cryogenic liquid by cooling gaseous H_2 below its typical boiling point of –253°C at atmospheric pressure [13]. Liquefying H_2 is attractive, due to the significant increase in the volumetric energy density, which allows for the storage of more H_2. This makes liquid H_2 particularly valuable for applications requiring large-scale storage or long-distance transportation. It is considered promising not only due to its gravimetric and volumetric densities but also due to the potential for low-pressure storage and high purification levels [38].

Liquid H_2 has a high specific energy and combusts cleanly, producing only water vapor as a byproduct, making it appealing for decarbonization. It plays a crucial role in several advanced energy and transportation applications, and its production is rising in tandem with the applications. Liquid H_2-based storage and transportation can supply the high purity of H_2 that is needed on a broad scale due to the rapidly expanding applications [39].

Liquid H_2 is often used in aerospace applications as a propellant for rockets and spacecraft [40]. They are best suited for launch vehicles where weight reduction is critical, due to the high gravimetric energy density. This is particularly ideal for the space companies, which have long relied on liquid H_2 in combination with liquid O_2 for high-efficiency propulsion missions [41].

Outside the aerospace industry, liquid H_2 facilitates the bulk transportation of H_2 between nations, similar to international commerce in liquefied natural gas. For long-distance H_2 transportation, dedicated liquid H_2 carriers have been developed, like the Suiso Frontier from Kawasaki Heavy Industries [42]. To effectively support the expansion of the global H_2 economy, this strategy is essential for linking areas with abundant renewable energy resources to those with a high demand for H_2.

Liquid H_2 is essential to H_2 stations in terms of retail and distribution levels, since it can be kept in cryogenic liquid form and then vaporized and compressed to high pressures before it is dispensed to fuel cell vehicles. These storage methods guarantee a consistent supply of H_2 for transport networks, in addition to minimizing the fuel station's physical size [43].

Also, in stationary and backup power systems, liquid H_2 offers long-term energy storage with a small geographical footprint [44]. This is especially valuable for critical infrastructure and remote locations where a steady power supply is essential, and H_2 may need to be stored for extended periods without significant degradation. Despite its numerous advantages, significant thermodynamic, safety, and economic issues still exist with the production, storage, and handling of liquid H_2.

H_2 liquefaction is a process fundamentally governed by challenging constraints. It is an energy-intensive process due to H_2's very low boiling point, which requires advanced cryogenic cooling technologies to cool it under atmospheric pressure and convert it into a liquid state. At critical pressure (~13.2 bar) and temperature (~ –240 K), it can no longer be liquefied [45]. This implies that H_2 liquefaction requires a substantial investment in energy and refrigeration.

The minimum theoretical work required for the H_2 liquefaction process at ideal thermodynamic conditions is approximately 2.7 to 3.0 kWh/kg [46]. However, in industrial applications, H_2 liquefaction systems require more work than the theoretical work (~10–13 kWh/kg). This amounts to approximately 30% of the H_2's energy content, which makes liquefaction a costly process.

3.2.1.1 Liquefaction Technologies

A H_2 liquefaction process generally comprises compression, pre-cooling, cryogenic expansion, and separation stages. The general process diagram is depicted in Figure 3.3. H_2 gas is fed to the compressor for compression, which is the first stage of the process, and is then optionally pre-cooled at the next stage for removing impurities that may have accumulated during cryogenic liquefaction [47]. The next stage involves cooling the H_2 in a cryogenic refrigeration cycle to convert the H_2 from ortho- to para-H_2 catalytically [48]. The cycle continues to the expansion stage and finally to the separation stage.

Three main types of H_2 liquefaction cycles exist. These are the Linde-Hampson cycle, Claude cycle, and the Brayton refrigeration cycle. The Linde-Hampson cycle has a basic system configuration. The system consists of components such as a compressor, a cooler, a heat exchanger, a Joule-Thomson valve, and a flash separator, as illustrated in Figure 3.4.

H_2 gas is fed to the compressor to raise both temperature and pressure of the gas at point 2, where work is done on the compressor. The compressed gas is precooled in the precooler, where heat is rejected to a fluid. Upon exiting the precooler (point 3), the gas is further cooled by exchanging heat with the uncondensed gas in the heat

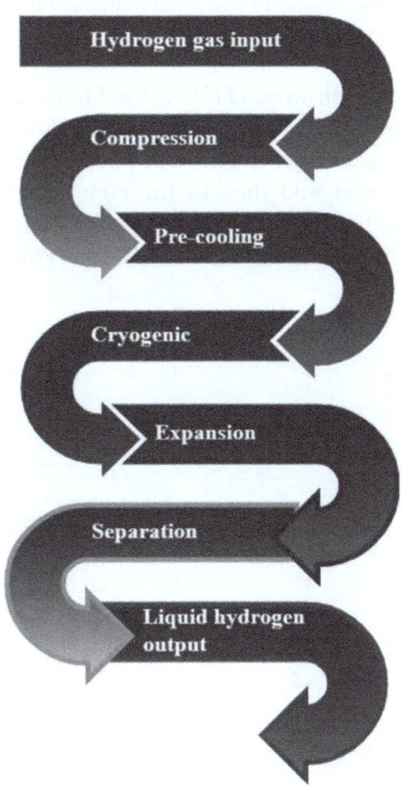

FIGURE 3.3 A general process diagram of hydrogen liquefaction.

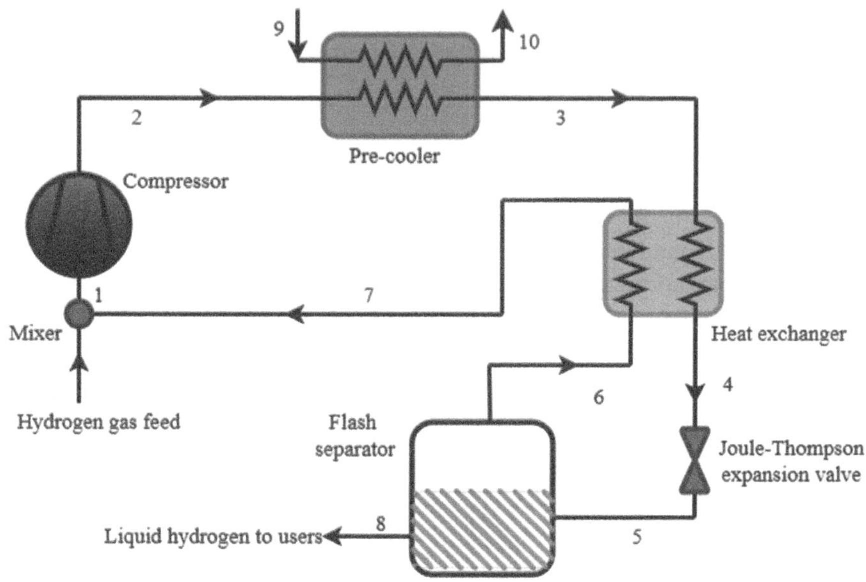

FIGURE 3.4 Linde-Thomson cycle for hydrogen liquefaction.

exchanger. A Joule-Thompson valve (JTV) is used to expand the cooled gas exiting the heat exchanger, resulting in a biphasic flow of the fluid.

A flash separator is used to collect the bi-phase fluid, which releases the gas back to the heat exchanger and then to the compressor, continuing the cycle. Liquid H_2 is now collected from the separator for consumption at point 8. The temperature-entropy diagram for the system's configuration during the cycle is shown in Figure 3.5.

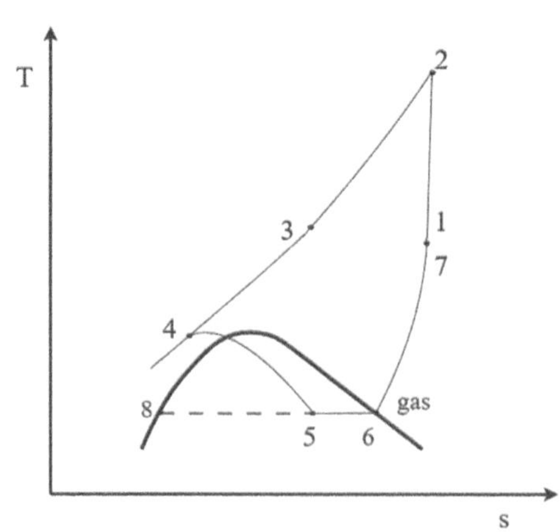

FIGURE 3.5 Linde-Thompson cycle temperature-entropy (T-s) diagram.

The energy and exergy balances for the components of the Linde-Thompson cycle are formulated as follows:

For the compressor,

$$\dot{W}_{com} = \left(\dot{\Xi}_2 - \dot{\Xi}_1\right) + \dot{\Xi}_{d,com} \tag{3.4}$$

where $\dot{\Xi}_1$ and $\dot{\Xi}_2$ are the exergy rates at the inlet (H_2 gas feed) and outlet (compressed H_2 gas), respectively. $\dot{\Xi}_{d,com}$ is the exergy destruction rate in the compressor. $\dot{W}_{com}$ is the compressor work done, computed as follows:

$$\dot{W}_{com} = \dot{m}_1(h_2 - h_1) \tag{3.5}$$

where $\dot{m}_1$ is the mass flow rate from point 1. h_2 and h_1 are the enthalpies at points 1 and 2, respectively, i.e.,

$$h = \int_{T_0}^{T} c_{p,H_2}(T)\,dT \tag{3.6}$$

c_{p,H_2} is the specific heat capacity for hydrogen, and T is the temperature. The isentropic temperature (T_{2s}) and temperature (T_2) at the compressor outlet are, respectively, obtained as:

$$T_{2s} = T_1\left(\frac{P_2}{P_1}\right)^{\frac{\gamma_{H_2-1}}{\gamma_{H_2}}} \tag{3.7}$$

$$T_2 = T_1\frac{T_{2s} - T_1}{\eta_{com}} \tag{3.8}$$

P_2/P_1 is the compression ratio of the compressor, and η_{com} is the isentropic efficiency of the compressor. γ_{H_2} is the ratio of specific heats for H_2 gas at constant pressure and volume.

The exergy balance for the pre-cooler is formulated as:

$$\dot{\Xi}_{10} - \dot{\Xi}_9 = \left(\dot{\Xi}_2 - \dot{\Xi}_3\right) + \dot{\Xi}_{d,prec} \tag{3.9}$$

where $\dot{\Xi}_9$ and $\dot{\Xi}_{10}$ are the exergy rates of the cooling at the inlet and outlet of the pre-cooler, respectively. $\dot{\Xi}_3$ is the exergy rate of the cooled H_2 gas exiting the pre-cooler. $\dot{\Xi}_{d,prec}$ is the exergy destruction rate in the pre-cooler.

For the heat exchanger, the exergy balance is formulated as:

$$\dot{\Xi}_7 - \dot{\Xi}_6 = \left(\dot{\Xi}_3 - \dot{\Xi}_4\right) + \dot{\Xi}_{d,HX} \tag{3.10}$$

$\dot{\Xi}_6$ and $\dot{\Xi}_7$ are the exergy rates of H_2 gas from the flash separator entering and exiting the heat exchanger, respectively. $\dot{\Xi}_4$ is the exergy rate of the cooled H_2 gas exiting the heat exchanger. $\dot{\Xi}_{d,HX}$ is the exergy destruction rate in the heat exchanger.

For the JTV, the exergy balance is formulated as:

$$\dot{\Xi}_4 = \dot{\Xi}_5 + \dot{\Xi}_{d,JTV} \tag{3.11}$$

where $\dot{\Xi}_4$ and $\dot{\Xi}_5$ are the exergy rates entering and exiting the valve, respectively. $\dot{\Xi}_{d,JTV}$ is the exergy destruction rate in the valve. Using the isenthalpic property of throttling, i.e., $h_4 = h_5$, the $\dot{\Xi}_{d,JTV}$ is computed as:

$$\dot{\Xi}_{d,JTV} = \dot{\Xi}_4 - \dot{\Xi}_5 = \dot{m}_4\left[T_0\left(s_5 - s_4\right)\right] \tag{3.12}$$

s_4 and s_5 are the entropies at points 4 and 5, respectively. $\dot{m}_4$ is the mass flow rate entering the valve.

For the flash separator, the exergy balance is formulated as:

$$\dot{\Xi}_5 = \dot{\Xi}_8 + \dot{\Xi}_6 + \dot{\Xi}_{d,FS} \tag{3.12}$$

where $\dot{\Xi}_5$ is the exergy rate of bi-phase H_2 exiting the JTV, $\dot{\Xi}_8$ is the exergy rate of the liquid H_2 exiting the flash separator, and $\dot{\Xi}_6$ is the exergy rate of H_2 gas exiting the flash separator. $\dot{\Xi}_{d,FS}$ is the exergy destruction in the flash separator.

Despite the Linde-Hampson cycle having a simple configuration, it requires high pressure from the compressor, and the high pressure requires more work from the compressor to raise both the pressure and temperature of the H_2 gas. Also, there is a low H_2 liquid yield from the system, indicating low system efficiency [49].

The Claude cycle is more complex compared to the Linde-Thompson cycle. Two more heat exchangers (HX2 and HX3) and an expander are added in the Claude cycle, thus yielding high efficiency when compared to that of Linde-Thompson [50].

In this cycle, after H_2 gas exits heat exchanger 1 (HX1), a splitter is used to split the gas into two parts: one portion is directed to the expander, and the other is directed to heat exchanger 2 (HX2). The expanded gas exiting the expander is mixed in mixer 2 (M2) with the gas exiting heat exchanger 3 (HX3), which is then used to fuel both heat exchangers 2 (HX2) and 1 (HX1) to complete the cycle. With this configuration, H_2 gas is further cooled before undergoing an isenthalpic process of throttling in the JTV. The Claude cycle is depicted in Figure 3.6, and the T-s diagram is shown in Figure 3.7.

The Kapitza cycle is similar to the Claude cycle. However, it has fewer components compared to the Claude cycle. In this configuration, the HX3 is removed. H_2 gas from the flash separator is mixed with the expanded gas from the expander, which is used to fuel HX2 and HX1. The Kapitza cycle is presented in Figure 3.8, and the T-s diagram is shown in Figure 3.9.

A further modification to both the Claude and Kapitza cycles is undertaken by the Heylandt cycle. In this configuration, the HX1 (Claude and Kapitza cycles) is removed. H_2 gas exiting the pre-cooled is split to the expander and HX1, from which the cycle is similar to the previous cycles. The Heylandt cycle and the T-s diagram are shown in Figures 3.10 and 3.11, respectively. These cycles are commercially used in H_2 liquefaction systems.

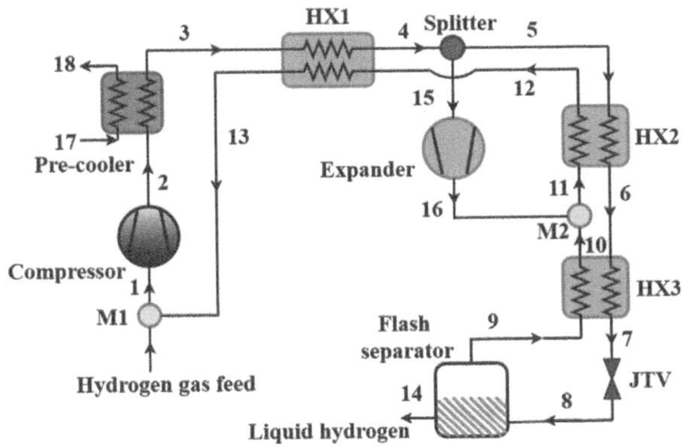

FIGURE 3.6 Claude cycle for hydrogen liquefaction. M1—mixer 1, M2—mixer 2, HX1—heat exchanger 1, HX2—heat exchanger 2, HX3—heat exchanger 3, and JTV—Joule-Thompson valve.

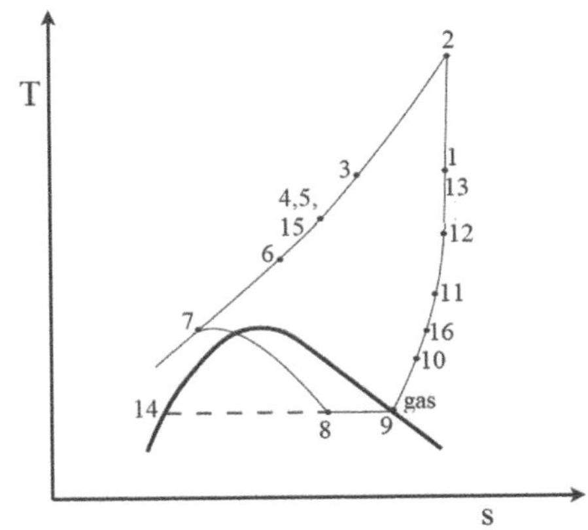

FIGURE 3.7 Claude cycle T-s diagram.

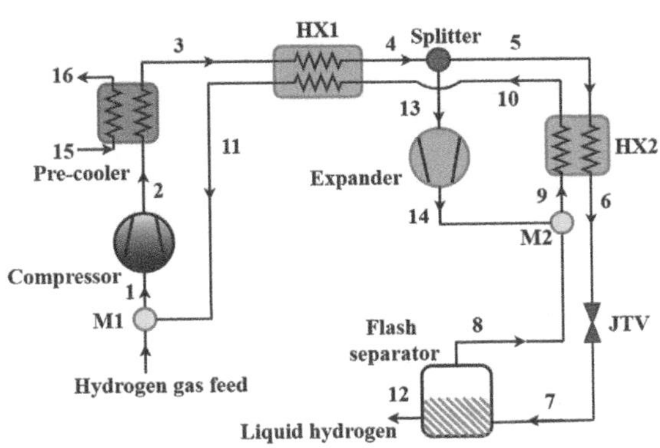

FIGURE 3.8 Kapitza cycle for hydrogen liquefaction. M1—mixer 1, M2—mixer 2, HX1—heat exchanger 1, HX2—heat exchanger 2, and JTV—Joule-Thompson valve.

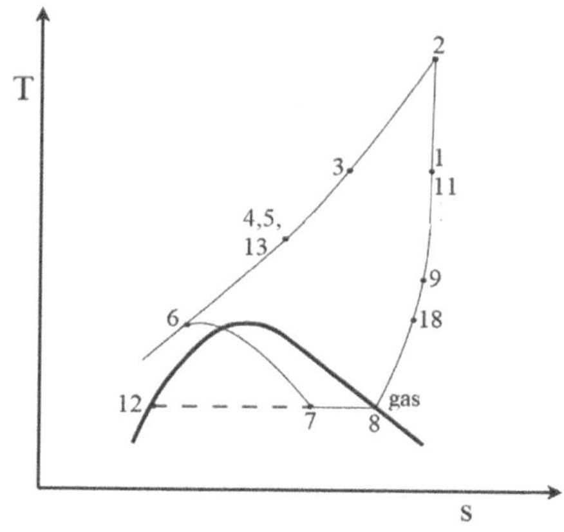

FIGURE 3.9 Kapitza cycle T-s diagram.

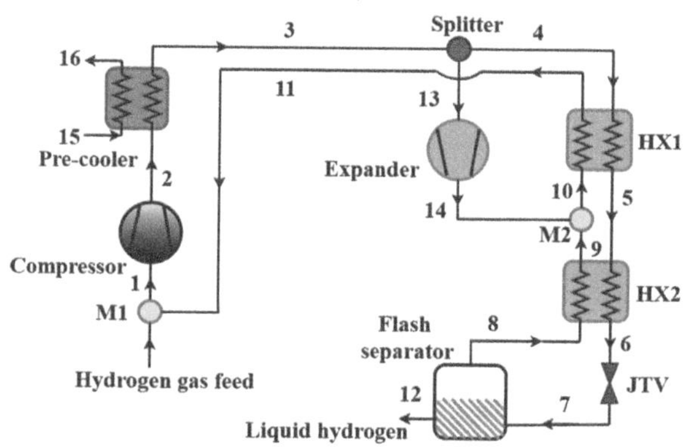

FIGURE 3.10 Heylandt cycle for hydrogen liquefaction. M1—mixer 1, M2—mixer 2, HX1—heat exchanger 1, HX2—heat exchanger 2, and JTV—Joule-Thompson valve.

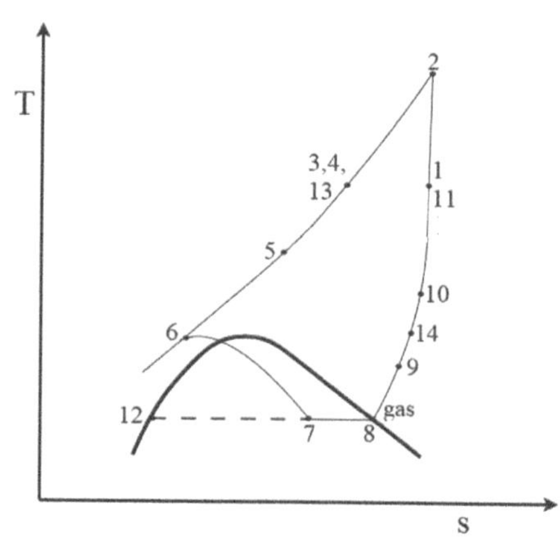

FIGURE 3.11 Heylandt cycle T-s diagram.

Liquid nitrogen pre-cooling in hybrid systems lowers the need for refrigeration, and advanced cycles and magnetic refrigeration are being developed to increase efficiency. A hybrid refrigeration cycle that includes pre-cooling and magnetic active regenerator cycles has been reported to increase liquefaction efficiency (i.e., >50%) [51, 52].

All of these processes, however, have significant upfront expenses (such as compressors, expanders, and cryogenic heat exchangers), call for advanced materials for containment and insulation, and present safety hazards due to their high purity and extremely low temperatures. Operationally, the financial load is increased by boil-off, parasitic losses, and maintenance.

3.2.1.2 Cryogenic Tanks

After H_2 is liquefied, it needs to be stored in specialized cryogenic tanks that are made to reduce heat loss and boil-off losses [53]. Liquid H_2 tanks can be classified as: (i) Dewar flasks (vacuum-insulated double-walled laboratory-scale applications); (ii) industrial double-walled tanks (large-scale tanks for H_2 plants and refueling stations); and (iii) spherical liquid H_2 tanks (reduced surface area for less heat transfer).

Even the best-insulated tanks experience heat intrusion, leading to partial vaporization of liquid H_2. Small vapor losses occur continuously during boil-off gas generation and are unavoidable due to heat transfer from the environment [36]. If left unchecked, vaporized H_2 can raise internal tank pressure, which could be dangerous.

However, there are mitigation strategies, such as venting, reliquefaction, and burn-off of flaring [53]. In venting, H_2 is safely released to the atmosphere. Reliquefaction involves recondensing boil-off gas using external cooling systems. The burn-off or flaring involves a controlled combustion of excess gas in industrial plants. The merits and demerits of H_2 liquefaction are summarized in Table 3.2.

3.2.2 Compressed Gas

Compressed gaseous H_2 is the most widely used method of physical storage. It is a high-pressure storage particularly for onboard storage in fuel cell electric vehicles and H_2 refueling stations. It is also a vehicle-mounted system which has a rapid

TABLE 3.2
Merits and Demerits of Hydrogen Liquefaction

Merits	Demerits
Essential for aerospace and bulk shipping	Boil-off gas losses over time
High volumetric energy density	Very high energy cost for liquefaction (30–35% of hydrogen's energy content)
Established industrial experience	Requires advanced cryogenic infrastructure
Enables centralized production and distribution	Safety risks from low temperatures and pressure buildup

speed of H_2 absorption and release. It also has a simple component structure and low energy consumption when compared to other methods of hydrogen storage [54].

Compressed H_2 storage stores high-pressure gaseous H_2 in a tank, which usually requires a typical storage pressure of 50 MPa for commercial vehicle fleets such as buses and trucks, and 70 MPa for light-duty passenger cars for extended driving range [55, 56]. Compressed gaseous H_2 offers simplicity and scalability, but suffers from low volumetric energy density and the need for costly, heavy tanks.

3.2.2.1 Types of Storage Cylinders

Tank technology is key to the widespread usage of compressed gaseous H_2. The tank's material must be cost-effective, meet safety standards, and be reliable [38]. There are four primary categories of high-pressure H_2 storage tanks: Type I, Type II, Type III, and Type IV [57].

The Type I category consists of all-metal tanks (made of steel or aluminum). The Type II tanks are lined with a metal hoop and a composite wrap. Type III tanks are fully wrapped composite with a metal liner. The Type IV tanks are fully composite, featuring a polymer liner.

Type I and Type II are suitable for H_2 filling stations but not for a tank mounted on a vehicle, as they are heavier, which makes it more challenging to achieve a high H_2 storage density [58]. Type III and Type IV use light, high-intensity fibers, which minimize the mass of the H_2 tank and can withstand higher pressure. As a result, they are commonly chosen for the H_2 storage tank installed on modern fuel cell electric vehicles. Table 3.3 compares these types of compressed gaseous tanks used.

A typical off-site process diagram of a compressed H_2 gas system dispensing compressed H_2 gas is shown in Figure 3.12. The system consists of a tube-trailer, compressor, high-pressure tank, buffer tank, and dispenser. High-pressure H_2 gas flows from a tube trailer via a pressure release valve (PRV1) to the compressor. The compressor increases the pressure of the gas, which is stored in a high-pressure

TABLE 3.3
Comparisons of Different Compressed Hydrogen Gas Tanks

Types	Pressure Capability	Materials Used	Merits/demerits	Applications
Type I	~20 MPa	All-metal (aluminum or steel)	High weight, low storage density, and corrosion in the internal tank	H_2 filling stations
Type II	~35 MPa	Metal liner with hoop-wound composite wrap	High weight, high issues with H_2 embrittlement, short lifespan	Industrial usage
Type III	~70 MPa	Fully wrapped composite with metal liner	High burst pressure, low weight, and impermeability	Automobiles
Type IV	~70 MPa	Fully composite with polymer liner	Low burst pressure, low weight	Automobiles

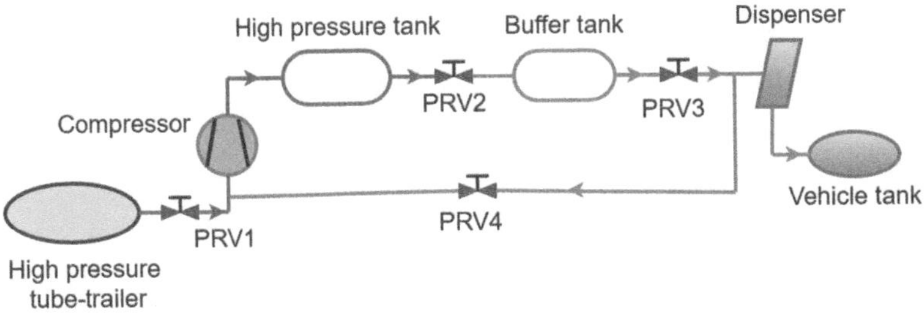

FIGURE 3.12 Off-site process diagram of compressed hydrogen gas system.

tank upon exiting the compressor. The gas is released to the buffer tank via another pressure release valve (PRV2), from which it is then released to the dispenser for filling. Table 3.4 summarizes the relevant thermodynamic equations utilized in modeling a typical compressed gas system. The table presents assumptions based on the equations.

TABLE 3.4
Equations for Modeling a Compressed Gas System

Parameter	Equations	Usage and Assumptions	Note
Estimated stored mass (ideal gas)	$m = \frac{PV}{RT}$	For flow to moderate pressures (1–2 MPa)	m = mass, p = pressure, v = volume, R = gas constant
Estimated stored mass (real gas)	$m = \frac{PV}{Z(p,T)RT}$	For high pressure (≤ 70 MPa)	Z = compressibility
Isentropic compression (ideal gas)	$T_2 = T_1\left(\frac{p_2}{p_1}\right)^{\frac{\gamma-1}{\gamma}}$, $W_s = \frac{R}{\gamma-1}T_1\left[\left(\frac{p_2}{p_1}\right)^{\frac{\gamma-1}{\gamma}} - 1\right]$	For adiabatic, reversible compression	γ = ratio of specific heats
Polytropic compression	$W_p = \frac{n}{n-1}RT_1\left[\left(\frac{p_2}{p_1}\right)^{\frac{n-1}{n}} - 1\right]$	Real compression with finite heat transfer	n = polytropic exponent (1.1 to 1.35 depending on intercooling)
Real gas reversible work	$W = \int_{v_2}^{v_1} pdv$	Accurate high-pressure work	
Thin-walled cylinder stresses	$\sigma_\theta = \frac{pr}{t}$, $\sigma_L = \frac{pr}{2t}$	For preliminary sizing and burst checks ($t \ll r$ (thin wall))	T = wall thickness, r = radius, p = internal pressure, σ_θ = hoop stress, σ_L= longitudinal stress

(Continued)

TABLE 3.4 (*Continued*)
Equations for Modeling a Compressed Gas System

Parameter	Equations	Usage and Assumptions	Note
Thick-walled cylinder	$\sigma_r = A - \frac{B}{r}$, $\sigma_\theta = A - \frac{B}{r^2}$	t is not $\ll r$	A and B are constants. σ_r= radial stress
Burst pressure	$p_{burst} \approx \frac{\sigma_u t}{r}$	Rough estimate	σ_u = ultimate stress
Fickian permeation flux	$J = -D\frac{dC}{dx}$, $J = \frac{P}{l}\Delta p(\text{Steady state})$	Permeation through polymers	D = diffusion coefficient, C = concentration of H_2 P = permeability, J = Fickian flux, l = thickness

3.2.3 Hydrogen Storage Safety and Challenges

Hydrogen storage systems must undergo intense testing to ensure compliance with regulations, safety, and reliability. SAE J2579, ISO/TC 197, ISO 11114-4, and NFPA 2 are among the standards that specify the procedures and requirements for acceptance for these tests [18, 19, 27, 28]. Common assessments consist of:

i. Leak and permeation testing: This determines the rate of H_2 leaking out from the system. H_2 sensors and helium leak detection are the primary methods for measuring permeation losses and ensuring compliance with permitted limits.
ii. Burst pressure testing: This establishes the storage vessel's maximum failure pressure. Catastrophic failure may be avoided by the H_2 tanks, which demonstrate that the burst pressure is significantly higher than the system's operating pressure.
iii. Fire exposure testing: This assesses how the system behaves in fire conditions, particularly the performance of pressure relief valves and other devices to keep tanks from rupturing and preserve structural integrity.
iv. Drop and impact testing: This simulates handling and accident situations in the real world and evaluates the storage system's mechanical integrity after impacts or drops.
v. Fatigue and cyclic tests: This verifies the vessel's capacity to withstand repeated pressure cycles, which are indicative of several refueling and discharging events throughout its life cycle.

Type IV composite H_2 sensors and helium leak detection storage tanks are bulkier than the conventional gasoline tanks, even with design improvements. Their larger volume constraints restrict the amount of space available for goods and passengers, limiting the design flexibility of vehicles. This space constraint may be a significant challenge for airspace and automobile applications.

Also, these storage tanks are far more expensive to produce than conventional fuel storage systems due to their reliance on high-strength composite materials [59]. Storing H_2 at high pressures increases the risk of leakage when damaged, as well as rapid decompression and explosion [60].

3.3 STORAGE BASED ON MATERIALS

Material H_2 storage is categorized into two types: chemical H_2 and metal hydride storage-based.

3.3.1 Storage of Chemical Hydrogen

There is also another potential method of H_2 storage, in addition to physical storage, which involves liquefied and compressed H_2 storage. As discussed earlier, these storage systems have practical and safety issues. Chemical H_2 storage offers an additional method for storing H_2. It involves storing H_2 chemically bound in compounds, which can be released when needed [61].

There are thermodynamic and kinetic requirements for the H_2 carriers to possess. The main thermodynamic requirements are that the carrier's bond formation and breaking reaction enthalpies must not be so high as to necessitate excessive heat input and yet be stable enough to safely store H_2 at ambient conditions. The kinetics of the carriers are equally important. It reveals the rate at which H_2 is dehydrogenated, i.e., the rate at which H_2 is being released from them. Additionally, the kinetics reveal the catalysts required and the optimal operating temperature.

These carriers are grouped into liquid-state and solid-state H_2 storage materials. Some of these liquid materials are ammonia, formic acid, methanol, liquid organic hydrogen carriers, and boron-based compounds. Ammonia (NH_3) has a high H_2 content by weight (~18 wt%), and it is easily liquefied under moderate conditions. It plays a crucial role in H_2 storage and transport [62, 63].

However, there are challenges associated with NH_3 in terms of energy required for production (NH_3 synthesis using the Haber-Bosch process). The production of ammonia requires approximately 95% of the total electricity used in the NH_3 synthesis process, with the remaining 5% allocated to the Haber–Bosch synthesis unit and air separation [64]. Also, NH_3 cracking process is an endothermic process, i.e.,

$$2NH_3 \leftrightharpoons N_2 + 3H_2 \left(\Delta H^0 = +46.1 \frac{\text{kJ}}{\text{mol}} \text{ of } NH_3 \right) \tag{3.13}$$

It has high-temperature requirements for the cracking, typically in the range of 400–900°C, depending on the catalysts used [65]. Catalyst cost and safety due to NH_3 toxicity still pose challenges.

TABLE 3.5
Merits and Demerits of Liquid-State Hydrogen Storage Materials

Merits	Demerits
High volumetric densities	High energy penalty for both hydrogenation and dehydrogenation
Compatible with existing liquid fuel infrastructure	There is a mass penalty as the carrier molecules themselves add weight
Safe long-term storage in many cases	Hydrogen must be purified after dehydrogenation due to the presence of catalyst residues
	Degradation of carrier molecules and catalysts over multiple cycles

Liquid organic H_2 carriers, such as methylcyclohexane and dibenzyltoluene systems, can undergo reversible hydrogenation and dehydrogenation in chemical reactions. They show promise for long-cycle H_2 storage and safe, cost-effective, large-scale transoceanic transportation, as well as stable and efficient H_2 storage [66]. However, techno-economic evaluations and life cycle analysis demonstrated that the systems are currently costly and unsustainable. Table 3.5 summarizes the merits and demerits of the liquid-state hydrogen storage materials.

3.3.2 Metal Hydrides

Metal hydrides store H_2 within metal or metal alloy materials that form metal-H_2 compounds, such as hydrides, through a reaction typically like:

$$M + \frac{y}{2} \leftrightharpoons MH_y \tag{3.14}$$

In practical applications, metal hydrides are useful for H_2 storage because they provide two sources of H_2: the H_2 produced through oxidation of the metal and the H_2 physically bonded within the hydride compound [61]. This dual mechanism enables relatively high gravimetric storage capacities. This is demonstrated in the reaction of lithium borohydride ($LiBH_4$) with water (H_2O), which may store H_2 to ~8.4 wt%, making it a potential material for efficient and compact H_2 storage systems. This reaction is demonstrated in Eq. (3.15):

$$LiBH_4 + 2H_2O \rightarrow LiBO_2 + 4H_2 \tag{3.15}$$

Despite such a hydrolytic system producing large amounts of H_2, it has several disadvantages. One of these is irreversibility, in which they are unable to regenerate expended material on board. After the release of H_2, the byproduct, which is typically a solid oxide, hydroxide, or salt, must be physically removed and brought back for regeneration. There, it undergoes an expensive and energy-intensive industrial process to be transformed back into the original hydride. This regeneration often requires high temperatures, reducing agents, or complex chemical processing [61].

Generally, interstitial hydrides, such as palladium hydride, are metals in which hydrogen atoms occupy interstitial sites. Although they frequently have lesser storage capacities and are more expensive, these tend to have excellent kinetics and are reversible [67]. For complex hydrides, such as the boron hydrides, a complex anion is created when hydrogen is covalently bonded to the metal atom. They have a higher hydrogen content, but often exhibit slow or poor kinetics, and sometimes face challenges to cycling stability.

The focus of research has been on lightweight complex aluminum hydrides. The sodium compound $NaALH_4$ is the most thoroughly researched reversible complex of aluminum hydrides. The $NaALH_4$ has a H_2 content of 5.5 wt%, good kinetics and reversibility, and has a thermal decomposition temperature of 125°C [67]. They undergo three decomposition steps [61], i.e.,

$$3NaAlH_4 \rightarrow Na_3AlH_6 + 2Al + 3H_2 \tag{3.16}$$

$$Na_3AlH_6 \rightarrow 3NaH + Al + \frac{3}{2}H_2 \tag{3.17}$$

$$3NaH \rightarrow 3Na + \frac{3}{2}H_2 \tag{3.18}$$

The first decomposition step (Eq. (3.16)) occurs at approximately 30°C with an equilibrium pressure of 10 bar, releasing 3.7 wt% H_2, about twice that of conventional low-temperature metal hydrides. The second decomposition step occurs at approximately 130°C with the same equilibrium pressure, releasing 1.9 wt% % H_2, which is typical of medium-temperature hydrides. The third decomposition step requires excessively high temperatures, making it unsuitable for practical hydrogen storage applications.

Metal hydrides are utilized in both portable and stationary storage applications. In portable storage applications, small metal hydride canisters are appropriate for powering portable gadgets, small electronics, and backup power systems due to their compact size and safety. In stationary storage applications, metal hydrides can be utilized for H_2 buffering, grid-scale energy storage, and thermal energy storage, as the exothermic and endothermic nature of hydride reactions enables effective heat control.

3.4 HYDROGEN TRANSPORTATION SYSTEMS

An essential component of the H_2 economy is transportation. It connects production sites, such as electrolysis plants, and end users, including power generation plants, refueling stations, and industrial plants. In contrast to conventional fossil fuels, H_2 presents unique transportation challenges due to its high diffusivity, low volumetric energy density, and the tendency to embrittle materials, as previously discussed.

Effective H_2 logistics are essential to minimize the levelized cost of H_2 and facilitate scaling of H_2 applications in energy storage, industrial decarbonization, and

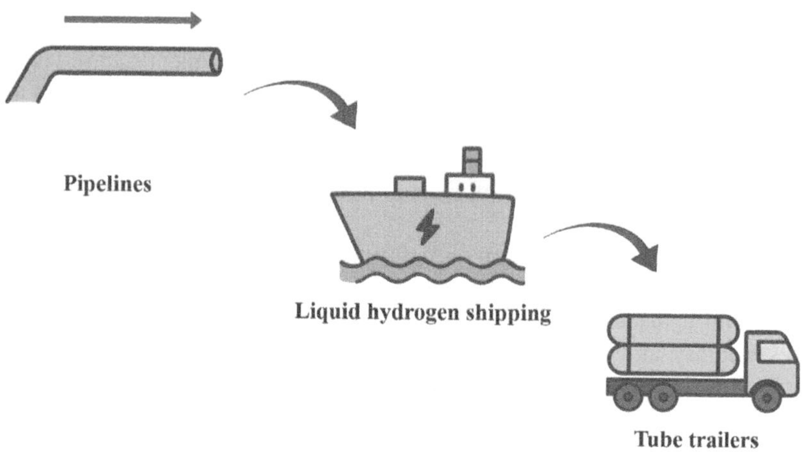

FIGURE 3.13 Hydrogen transportation options.

mobility [68]. Several transportation alternatives exist, including compressed gas trucking, pipeline transport, liquid hydrogen shipping, and chemical carrier transport. The decision on the type of transportation options depends on scalability, costs, distance, and infrastructure availability.

Transportation methods are broadly classified into three categories. They include pipeline transport, liquid H_2 shipping, and truck transport (using tube trailers). Figure 3.13 illustrates these H_2 transportation options.

Pipelines similar to those used in natural gas networks are the most efficient and cost-effective option for transporting H_2 on a large scale and over long distances. Their main benefits include scalability for industrial use, continuous delivery with reduced operating costs, and integration with the existing natural gas infrastructure through fuel blending. However, there are challenges, such as H_2 embrittlement of pipeline steel due to its high diffusivity, leakage through seals and joints due to the small molecular size, and a high initial cost for pipelines [69, 70].

One of the main design equations for H_2 pipeline transport involves the pressure drop over the length of the pipe, which is similar to natural gas transport, i.e.,

$$\Delta P = \frac{fL\rho\upsilon^2}{2D} \tag{3.19}$$

where L is the length of the pipeline, f is the friction factor, ρ is hydrogen density, υ is the flow velocity, and D is the pipe diameter.

Austenite stainless steels and composite pipes are preferred materials for pipelines due to their resistance to embrittlement [71]. To decarbonize the current grids without requiring a major infrastructure, hydrogen blending is also being explored.

Similar to liquefied natural gas, the transportation of liquid H_2 is crucial to the global trade of H_2. To liquefy, H_2 must first be cooled to –253°C, which minimizes its volume by a factor of approximately 800, allowing for long-distance shipping [72]. The main components of the liquid H_2 shipping systems are the liquefaction plants (consume between 25% and 35% of the energy content in hydrogen), cryogenic tanks (properly insulated to minimize boil-off gas), and liquid hydrogen carriers (specialized vessels equipped with pressurized cryogenic storage, such as the Suiso Frontier).

Compressed H_2 tube trailers store H_2 in Type IV composite tanks mounted on trucks and compress it from 35 to 70 MPa [33]. They are used for medium-distance distribution, especially where pipelines are not available. Benefits of this transportation method include flexibility and rapid deployment, as well as the elimination of the need for liquefaction infrastructure and utilities in early market stages. However, their drawbacks include high transportation costs and limited payload due to weight and space constraints [73].

3.5 CODES, STANDARDS, AND RISK MITIGATION STRATEGIES

As a continuation of Sections 3.1.1 and 3.2.3 of H_2 safety, this section focuses on the international codes, standards, and mitigation strategies. As previously discussed, H_2 has high diffusivity and low molecular weight, allowing it to quickly leak through microscopic gaps. Its low ignition energy (0.02 mJ) and wide flammability range (4–75% by volume in air) make it prone to accidental ignition and combustion, posing a greater danger compared to conventional fuels [74, 75].

H_2 combustion in air can have an adiabatic flame temperature up to 2318 K, leading to significant heat release [76]. In addition, the near-invisible nature of the H_2 flames makes detection and response more difficult [77]. Due to these features, the safety of H_2 needs to be considered across the entire value chain, including generation, transportation, storage, distribution, and final-use applications.

H_2 leakage is one of the most serious risks due to its small molecular size, as it escapes easily through fittings, seals, and joints. When H_2 is mixed with air for combustion, leaked H_2 can be highly flammable. The LFL and upper flammability limit (UFL) are utilized in quantifying H_2 risks [78], i.e.,

$$\text{(LFL) } 4\% \leq \text{hydrogen concentration} \leq 75\% \text{ (UFL)} \tag{3.20}$$

Equation (3.20) indicates that at LFL, i.e., 4% H_2 by volume in air, the mixture has sufficient H_2 to combust; below 4%, the H_2 is insufficient for a flame to propagate. At UFL, i.e., 75% H_2 by volume in air, there is too little O_2 for continuous combustion and above the mixture is too rich for combustion. The adiabatic flame speed, which is much larger for H_2 (~100–120 m/s) than for NH_3 (~30–40 m/s), is correlated with the severity of the explosion [79–81].

The Hindenburg tragedy that occurred in 1937, in which a H_2-filled airship combusted during landing, is the most famous historical H_2 incident [82]. The incident drew attention to the dangers of widespread H_2 use; even later analysis suggested that flammable coating materials may have contributed to the fire. The recent H_2 incident, exemplified by the 2019 explosion at the Norway H_2 station, which was caused by defective H_2 storage equipment [83], underscores the dangers associated

TABLE 3.6
International Standards and Their Focus Areas

International Standards	Focus Areas
ISO/TC 197	Deals with hydrogen production, storage, transport, and safety
ISO 11114-4	Deals with metallic materials' resistance to hydrogen embrittlement
SAE J2579	Deals with vehicle fuel system integrity tests
NFPA 2	Deals with the hydrogen technologies code (fire safety and venting)

with cryogenic storage. These instances underscore the importance of adhering to stringent codes and standards.

Hydrogen safety is governed by a variety of international standards developed by organizations such as ISO, SAE, and NFPA. These standards ensure safe design, testing, and operation of H_2 systems. Table 3.6 highlights the international standards and their focus area as they relate to H_2.

Common risk mitigation techniques in H_2 storage include leak detection, ventilation, passive safety measures, and emergency response procedures. The migration strategies are highlighted in the following [84–86]:

i. Leak detection technologies include optical sensors such as fiber-optic sensors that enable quick and real-time monitoring, and electrochemical sensors that can detect H_2 leaks at the parts per million (ppm) level.
ii. Both ventilation and passive safety measures are included in the design facilities so that H_2 cannot accumulate. These could consist of inlet vents at floor level, roof vents, or exhaust openings, utilizing either active or passive ventilation. Also, physical barriers are required.
iii. Emergency response protocols include automatic shutdown systems, fire suppression systems for H_2 flames, and frequent staff training combined with safety exercises to ensure that employees are prepared for leaks, fires, or system failures.

REFERENCES

1. B. W. Ang, W. L. Choong, and T. S. Ng, 'Energy security: Definitions, dimensions and indexes', *Renewable and Sustainable Energy Reviews*, vol. 42, pp. 1077–1093, 2015. doi: 10.1016/j.rser.2014.10.064.
2. T. Amirthan, and M. S. A. Perera, 'The role of storage systems in hydrogen economy: A review', *Journal of Natural Gas Science and Engineering*, vol. 108, p. 104843, 2022. doi: 10.1016/j.jngse.2022.104843.
3. A. S. Mehr, A. D. Phillips, M. P. Brandon, M. T. Pryce, and J. G. Carton, 'Recent challenges and development of technical and technoeconomic aspects for hydrogen storage, insights at different scales; A state of art review', *International Journal of Hydrogen Energy*, vol. 70, pp. 786–815, 2024. doi: 10.1016/j.ijhydene.2024.05.182.

4. L. Mulky, S. Srivastava, T. Lakshmi, E. R. Sandadi, S. Gour, N. A. Thomas, S. S. Priya, 'An overview of hydrogen storage technologies – Key challenges and opportunities', *Materials Chemistry and Physics*, vol. 325, p. 129710, 2024. doi: 10.1016/j.matchemphys.2024.129710.
5. A. Mehrabianbardar, M. Shirinbayan, Z. Jendli, S. Gillet, S. Nouira, and J. Fitoussi, 'A review: Challenges, processes, and innovations in high-pressure hydrogen storage technologies', *International Journal of Material Forming*, vol. 18, no. 3, p. 77, 2025. doi: 10.1007/s12289-025-01934-3.
6. 'Hydrogen Analysis Resource Center', Hydrogen Tools, 2025. [Online]. Available: https://www.h2tools.org/hyarc/hydrogen-data/comparative-properties-hydrogen-and-other-fuels
7. F. Dawood, M. Anda, and G. M. Shafiullah, 'Hydrogen production for energy: An overview', *International Journal of Hydrogen Energy*, vol. 45, no. 7, pp. 3847–3869, 2020. doi: 10.1016/j.ijhydene.2019.12.059.
8. M. Krebsz, T. Pasinszki, S. Sreenath, and V. P. Ting, 'Advances in catalysing the hydrogen storage in main group metals and their tetrahydroborates and tetrahydroaluminates', *EES Catalysis*, vol. 3, no. 6, pp. 1196–1228, 2025. doi: 10.1039/D5EY00134J.
9. J. Zheng, X. Liu, P. Xu, P. Liu, Y. Zhao, and J. Yang, 'Development of high pressure gaseous hydrogen storage technologies', *International Journal of Hydrogen Energy*, vol. 37, no. 1, pp. 1048–1057, 2012. doi: 10.1016/j.ijhydene.2011.02.125.
10. S. S. Samantaray, S. T. Putnam, and N. P. Stadie, 'Volumetrics of hydrogen storage by physical adsorption', *Inorganics*, vol. 9, no. 6, p. 45, 2021. doi: 10.3390/inorganics9060045.
11. M. Kunowsky, J. P. Marco-Lózar, and A. Linares-Solano, 'Material demands for storage technologies in a hydrogen economy', *Journal of Renewable Energy*, vol. 2013, pp. 1–16, 2013. doi: 10.1155/2013/878329.
12. 'Heat values of various fuels', World Nuclear Association, 2025. [Online]. Available: https://world-nuclear.org/information-library/facts-and-figures/heat-values-of-various-fuels
13. M. Aziz, 'Liquid hydrogen: A review on liquefaction, storage, transportation, and safety', *Energies*, vol. 14, no. 18, p. 5917, 2021. doi: 10.3390/en14185917.
14. J. Jeon, and S. J. Kim, 'Recent progress in hydrogen flammability prediction for the safe energy systems', *Energies*, vol. 13, no. 23, p. 6263, 2020. doi: 10.3390/en13236263.
15. C. Bauer, 'Effect of hydrogen addition on the performance of methane-fueled vehicles. Part I: Effect on S.I. engine performance', *International Journal of Hydrogen Energy*, vol. 26, no. 1, pp. 55–70, 2001. doi: 10.1016/S0360-3199(00)00067-7.
16. I. Wierzba, and V. Kilchyk, 'Flammability limits of hydrogen–carbon monoxide mixtures at moderately elevated temperatures', *International Journal of Hydrogen Energy*, vol. 26, no. 6, pp. 639–643, 2001. doi: 10.1016/S0360-3199(00)00114-2.
17. R. Ono, M. Nifuku, S. Fujiwara, S. Horiguchi, and T. Oda, 'Minimum ignition energy of hydrogen–air mixture: Effects of humidity and spark duration', *Journal of Electrostatics*, vol. 65, no. 2, pp. 87–93, 2007. doi: 10.1016/j.elstat.2006.07.004.
18. SAE J2579, 'Standard for Fuel Systems in Fuel Cell and Other Hydrogen Vehicles J2579_202301', 2023. [Online]. Available: https://www.sae.org/standards/content/j2579_202301/
19. ISO 11114-4, 'Transportable gas cylinders—Compatibility of cylinder and valve materials with gas contents. Part 4: Test methods for selecting steels resistant to hydrogen embrittlement', 2017. [Online]. Available: https://www.iso.org/standard/64587.html
20. O. Ferreira, A. Fonseca, C. F. Adame, N. Bundaleski, R. Robinson, and O. M. N. D. Teodoro, 'Advancing hydrogen leak detection: Design and calibration of reference leaks', *International Journal of Hydrogen Energy*, vol. 68, pp. 1090–1096, 2024. doi: 10.1016/j.ijhydene.2024.04.328.

21. Marcogaz, 'Leak detection and repair LDAR technical recommendations based on best practices applied by EUROPEAN gas system operators', 2025. [Online]. Available: https://www.marcogaz.org/
22. M. Wetegrove *et al.*, 'Preventing hydrogen embrittlement: The role of barrier coatings for the hydrogen economy', *Hydrogen*, vol. 4, no. 2, pp. 307–322, 2023. doi: 10.3390/hydrogen4020022.
23. S. Lynch, 'Hydrogen embrittlement phenomena and mechanisms', *Corrosion Reviews*, vol. 30, no. 3–4, pp. 105–123, 2012. doi: 10.1515/corrrev-2012-0502.
24. I. M. Robertson *et al.*, 'Hydrogen embrittlement understood', *Metallurgical and Materials Transactions B*, vol. 46, no. 3, pp. 1085–1103, 2015. doi: 10.1007/s11663-015-0325-y.
25. ASTM G142-98(2022), 'Standard Test Method for Determination of Susceptibility of Metals to Embrittlement in Hydrogen Containing Environments at High Pressure, High Temperature, or Both', 2022. [Online]. Available: https://store.astm.org/standards/g142
26. A. Schiaroli *et al.*, 'A comprehensive review on liquid hydrogen transfer operations and safety considerations for mobile applications', *International Journal of Hydrogen Energy*, vol. 107, pp. 164–182, 2025. doi: 10.1016/j.ijhydene.2024.12.266.
27. ISO/TC 197, 'Hydrogen technologies', 2025. [Online]. Available: https://www.iso.org/committee/54560.html
28. NFPA 2, 'Hydrogen Technologies Code', 2025. [Online]. Available: https://www.nfpa.org/codes-and-standards/nfpa-2-standard-development/2
29. M. M. Hossain Bhuiyan, and Z. Siddique, 'Hydrogen as an alternative fuel: A comprehensive review of challenges and opportunities in production, storage, and transportation', *International Journal of Hydrogen Energy*, vol. 102, pp. 1026–1044, 2025. doi: 10.1016/j.ijhydene.2025.01.033.
30. O. Ayakdas, and L. Aydin, '*Researches on Science and Art in 21st Century Turkey*', Gece Kitapligi, 2018.
31. A. Basile, and A. Iulianelli, *Advances in Hydrogen Production, Storage and Distribution*, in Woodhead Publishing series in energy, no. 63, Woodhead Publishing, 2014.
32. M. M. Whiston, I. M. Lima Azevedo, S. Litster, C. Samaras, K. S. Whitefoot, and J. F. Whitacre, 'Hydrogen storage for fuel cell electric vehicles: Expert elicitation and a levelized cost of driving model', *Environmental Science & Technology*, vol. 55, no. 1, pp. 553–562, 2021. doi: 10.1021/acs.est.0c04145.
33. H. K. Shin, and S. K. Ha, 'A review on the cost analysis of hydrogen gas storage tanks for fuel cell vehicles', *Energies*, vol. 16, no. 13, p. 5233, 2023. doi: 10.3390/en16135233.
34. H. Luo, J. Xiao, P. Bénard, R. Chahine, and T. Yang, 'Effects of filling strategies on hydrogen refueling performance', *International Journal of Hydrogen Energy*, vol. 51, pp. 664–675, 2024. doi: 10.1016/j.ijhydene.2023.09.029.
35. V. K. Kukkapalli, S. Kim, and S. A. Thomas, 'Thermal management techniques in metal hydrides for hydrogen storage applications: A review', *Energies*, vol. 16, no. 8, p. 3444, 2023. doi: 10.3390/en16083444.
36. R. Morales-Ospino, A. Celzard, and V. Fierro, 'Strategies to recover and minimize boil-off losses during liquid hydrogen storage', *Renewable and Sustainable Energy Reviews*, vol. 182, p. 113360, 2023. doi: 10.1016/j.rser.2023.113360.
37. I. Staffell *et al.*, 'The role of hydrogen and fuel cells in the global energy system', *Energy & Environmental Science*, vol. 12, no. 2, pp. 463–491, 2019. doi: 10.1039/C8EE01157E.
38. M. Li *et al.*, 'Review on the research of hydrogen storage system fast refueling in fuel cell vehicle', *International Journal of Hydrogen Energy*, vol. 44, no. 21, pp. 10677–10693, 2019. doi: 10.1016/j.ijhydene.2019.02.208.

39. L. Yin, and Y. Ju, 'Review on the design and optimization of hydrogen liquefaction processes', *Frontiers in Energy*, vol. 14, no. 3, pp. 530–544, 2020. doi: 10.1007/s11708-019-0657-4.
40. S. Tiwari, M. J. Pekris, and J. J. Doherty, 'A review of liquid hydrogen aircraft and propulsion technologies', *International Journal of Hydrogen Energy*, vol. 57, pp. 1174–1196, 2024. doi: 10.1016/j.ijhydene.2023.12.263.
41. F. Xie, and Q. Sun, 'Comprehensive performance evaluation of densified liquid hydrogen/liquid oxygen as propulsion fuel', *Energies*, vol. 15, no. 4, p. 1365, 2022. doi: 10.3390/en15041365.
42. Kawasaki, 'World's First Liquefied Hydrogen Carrier SUISO FRONTIER Launches Building an International Hydrogen Energy Supply Chain Aimed at Carbon-free Society', 2019. [Online]. Available: https://global.kawasaki.com/en/corp/newsroom/news/detail/?f=20191211_3487
43. D. Kang, H. Mun, J. Park, and I. Lee, 'System design and economic evaluation of a liquid hydrogen superstation', *Korean Journal of Chemical Engineering*, vol. 42, no. 2, pp. 233–255, 2025. doi: 10.1007/s11814-024-00351-7.
44. A. Backurs, L. Jansons, L. Zemite, and A. Laizans, 'The practical implementation of hydrogen-based sustainable power generation backup', *Latvian Journal of Physics and Technical Sciences*, vol. 61, no. 6, pp. 69–79, 2024. doi: 10.2478/lpts-2024-0044.
45. A. A. Levikhin, and A. A. Boryaev, 'Physical properties and thermodynamic characteristics of hydrogen', *Heliyon*, vol. 10, no. 17, p. e36414, 2024. doi: 10.1016/j.heliyon.2024.e36414.
46. Ø Wilhelmsen, D. Berstad, A. Aasen, P. Nekså, and G. Skaugen, 'Reducing the exergy destruction in the cryogenic heat exchangers of hydrogen liquefaction processes', *International Journal of Hydrogen Energy*, vol. 43, no. 10, pp. 5033–5047, 2018. doi: 10.1016/j.ijhydene.2018.01.094.
47. A. Léon, *Hydrogen Technology: Mobile and Portable Applications*, Springer, 2008.
48. J. Essler, C. Haberstroh, and H. Quack, 'Report on Technology Overview and Barriers to Energy- and Cost-Efficient Large Scale Hydrogen Liquefaction', 2012. [Online]. Available: https://www.idealhy.eu/uploads/documents/IDEALHY_D1-1_Report_Tech_Overview_and_Barriers_web2.pdf
49. M. Kılıç, and A. F. Altun, 'Comprehensive thermodynamic performance evaluation of various gas liquefaction cycles for cryogenic energy storage', *Sustainability*, vol. 15, no. 24, p. 16906, 2023. doi: 10.3390/su152416906.
50. S. Hamdy, F. Moser, T. Morosuk, and G. Tsatsaronis, 'Exergy-based and economic evaluation of liquefaction processes for cryogenics energy storage', *Energies*, vol. 12, no. 3, p. 493, 2019. doi: 10.3390/en12030493.
51. T. Numazawa, and K. Matsumoto, 'Development of advanced hydrogen liquefaction system by using magnetic refrigeration technology for JST-MIRAI large-scale project', *TEION KOGAKU (Journal of Cryogenics and Superconductivity Society in Japan)*, vol. 58, no. 2, pp. 58–62, 2023. doi: 10.2221/jcsj.58.58.
52. T. Numazawa, K. Kamiya, T. Utaki, and K. Matsumoto, 'Magnetic refrigerator for hydrogen liquefaction', *Cryogenics*, vol. 62, pp. 185–192, 2014. doi: 10.1016/j.cryogenics.2014.03.016.
53. A. Naquash, N. Agarwal, and M. Lee, 'A review on liquid hydrogen storage: Current status, challenges and future directions', *Sustainability*, vol. 16, no. 18, p. 8270, 2024. doi: 10.3390/su16188270.
54. Y. Wang, X. Dai, H. You, and M. Gao, 'Research on the design of hydrogen supply system of 70 MPa hydrogen storage cylinder for vehicles', *International Journal of Hydrogen Energy*, vol. 43, no. 41, pp. 19189–19195, 2018. doi: 10.1016/j.ijhydene.2018.08.138.

55. T. Q. Hua, H.-S. Roh, and R. K. Ahluwalia, 'Performance assessment of 700-bar compressed hydrogen storage for light duty fuel cell vehicles', *International Journal of Hydrogen Energy*, vol. 42, no. 40, pp. 25121–25129, 2017. doi: 10.1016/j.ijhydene.2017.08.123.
56. R. K. Ahluwalia, J. K. Peng, H. S. Roh, T. Q. Hua, C. Houchins, and B. D. James, 'Supercritical cryo-compressed hydrogen storage for fuel cell electric buses', *International Journal of Hydrogen Energy*, vol. 43, no. 22, pp. 10215–10231, 2018. doi: 10.1016/j.ijhydene.2018.04.113.
57. 'ANSI HGV 2-2014 – Compressed hydrogen gas vehicle fuel containers', American National Standards Institute, 2014. [Online]. Available: https://webstore.ansi.org/standards/csa/csaansihgv2014r2019
58. J.-Q. Li, J.-C. Li, K. Park, S.-J. Jang, and J.-T. Kwon, 'An analysis on the compressed hydrogen storage system for the fast-filling process of hydrogen gas at the pressure of 82 MPa', *Energies*, vol. 14, no. 9, p. 2635, 2021. doi: 10.3390/en14092635.
59. A. Magliano, C. Perez Carrera, C. M. Pappalardo, D. Guida, and V. P. Berardi, 'A comprehensive literature review on hydrogen tanks: Storage, safety, and structural integrity', *Applied Sciences*, vol. 14, no. 20, p. 9348, 2024. doi: 10.3390/app14209348.
60. X. Li *et al.*, 'Exploring the gas permeability of type IV hydrogen storage cylinder liners: Research and applications', *Materials*, vol. 18, no. 13, p. 3127, 2025. doi: 10.3390/ma18133127.
61. U. Eberle, M. Felderhoff, and F. Schüth, 'Chemical and physical solutions for hydrogen storage', *Angewandte Chemie International Edition*, vol. 48, no. 36, pp. 6608–6630, 2009. doi: 10.1002/anie.200806293.
62. J. W. Makepeace, T. J. Wood, H. M. A. Hunter, M. O. Jones, and W. I. F. David, 'Ammonia decomposition catalysis using non-stoichiometric lithium imide', *Chemical Science*, vol. 6, no. 7, pp. 3805–3815, 2015. doi: 10.1039/C5SC00205B.
63. V. Negro, M. Noussan, and D. Chiaramonti, 'The potential role of ammonia for hydrogen storage and transport: A critical review of challenges and opportunities', *Energies*, vol. 16, no. 17, p. 6192, 2023. doi: 10.3390/en16176192.
64. International Energy Agency, 'Ammonia Technology Roadmap', 2021. [Online]. Available: https://www.iea.org/reports/ammonia-technology-roadmap
65. Y. Wang *et al.*, 'Low–temperature ammonia decomposition catalysts for direct ammonia solid oxide fuel cells', *Journal of the Electrochemical Society*, vol. 167, no. 6, p. 064501, 2020. doi: https://doi.org/10.1149/1945-7111/ab7b5b.
66. E. B. Agyekum, B. Tarawneh, and F. L. Rashid, 'State-of-the-art review of liquid organic hydrogen carriers for hydrogen storage and transport: Challenges and perspectives', *Applied Energy*, vol. 400, p. 126555, 2025. doi: 10.1016/j.apenergy.2025.126555.
67. W. Grochala, and P. P. Edwards, 'Thermal decomposition of the non-interstitial hydrides for the storage and production of hydrogen', *Chemical Reviews*, vol. 104, no. 3, pp. 1283–1316, 2004. doi: 10.1021/cr030691s.
68. H. Miao, Y. Yu, Y. Wan, Y. Zhang, and T. Ma, 'Levelized cost of long-distance large-scale transportation of hydrogen in China', *Energy*, vol. 310, p. 133201, 2024. doi: 10.1016/j.energy.2024.133201.
69. X. Fan, and Y. F. Cheng, 'Hydrogen pipelines and embrittlement in gaseous environments: An up-to-date review', *Applied Energy*, vol. 387, p. 125636, 2025. doi: 10.1016/j.apenergy.2025.125636.
70. A. Raj *et al.*, 'Evaluating hydrogen gas transport in pipelines: Current state of numerical and experimental methodologies', *International Journal of Hydrogen Energy*, vol. 67, pp. 136–149, May 2024. doi: 10.1016/j.ijhydene.2024.04.140.
71. K. M. Bertsch, A. Nagao, B. Rankouhi, B. Kuehl, and D. J. Thoma, 'Hydrogen embrittlement of additively manufactured austenitic stainless steel 316 L', *Corrosion Science*, vol. 192, p. 109790, 2021. doi: 10.1016/j.corsci.2021.109790.

72. K. Isano, T. Komiya, Y. Matsuda, D. Kariya, and S. Kamiya, 'Hydrogen Production—Development of Hydrogen Liquefaction Systems', 2021. [Online]. Available: https://global.kawasaki.com/en/corp/rd/magazine/182/pdf/n182en05.pdf
73. A. Burke, J. Ogden, and L. Fulton, 'Hydrogen Storage and Transport: Technologies and Costs', Institute of Transportation Studies, UC Davis, 2024. [Online]. Available: https://escholarship.org/content/qt83p5k54m/qt83p5k54m_noSplash_8bb1326c13cfb9aa3d0d376ec26d3e06.pdf
74. K. van Wingerden, 'Hydrogen – Overcoming safety challenges to achieve a cleaner future', 2022. [Online]. Available: https://www.vysusgroup.com/articles/hydrogen-safety-challenges
75. A. Kumamoto, H. Iseki, R. Ono, and T. Oda, 'Measurement of minimum ignition energy in hydrogen-oxygen-nitrogen premixed gas by spark discharge', *Journal of Physics: Conference Series*, vol. 301, p. 012039, 2011. doi: 10.1088/1742-6596/301/1/012039.
76. F. Rigas, and P. R. Amyotte, 'Myths and facts about hydrogen hazards', *Chemical Engineering Transactions*, vol. 31, pp. 913–918, 2013. doi: 10.3303/CET1331153.
77. R. W. Schefer, W. D. Kulatilaka, B. D. Patterson, and T. B. Settersten, 'Visible emission of hydrogen flames', *Combustion and Flame*, vol. 156, no. 6, pp. 1234–1241, 2009. doi: 10.1016/j.combustflame.2009.01.011.
78. H. Le, S. Nayak, and M. S. Mannan, 'Upper flammability limits of hydrogen and light hydrocarbons in air at subatmospheric pressures', *Industrial & Engineering Chemistry Research*, vol. 51, no. 27, pp. 9396–9402, 2012. doi: 10.1021/ie300268x.
79. E. A. Shchepakina *et al.*, 'Physical and chemical features of hydrogen combustion and their influence on the characteristics of gas turbine combustion chambers', *Applied Sciences*, vol. 13, no. 6, p. 3754, 2023. doi: 10.3390/app13063754.
80. L. Giannini, N. Razavi, A. Alvaro, and N. Paltrinieri, 'Embrittlement, degradation, and loss prevention of hydrogen pipelines', *MRS Bulletin*, vol. 49, no. 5, pp. 464–477, 2024. doi: 10.1557/s43577-024-00695-9.
81. B. E. Okpeke, C. Ait Aider, L. Baetcke, and S. Ehlers, 'Assessment of boil-off losses and their cost implication during liquid hydrogen tank filling with and without precooling', *Energies*, vol. 17, no. 16, p. 4092, 2024. doi: 10.3390/en17164092.
82. 'The Hindenburg Disaster', 2025. [Online]. Available: https://www.airships.net/hindenburg/disaster/
83. 'Norway fines Nel units $3 million over 2019 blast at hydrogen fuel station', Reuters, 2019. [Online]. Available: https://www.reuters.com/article/business/energy/norway-fines-nel-units-3-million-over-2019-blast-at-hydrogen-fuel-station-idUSKBN2AG2MF/
84. N. Moussa *et al.*, 'Hydrogen sensing technologies for the safe and reliable decarbonization of electric power: A review', *Energies*, vol. 17, no. 18, p. 4532, 2024. doi: 10.3390/en17184532.
85. M. Yang, and J. Dai, 'Fiber optic hydrogen sensors: A review', *Photonic Sensors*, vol. 4, no. 4, pp. 300–324, 2014. doi: 10.1007/s13320-014-0215-y.
86. 'Hydrogen Leak Detection', Hydrogen Tools, 2025. [Online]. Available: https://h2tools.org/hydrogen-leak-detection

4 Green Hydrogen Applications in Energy Systems

4.1 PREAMBLE

Green hydrogen is emerging as a cornerstone of the global transition to a carbon-neutral energy system, valued for its unique versatility as an energy vector [1, 2]. Produced through the electrolysis of water using renewable electricity, green hydrogen is a zero-emission alternative that can be utilized across a wide spectrum of energy-consuming sectors. A typical depiction of electrolysis is shown in Figure 4.1. The reaction process at the anode and cathode that resulted in the production of hydrogen is shown in Eqs. (4.1) and (4.2).

$$X^- \rightarrow X + e^- \tag{4.1}$$

$$M^+ + e^- \rightarrow M \tag{4.2}$$

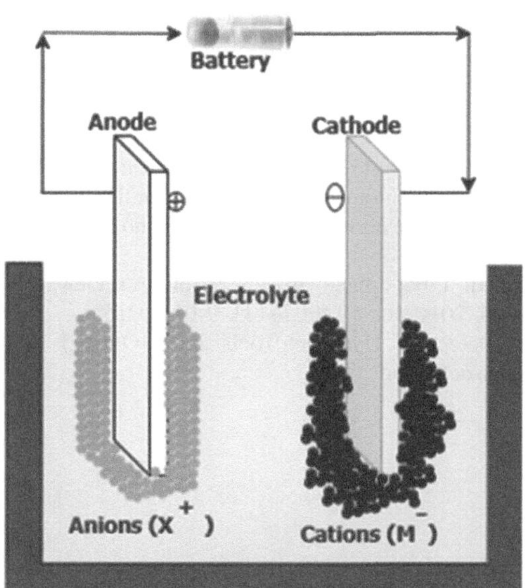

FIGURE 4.1 Schematic of electrolysis process.

DOI: 10.1201/9781003727194-4

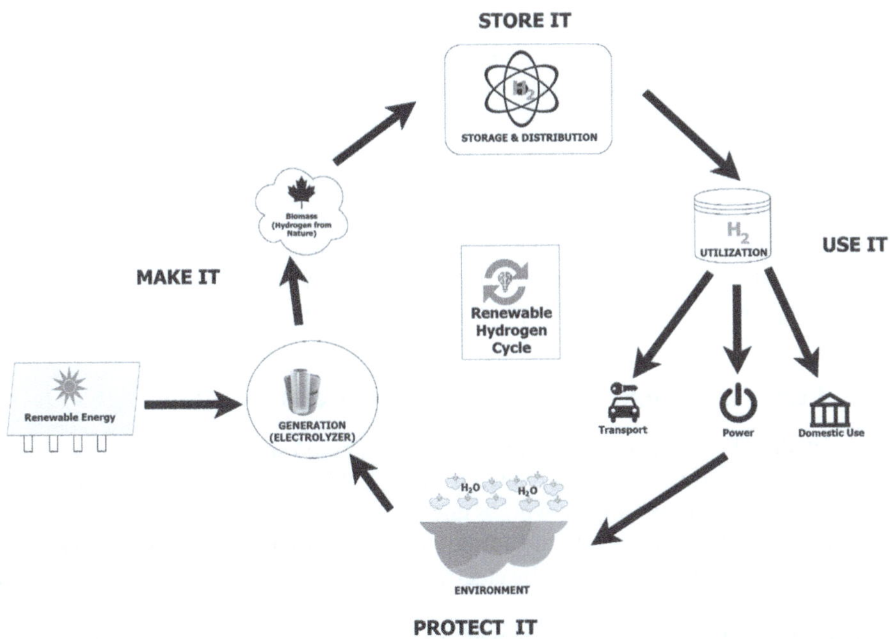

FIGURE 4.2 Schematic of hydrogen process.

Unlike electricity, which is limited in its direct applicability to some processes and sectors, hydrogen can be stored, transported, and converted into multiple energy forms, making it exceptionally adaptable, as shown in Figure 4.2.

It can be combusted for heat, used in fuel cells to generate electricity, or serve as a feedstock in industrial chemical reactions. These characteristics allow green hydrogen to complement renewable energy sources, providing solutions to intermittency and storage challenges while facilitating energy use in applications where direct electrification is impractical or inefficient [3, 4].

To achieve deep decarbonization, green hydrogen must be widely applicable. Numerous industries are now highly dependent on fossil fuels and present substantial obstacles to direct electrification, such as heavy industry, long-distance transportation, and high-temperature heat production [5, 6]. Similarly, hydrogen can replace the role of carbon-intensive fuels and chemical inputs in sectors such as steelmaking and refining. Furthermore, by collecting surplus renewable energy during times of oversupply and releasing it when demand outpaces generation, green hydrogen can be a crucial component in electrical grid balancing [7, 8]. The concept of "sector coupling," which connects the power, heat, transportation, and industrial sectors, is made possible by the diverse applications of green hydrogen. Along with improving resilience and energy efficiency, this integration also helps to create synergies that lower system emissions as a whole [9, 10]. Furthermore, green hydrogen helps enable the widespread use of intermittent renewable energy sources, such as solar and wind, by providing long-duration and seasonal energy storage.

In order to fully realize the potential of green hydrogen in these diverse applications, a concerted effort encompassing infrastructure development, innovation, and supportive legislative frameworks is needed. The total energy shift can be accelerated by the successful application of hydrogen technologies in one industry, which can spur improvements and cost savings in other industries. The variety of applications is a strategic necessity in this situation rather than just a technological benefit. It guarantees that green hydrogen can serve as a systemic facilitator of clean energy transitions globally as well as a tool for decarbonization. Accordingly, reaching global climate goals and ensuring a sustainable energy future depend on investigating and expanding the range of applications for green hydrogen [11, 12].

4.2 POWER GENERATION AND GRID STABILIZATION

Green hydrogen's incorporation into power generation systems is improving grid stability and tackling the issues related to renewable energy sources' unpredictability, which is revolutionizing the world's energy landscape. An adaptable, low-carbon energy source that can greatly aid in the decarbonization of the power industry is green hydrogen, which is created by electrolyzing solar or wind-generated electricity [13, 14].

The intermittent nature of renewable energy sources, particularly wind and solar, is one of the primary obstacles to their widespread adoption. Power shortages during times of low resource availability and surpluses during times of peak output can be caused by generation variability. As a temporal buffer, green hydrogen presents a viable remedy. Water electrolysis, which separates water into hydrogen and oxygen, can be powered by excess renewable electricity when it is available. This process, called power-to-hydrogen (P2H), stores energy in chemical form and helps prevent the curtailment of excess electricity as shown in Eqs. (4.3) and (4.4).

4.2.1 Electrolysis (Power to Hydrogen)

Electricity from renewables drives the electrolysis of water:

$$2H_2O_{(l)} \xrightarrow{\text{Electricity}} 2H_{2(g)} + O_{2(g)} \tag{4.3}$$

This reaction stores energy in the chemical bonds of hydrogen gas.

4.2.2 Hydrogen Fuel Cell (Hydrogen to Power)

Stored hydrogen is converted back to electricity using a fuel cell:

$$2H_{2(g)} + O_{2(g)} \rightarrow 2H_2O_{(l)} + \text{Electricity} + \text{Heat} \tag{4.4}$$

This reversible P2H2P cycle helps manage intermittency in solar and wind power systems by shifting energy from periods of excess to periods of deficit.

Fuel cells or turbines compatible with hydrogen can be used to convert this stored hydrogen back into power through power-to-hydrogen-to-power (P2H2P) cycles. These cycles improve grid resilience by supplying dispatchable power during times of high electricity demand or low renewable generation. P2H2P has a lower round-trip efficiency than battery storage, but its long-duration and seasonal storage possibilities make it valuable for energy systems that rely heavily on variable renewables for balance.

Additionally, grid balancing and the provision of auxiliary services—both of which are critical to preserving the stability and dependability of electrical networks—are made possible by green hydrogen. Hydrogen systems are capable of load shifting through dynamic electrolyzer operation, which modifies energy demand in real-time to match supply availability as shown in Eqs. (4.5) and (4.6) [15, 16]. This lessens dependency on peaking fossil fuel plants and helps alleviate grid congestion.

Off-peak (electrolysis):

$$2H_2O_{(l)} \xrightarrow{\text{Excess Electricity}} 2H_{2(g)} + O_{2(g)} \tag{4.5}$$

Peak-time (fuel cell or turbine):

$$2H_{2(g)} + O_{2(g)} \rightarrow 2H_2O_{(l)} + \text{Electricity} \tag{4.6}$$

Additionally, by providing stored energy at times of peak demand, hydrogen-based systems can help with peak shaving by flattening demand curves and improving the use of current grid infrastructure. Hydrogen turbines and fuel cells are particularly valuable for frequency regulation, as they can rapidly respond to grid imbalances caused by sudden changes in load or generation.

Because hydrogen is dispatchable, it can serve as a backup energy source in the event of grid outages or disruptions, enhancing system resilience overall. This function is especially important as energy systems shift to a high proportion of renewable energy sources and phase out conventional generating sources that provide inertia. Hydrogen allows for medium- to long-term and seasonal energy storage, in contrast to traditional batteries, which are more appropriate for short-duration energy storage. In areas where energy consumption and the supply of renewable energy exhibit clear seasonal trends, this skill is particularly significant. Furthermore, hydrogen storage can be used at many scales, from deep salt caverns for large-scale, grid-level storage to compressed hydrogen tanks for local energy systems [17, 18]. Power systems are more adaptable in a variety of settings thanks to the scalability and geographic flexibility of hydrogen storage options. Hydrogen systems can be progressively included in intelligent control and management frameworks with the introduction of smart grid technology. Electrolyzers, storage units, and hydrogen-fueled production assets can operate more efficiently thanks to real-time monitoring and predictive analytics made possible by IoT and AI. This digitization guarantees that hydrogen plays a role in modernizing power infrastructure as well as energy security and decarbonization.

With its vital functions in energy storage as shown in Eqs. (4.7) and (4.8), grid stabilization, and the integration of renewable energy, green hydrogen is set to become a key component of next-generation power systems. It is positioned as a strategic enabler of robust, low-carbon power networks due to its role in regulating intermittency, providing seasonal storage, and supplying ancillary services. Its full promise can only be achieved with sustained investment in infrastructure, technology development, and enabling legislative frameworks.

Hydrogen produced during high renewable generation periods (e.g., summer) can be stored and used months later (e.g., winter):

Summer (storage):

$$2H_2O \xrightarrow{\text{Renewable Electricity}} 2H_2 + O_2 \tag{4.7}$$

Winter (usage):

$$2H_2 + O_2 \rightarrow 2H_2O + \text{Electricity} \tag{4.8}$$

4.3 APPLICATIONS IN INDUSTRIES

Figure 4.3 illustrates how green hydrogen, which is created by electrolyzing water with renewable energy, is becoming a game-changer in industrial decarbonization. Green hydrogen is a sustainable energy source for challenging industrial sectors since it produces no greenhouse gases (GHGs), in contrast to gray hydrogen, which comes from fossil fuels. Because of its adaptability, it may be used for direct combustion, feedstock substitution, and incorporation into chemical synthesis and high-temperature processes, greatly lowering carbon footprints in a variety of industrial sectors [19, 20].

Green hydrogen has a lot of potential for use in high-temperature industrial processes where electrification is difficult or ineffective. Burning fossil fuels has historically produced the constant, intense heat needed by industries like steel, cement, and glassmaking, which frequently reach temperatures of above 1,000°C.

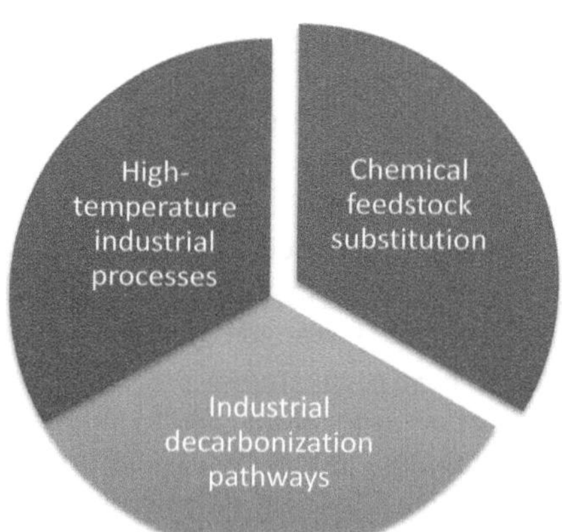

FIGURE 4.3 Applications in industries.

Hydrogen can be used as a reducing agent in Direct Reduced Iron (DRI) operations in the steel industry, replacing carbon-based sources such as coke. By emitting just water vapor rather than carbon dioxide, hydrogen-based DRI makes it possible to produce steel with almost zero emissions. Pilot projects showcasing this potential have already been started by companies in Germany and Sweden (such as HYBRIT). Hydrogen can be used to decarbonize the fuel combustion and thermal process (calcining limestone) in the cement industry, another carbon-intensive industry. Similar to this, hydrogen combustion can be used in place of natural gas burners in the glassmaking industry to melt raw materials, producing high-quality heat with less pollutants and particulate matter.

Many businesses use hydrogen as a basic chemical feedstock, and switching from gray to green hydrogen offers a clear path to lowering emissions. The production of methanol and ammonia is one of the most important uses. Under high pressure, hydrogen and nitrogen react to produce ammonia, mostly using the Haber-Bosch process. Significant amounts of hydrogen, which is now produced nearly exclusively from fossil fuels, are used in this process. A significant contributor to agricultural carbon footprints, fertilizer production emissions can be significantly reduced by switching to green hydrogen [21, 22].

Similar to this, green hydrogen and captured carbon dioxide can be used to create methanol, a solvent and precursor for polymers and other compounds. When paired with biogenic CO_2 sources, this allows for carbon-neutral or even carbon-negative routes. Hydrogen is also essential for hydrocracking and desulfurization in the petrochemical and refining sectors. Without requiring significant modification, green hydrogen may easily replace gray hydrogen in these operations, lowering Scope 1 and Scope 2 emissions. This is especially beneficial as the demand for low-carbon chemicals and fuels rises globally.

The integration of green hydrogen into industrial applications supports sector-specific decarbonization roadmaps that align with national and international climate goals. Case studies from various regions demonstrate how green hydrogen adoption is advancing.

For example, the German chemical industry is implementing hydrogen hubs linked with offshore wind farms to provide a continuous green hydrogen supply. In Japan, hydrogen is being blended into municipal gas grids to supply both industrial and residential users, creating a hybrid low-emission energy system [23, 24].

With its abundant solar resources, the Middle East is investing in large-scale green hydrogen production for exporting ammonia, combining chemical manufacturing and renewable energy to promote decarbonized international trade. These programs showcase best practices include government-industry collaborations, the colocation of industrial clusters and renewable energy sources, and circular economy models that make use of byproducts like oxygen.

Furthermore, by analyzing life-cycle emissions, costs, and return on investment, digital tools and techno-economic assessment models are directing the adoption of hydrogen. Adopting green hydrogen certification programs promotes market development for low-emission industrial goods and guarantees transparency.

Through its use in high-temperature processes, the replacement of chemical feedstocks, and its strategic integration into current systems, green hydrogen offers a feasible route for decarbonizing high-emission industrial sectors. Green hydrogen may drastically cut industrial emissions by taking the place of gray hydrogen, facilitating clean combustion, and encouraging innovation in production and storage. But widespread adoption necessitates coordinated R&D, legislative backing, and infrastructure development efforts. Green hydrogen has the potential to become a key component of sustainable industrial transformation as costs come down and technological maturity increases [25, 26].

4.4 INCORPORATION OF FUEL CELLS IN TRANSPORTATION

Particularly for forms of mobility that are less suitable for battery-electric solutions because of limitations in payload, range, or recharging, green hydrogen is becoming more widely acknowledged as a key component for decarbonizing the transportation industry. A viable substitute is the hydrogen fuel cell electric vehicle (FCEV), which uses electrochemical reactions to turn hydrogen into energy with just water vapor as a byproduct [27, 28]. They are ideal for a wide range of applications, from light-duty vehicles to heavy-duty, long-haul transport modes, due to their high energy density and quick refilling capability.

Commercial trucks, buses, and passenger cars are increasingly using hydrogen FCEVs, which provide a low-emission substitute for internal combustion engines. Large automakers like Hyundai, Honda, and Toyota have created hydrogen-powered vehicles (such as the Hyundai NEXO and Toyota Mirai), demonstrating the practicality of hydrogen for personal transportation. Cities in North America, Asia, and Europe have implemented hydrogen fuel cell buses for urban transportation. These buses are especially useful in places with high transportation demands and little time for recharging because they operate silently, produce no local emissions, and have longer ranges than battery-electric buses.

The use of FCEVs has also advanced in commercial trucking, which requires significant energy production and long-range capabilities. As interest in fleet deployment for logistics and freight transport grows, companies such as Hyundai and Nikola Motors are leading the way in developing fuel cell trucks with ranges surpassing 600 miles. Fuel cell trucks provide quicker refilling than battery-electric vehicles and are less susceptible to cold weather, which can impair battery performance.

The advantages of hydrogen go beyond transportation by road to industries where endurance and weight are crucial. To meet decarbonization targets, the rail, maritime, and aviation sectors are actively investigating fuel cell integration [29, 30]. Fuel cell trains are being tested and implemented in non-electrified rail systems, especially in Europe. The world's first hydrogen-powered passenger train, the Alstom Coradia iLint, runs in Germany and serves as an example of how hydrogen might take the role of diesel engines in regional lines.

There is tremendous pressure on the maritime industry to cut emissions, particularly in light of global laws like the IMO 2020 sulfur cap. A clean propulsion option for ferries, cargo ships, and port operations is provided by hydrogen fuel cells, which may be used with onboard hydrogen or ammonia (a hydrogen carrier). Initiatives such as the ZEROe and HYSEAS III concepts investigate hybrid systems that combine batteries and hydrogen fuel cells. Although it is still in its infancy, hydrogen is becoming more and more popular in aviation as a sustainable substitute for jet fuel. Fuel cell-powered regional aircraft are now being tested by startups like ZeroAvia and Universal Hydrogen, while Airbus has put forward hydrogen-powered aircraft proposals aimed toward 2035. Although hydrogen has a high energy density per unit weight, storage and safety issues necessitate continued research and development.

The establishment of a strong network of hydrogen refueling stations (HRS) and distribution infrastructure is essential to the growth of hydrogen mobility. Electric vehicles can be charged using readily accessible electrical grids, but FCEVs need specific refueling stations that can safely compress, store, and dispense hydrogen at high pressures (usually 350 or 700 bar). Leading the way in HRS implementation are nations like South Korea, Japan, and Germany, aided by public-private collaborations and strategic government initiatives. But in some areas, the pace of infrastructure development has been hampered by high capital expenditures, site needs, and complicated approval procedures.

Furthermore, distributing hydrogen through pipes, high-pressure tube trailers, or tankers carrying liquefied hydrogen is still a logistical difficulty. To maximize cost and availability, two options are being considered: implementing decentralized electrolyzer-based refueling stations or establishing centralized production with regional distribution hubs [31, 32]. A strong path to decarbonization is provided by the integration of hydrogen fuel cells into the transportation industry, especially in applications where batteries are limited. FCEVs offer adaptable, effective, and environmentally friendly substitutes for anything from passenger cars to large trucks, long-distance rail and sea transportation, and more. However, to fully realize this promise, infrastructure for refueling, hydrogen production, and accompanying regulatory frameworks must be developed simultaneously. Unlocking hydrogen's full potential across worldwide transportation networks would need sustained technological advancement, calculated financial investments, and cooperative international efforts.

4.5 APPLICATION IN RESIDENTIAL AND COMMERCIAL SECTORS

Green hydrogen, which is created by electrolyzing water with renewable power, has revolutionary potential to reduce the carbon footprint of both home and business energy use. Buildings, which have historically relied on fossil fuels for power and heating, are a major source of carbon emissions worldwide. A sustainable substitute is provided by incorporating hydrogen into home and business energy systems, especially for cogeneration, decentralized energy systems, and heating applications. As shown in Figure 4.4, recent technological advancements and pilot projects indicate that hydrogen can be essential to the creation of low-carbon communities and buildings [33, 34].

Heating with hydrogen-compatible boilers and combined heat and power (CHP) systems is one of the most obvious uses of hydrogen in homes and businesses. Hydrogen-ready boilers, which cleanly burn hydrogen to generate heat and hot water, can be modified or used in place of traditional natural gas boilers. Prototype hydrogen boilers that can run on blended fuels or 100% hydrogen have already been created by manufacturers like Worcester Bosch and Baxi.

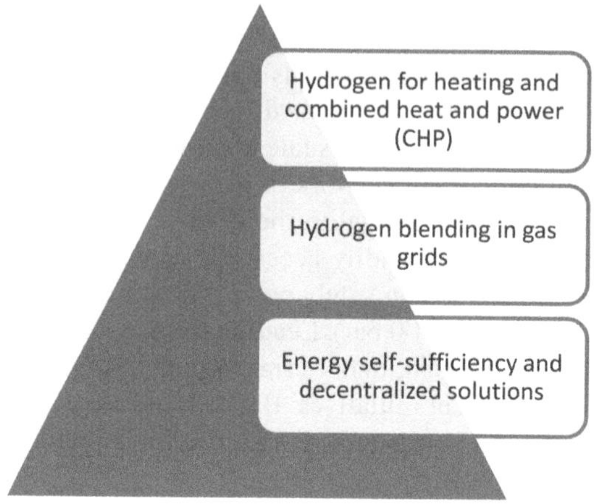

FIGURE 4.4 Application in residential and commercial sectors.

Furthermore, by co-generating heat and electricity at the place of use, fuel cell-based micro-CHP units—in particular, proton exchange membrane fuel cells, or PEMFCs—offer an effective option. Compared to traditional electricity generation, these systems can attain an overall efficiency of above 85%. The ENE-FARM project has placed more than 400,000 fuel cell devices in homes in nations like Japan, demonstrating the technology's feasibility for urban residential applications. Micro-CHP units improve local energy resilience and help balance grid load by using hydrogen for both heating and power.

Integrating hydrogen into current natural gas distribution networks is a crucial transitional approach to decarbonize heating without undergoing a significant infrastructure revamp. Several nations, including the UK, Germany, and the Netherlands, have explored blending ratios of up to 20% hydrogen by volume. Projects such as Germany's GRHYD and the UK's HyDeploy have shown that hydrogen-enriched gas mixes are both technically feasible and widely accepted.

Technically speaking, low-percentage blending is typically a cost-effective temporary fix that doesn't necessitate major modifications to end user appliances or pipes. However, methane and hydrogen have different combustion properties, safety levels, and compatibility with different materials. This calls for a thorough evaluation of burner performance, leak detection systems, and pipeline integrity. To guarantee the safe implementation of hydrogen mixes in the gas grid, regulatory frameworks and standards are being updated appropriately [35, 36].

Higher blending ratios or complete conversion to hydrogen networks may become possible as hydrogen availability rises, especially from local electrolysis driven by renewable energy sources. A scalable route for widespread residential decarbonization is provided by this strategy. As part of the larger shift to decentralized and sustainable energy sources, the idea of hydrogen-powered smart homes and energy communities is gaining popularity. Under these arrangements, homes with solar photovoltaic (PV) systems can use surplus electricity to power small-scale electrolyzers that create hydrogen. The hydrogen can subsequently be used in fuel cells or boilers after being stored on-site, serving as a long-term, emission-free energy storage medium.

This method lessens dependency on centralized power infrastructure while increasing energy self-sufficiency. The Hydrogen House Project in the United States and the H2Village in Japan, for example, have shown that hydrogen-based autonomous energy systems are technically feasible. Energy communities—groups of residential and commercial buildings connected by a shared hydrogen infrastructure—can balance energy loads, pool resources, and realize significant economies of scale in larger-scale applications.

Furthermore, real-time energy consumption optimization is made possible by connecting hydrogen systems with smart meters, demand response technologies, and building energy management systems (BEMS). This improves efficiency and reduces emissions. Green hydrogen, which offers adaptable, clean, and locally deployable solutions for heating, power generation, and energy storage, offers a strong chance to decarbonize the residential and commercial sectors. Hydrogen can improve energy resilience and complement current technologies, whether it is used in decentralized smart homes, gas grid blending, micro-CHP systems, or hydrogen boilers. Even though there are still issues with infrastructure, cost, and public opinion, continued innovation and legislative backing indicate that hydrogen will become more and more significant in the next few decades as zero-emission buildings and communities are developed [37, 38].

4.6 SECTOR INTEGRATION AND ENERGY STORAGE

Innovative strategies are required to control supply unpredictability, improve system flexibility, and accomplish deep decarbonization as a result of the growing share of renewable energy sources in global energy systems. Sector integration or sector coupling, which connects the historically distinct energy sectors of industry, transportation, heating, and power generation via shared energy carriers, is one possible approach. An important energy vector supporting this integration is green hydrogen, which is generated from renewable electricity [39]. Hydrogen makes energy systems more robust, effective, and sustainable when paired with cutting-edge energy storage technologies.

Sector coupling involves creating synergies between electricity, heating, transport, and industrial sectors by using hydrogen as a universal energy carrier. Unlike fossil fuels, hydrogen can be produced using excess renewable power and then converted back into electricity, heat, or mechanical energy, allowing surplus renewable generation to be stored and utilized across sectors.

Electrolysis is used in power-to-gas (P2G) applications to transform solar- and wind-generated electricity into hydrogen, which can then be stored or fed into natural gas infrastructures. By transferring energy across time and sectors, this connection aids in resolving the intermittent nature of renewable energy sources. The power-to-heat pathway provides decarbonized heating to homes and businesses by using hydrogen in fuel cells or boilers. In the meantime, power-to-mobility offers a clean substitute for fossil fuels by integrating hydrogen into FCEVs for rail, heavy-duty haulage, and passenger transportation.

Hydrogen takes the place of fossil fuels and feedstocks in industry, especially in high-temperature operations like chemical synthesis, steelmaking, and cement manufacture. By using green hydrogen instead of fossil fuels, this multi-sector connection lowers overall system emissions and allows for flexible load control to maximize the use of renewable energy. Multiple energy transports and storage technologies are used in integrated energy systems to produce a reliable and effective energy network. Along with batteries, bioenergy, and thermal storage, green hydrogen is essential to hybrid systems in order to increase flexibility and dependability [40].

Batteries' electrochemical storage allows for quick, short-term energy balancing, making them perfect for minute-to-hour-scale variations. For long-term or seasonal storage, they are constrained by cost and capacity. In this case, hydrogen shines because it makes long-term energy storage possible, allowing excess renewable electricity to be stored seasonally for months or more.

Through the sustainable conversion of biomass, bioenergy provides dispatchable renewable fuels for power and heating, complementing hydrogen. In integrated systems, hydrogen and bioenergy can be combined to help meet different energy needs while reducing carbon emissions.

Molten salts and phase-change materials are examples of thermal energy storage that can store heat for later use, improving heating and industrial processes' energy efficiency. When combined with fuel cells and hydrogen production, thermal storage allows for a comprehensive strategy to decarbonize the heat and power sectors. These hybrid arrangements save overall costs, increase system efficiency, and optimize energy flow patterns. For example, at peak generation, wind turbines and solar PV power an integrated microgrid. While batteries balance short-term swings and thermal storage provides heat, excess electricity powers electrolyzers that produce hydrogen, which is then stored and converted into fuel cells during periods of low renewable supply [41]. These systems lessen reliance on fossil fuels, guarantee a steady supply, and optimize the use of renewable resources.

Hydrogen is a very adaptable vector for energy systems due to its inherent qualities as an energy carrier, including its high energy density, adaptability in end use applications, and compatibility with current infrastructure. Under changing demand patterns and rising renewable penetration, this flexibility improves the energy networks' resilience and adaptability. Compressed gas, liquefied, or chemically bound in carriers are some of the ways that hydrogen can be kept, allowing for the temporal and spatial separation of production and consumption. This feature makes it possible to balance energy over both short and long durations, which is crucial for including the fluctuation of seasonal renewable power.

Moreover, integrating hydrogen into natural gas grids offers a low-carbon transition route while preserving the stability of the energy supply. Decarbonization of transportation and mobility is facilitated by dedicated hydrogen pipelines and filling facilities. Hydrogen turbines and fuel cells provide dispatchable power generating choices that aid in peak load management and grid frequency regulation. By diversifying energy sources and supply chains, hydrogen's adaptability also supports industry resilience. For instance, hydrogen-fueled microgrids or backup generators sustain vital services during power outages or supply interruptions, promoting energy security and disaster recovery.

Green hydrogen provides a revolutionary method to create robust, adaptable, and low-carbon energy systems when combined with complementary energy storage technologies in the power, heating, transportation, and industrial sectors. Sector coupling solutions promote cross-sector decarbonization, allow for long-duration energy storage, and optimize the use of renewable resources [42]. Hydrogen's function as a flexible energy vector will be essential to the sustainable transition as energy systems change. To fully realize hydrogen's potential in integrated energy solutions globally, further innovation, infrastructure development, and governmental assistance are required.

4.7 ENVIRONMENTAL IMPACTS AND TECHNO-ECONOMIC VIABILITIES

Figure 4.5 illustrates how green hydrogen has become a viable clean energy carrier as global efforts to decarbonize energy systems and slow down climate change increase. Assessing its potential as a sustainable substitute for fossil fuels requires an understanding of its techno-economic feasibility and environmental effects [43].

The environmental advantages of green hydrogen throughout its life cycle, resource and land-use concerns, and the changing economic environment influencing its implementation are all covered in this.

Compared to alternatives based on fossil fuels, green hydrogen production, mainly by water electrolysis driven by renewable energy, offers substantial environmental benefits. From feedstock extraction to end-use, the hydrogen value chain's

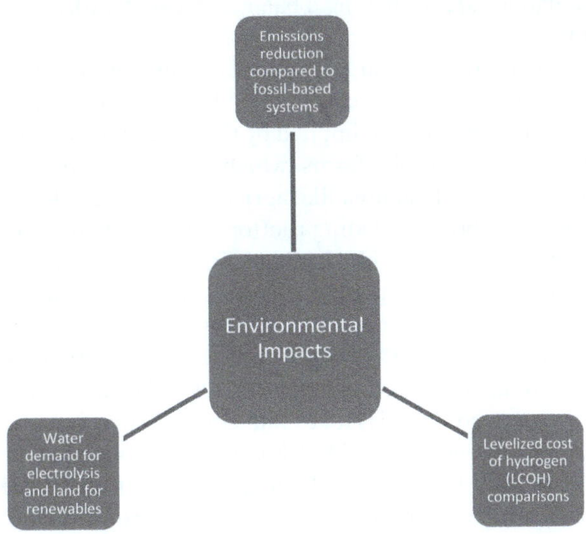

FIGURE 4.5 Environmental impacts and techno-economic viabilities.

GHG emissions, pollutant outputs, and resource consumption are all assessed using a life cycle assessment (LCA) methodology.

Green hydrogen significantly lowers carbon emissions as compared to gray hydrogen, which is made from natural gas by steam methane reforming. If it is fueled by renewable energy, its operational GHG footprints can frequently be almost zero. Research shows that using renewable electricity can reduce CO_2 emissions by 90–100% throughout the course of its life cycle, which significantly helps achieve climate mitigation targets. Green hydrogen replaces fossil fuels and their related emissions of CO_2, sulfur oxides (SO_x), nitrogen oxides (NO_x), and particulate matter when it is incorporated into industries including heavy manufacturing, transportation, and power generation. Public health and local air quality are improved by this decrease. Furthermore, green hydrogen's contribution to grid flexibility permits a greater penetration of renewable energy, which indirectly increases environmental benefits by reducing curtailment and the need for fossil fuel backup [44].

However, the electrical source utilized for electrolysis has a significant impact on the environmental benefits. Overall, carbon reductions are reduced if grid electricity has a substantial fossil fuel component. Therefore, to optimize environmental benefits, hydrogen production must be coupled with either surplus renewable electricity or dedicated renewable generating.

Electrolysis, which separates water molecules into hydrogen and oxygen, necessitates significant water resources for the creation of green hydrogen. Water consumption typically ranges around 9 l/kg of hydrogen produced, though it varies depending on the type of electrolyzer and operating conditions. In areas with limited water resources, this water footprint is concerning even though it is small in comparison to other industrial operations. Potential ways to lessen water-related limitations include integrating desalination or wastewater treatment, reusing electrolyzer cooling water, and procuring water sustainably. Life cycle studies stress how crucial site-specific evaluations are to striking a balance between hydrogen deployment and water availability.

Green hydrogen also requires large-scale solar, wind, or hydropower installations because it depends on renewable electricity. Land use consequences for the construction of renewable infrastructure are implied by this criterion. Significant surface area is needed for wind parks and solar farms, which can have an impact on biodiversity, ecosystems, and competing land uses like agriculture [45]. Land-use disputes can be lessened by creative methods, including rooftop solar, offshore wind, and agrivoltaics, which combine solar panels with farmland. Furthermore, transmission losses are reduced and resource efficiency is increased by strategically integrating renewable energy sources and hydrogen manufacturing plants.

One important indicator for assessing the economic viability of hydrogen production paths is the levelized cost of hydrogen (LCOH). The average lifetime cost per unit of hydrogen generated (LCOH) takes into consideration system efficiency, electricity costs, capital expenditures (CAPEX), and operating expenditures (OPEX). In the past, gray or blue hydrogen—which is made from fossil fuels with carbon capture—has been less expensive than green hydrogen. However, costs have significantly decreased due to economies of scale, rapid developments in electrolyzer technology, and falling prices for renewable electricity. According to recent estimations, in ideal areas with a wealth of inexpensive renewable energy sources, green hydrogen LCOH can approach $2 to $3 per kilogram, and in certain situations, it can even approach parity with hydrogen derived from fossil fuels.

In order to speed up market adoption, policy frameworks like carbon pricing, tax credits, and subsidies are essential. The relative appeal of green hydrogen is raised by carbon pricing schemes that absorb the environmental costs of fossil fuels. In a similar vein, direct financial incentives lower upfront investment barriers for the deployment of electrolyzers and the increase of renewable capacity. Scaling the hydrogen economy also requires investments in infrastructure, such as refueling stations, storage facilities, and hydrogen pipelines [46]. International collaboration and public-private partnerships can aid in raising the necessary funds.

By significantly lowering life cycle GHG emissions and local pollutants, green hydrogen offers strong environmental advantages and is positioned as a major facilitator of global decarbonization. However, sustainable resource planning and technical advancements are needed to address issues with land demand and water use. Green hydrogen's techno-economic prospects are quickly improving because of advancements in technology, falling prices for renewable energy, and favorable legislative frameworks. Unlocking hydrogen's promise as a pillar of a sustainable, low-carbon energy future will need sustained efforts to lower the LCOH and construct infrastructure. The key to the successful large-scale deployment of green hydrogen continues to be the integration of economic viability and environmental sustainability [47, 48].

4.8 CONCLUSION

An essential component of the global effort to create a resilient and decarbonized energy system is green hydrogen. Its diverse range of uses includes industries that are frequently challenging to decarbonize with direct electrification alone, such as transportation, industrial processing, power generation, home heating, and energy storage. Green hydrogen is a versatile, emission-free energy source that not only takes the place of fossil fuels in carbon-intensive operations but also improves the dependability and efficiency of renewable energy systems by facilitating grid stabilization and long-duration energy storage. Its usefulness as a systemic solution for energy transformation is highlighted by its ability to act as a bridge between sectors, powering fuel cells in automobiles, producing heat in industries, and storing excess electricity.

But more than just technological preparedness is needed for green hydrogen to be successfully incorporated into global energy infrastructure. In order to realize its full potential, integrated planning that takes supply chains, infrastructure, and end-use requirements into account holistically is necessary. In order to coordinate investments, align standards, and guarantee the interoperability of hydrogen technologies, cross-sector collaboration is crucial. To overcome present obstacles, such as high production costs, inadequate infrastructure, and unclear regulations, governments, businesses, and academia must collaborate.

Furthermore, scaling up the production and use of green hydrogen requires international cooperation. Globally equal access to hydrogen technology can be promoted and cost reductions accelerated by coordinated legislation, global funding sources, and cooperative research and development initiatives. In conclusion, even if green hydrogen is not a panacea, its strategic application in a variety of industries will be essential to creating an energy future that is secure, sustainable, and climate resilient [49, 50].

REFERENCES

1. J. L. Hinojosa, S. Villamizer, N. Gama (2023) Greeen Hydrogen Opportunities for the Caribbean. G. Jackson, J. Williams, G. Castellanos, R. G. Aiello, C. Gischler and A. B. Teixeira (Eds.). https://doi.org/10.18235/0004621
2. M. Lambert, and B. Fattouh, 'The role of hydrogen in the energy transition', in *Oxford Energy Forum: A Quarterly Journal for Debating Energy Issues and Policies* (No. 127), The Oxford Institute for Energy Studies, 2021.
3. A. H. Schrotenboer, A. A. Veenstra, M. A. uit het Broek, and E. Ursavas, 'A green hydrogen energy system: Optimal control strategies for integrated hydrogen storage and power generation with wind energy', *Renewable and Sustainable Energy Reviews*, vol. 168, p. 112744, 2022.
4. S. M. Tatar, H. Akulker, H. Sildir, and E. Aydin, 'Optimal design and operation of integrated microgrids under intermittent renewable energy sources coupled with green hydrogen and demand scenarios', *International Journal of Hydrogen Energy*, vol. 47, no. 65, pp. 27848–27865, 2022.
5. G. P. Thiel, and A. K. Stark, 'To decarbonize industry, we must decarbonize heat', *Joule*, vol. 5, no. 3, pp. 531–550, 2021.
6. Z. Y. Xu, R. Z. Wang, and C. Yang, 'Perspectives for low-temperature waste heat recovery', *Energy*, vol. 176, pp. 1037–1043, 2019.
7. S. Furfari, and A. Clerici, 'Green hydrogen: The crucial performance of electrolysers fed by variable and intermittent renewable electricity', *The European Physical Journal Plus*, vol. 136, no. 5, p. 509, 2021.
8. A. Rabiee, A. Keane, and A. Soroudi, 'Green hydrogen: A new flexibility source for security constrained scheduling of power systems with renewable energies', *International Journal of Hydrogen Energy*, vol. 46, no. 37, pp. 19270–19284, 2021.
9. A. E. H. Berjawi, S. L. Walker, C. Patsios, and S. H. R. Hosseini, 'An evaluation framework for future integrated energy systems: A whole energy systems approach', *Renewable and Sustainable Energy Reviews*, vol. 145, p. 111163, 2021.
10. A. N. Matheri, B. Mohamed, and J. C. Ngila, 'Smart climate resilient and efficient integrated waste to clean energy system in a developing country: Industry 4.0', in *African Handbook of Climate Change Adaptation*, Springer International Publishing, 2021, pp. 1053–1080.
11. F. Eljack, and M. K. Kazi, 'Prospects and challenges of green hydrogen economy via multi-sector global symbiosis in Qatar', *Frontiers in Sustainability*, vol. 1, p. 612762, 2021.
12. F. Qureshi, M. Yusuf, H. Kamyab, D. V. N. Vo, S. Chelliapan, S. W. Joo, and Y. Vasseghian, 'Latest eco-friendly avenues on hydrogen production towards a circular bioeconomy: Currents challenges, innovative insights, and future perspectives', *Renewable and Sustainable Energy Reviews*, vol. 168, p. 112916, 2022.
13. T. R. Ayodele, and J. L. Munda, 'Potential and economic viability of green hydrogen production by water electrolysis using wind energy resources in South Africa', *International Journal of Hydrogen Energy*, vol. 44, no. 33, pp. 17669–17687, 2019.
14. T. Demirdelen, F. Ekinci, B. D. Mert, Karasu, and M. Tümay, 'Green touch for hydrogen production via alkaline electrolysis: The semi-flexible PV panels mounted wind turbine design, production and performance analysis', *International Journal of Hydrogen Energy*, vol. 45, no. 18, pp. 10680–10695, 2020.

15. M. G. Dozein, A. Jalali, and P. Mancarella, 'Fast frequency response from utility-scale hydrogen electrolyzers', *IEEE Transactions on Sustainable Energy*, vol. 12, no. 3, pp. 1707–1717, 2021.
16. B. Flamm, C. Peter, F. N. Büchi, and J. Lygeros, 'Electrolyzer modeling and real-time control for optimized production of hydrogen gas', *Applied Energy*, vol. 281, p. 116031, 2021.
17. M. P. Alves, W. Gul, C. A. Cimini Junior, and S. K. Ha, 'A review on industrial perspectives and challenges on material, manufacturing, design and development of compressed hydrogen storage tanks for the transportation sector', *Energies*, vol. 15, no. 14, p. 5152, 2022.
18. A. G. Olabi, A. A. Abdelghafar, A. Baroutaji, E. T. Sayed, A. H. Alami, H. Rezk, and M. A. Abdelkareem, 'Large-scale hydrogen production and storage technologies: Current status and future directions', *International Journal of Hydrogen Energy*, vol. 46, no. 45, pp. 23498–23528, 2021.
19. E. Muh, F. Tabet, and S. Amara, 'Biomass conversion to fuels and value-added chemicals: A comprehensive review of the thermochemical processes', *Current Alternative Energy*, vol. 4, no. 1, pp. 3–25, 2021.
20. F. Ueckerdt, C. Bauer, A. Dirnaichner, J. Everall, R. Sacchi, and G. Luderer, 'Potential and risks of hydrogen-based e-fuels in climate change mitigation', *Nature Climate Change*, vol. 11, no. 5, pp. 384–393, 2021.
21. M. Farghali, A. I. Osman, K. Umetsu, and D. W. Rooney, 'Integration of biogas systems into a carbon zero and hydrogen economy: A review', *Environmental Chemistry Letters*, vol. 20, no. 5, pp. 2853–2927, 2022.
22. S. Mathur, H. Waswani, D. Singh, and R. Ranjan, 'Alternative fuels for agriculture sustainability: Carbon footprint and economic feasibility', *AgriEngineering*, vol. 4, no. 4, pp. 993–1015, 2022.
23. M. Cavana, A. Mazza, G. Chicco, and P. Leone, 'Electrical and gas networks coupling through hydrogen blending under increasing distributed photovoltaic generation', *Applied Energy*, vol. 290, p. 116764, 2021.
24. K. Topolski, E. P. Reznicek, B. C. Erdener, C. W. San Marchi, J. A. Ronevich, L. Fring, K. Simmons, O. J. G. Fernandez, B. M. Hodge, and M. Chung, 2022. Hydrogen blending into natural gas pipeline infrastructure: Review of the state of technology. (p. 82477). National Renewable Energy Laboratory.
25. A. Patonia, and R. Poudineh, 2022. *Cost-competitive green hydrogen: how to lower the cost of electrolysers?* (No. 47). OIES Paper: EL. https://www.econstor.eu/bitstream/10419/253279/1/1787058085.pdf
26. E. Taibi, R. Miranda, M. Carmo, and H. Blanco, 2020. Green hydrogen cost reduction. https://isk.nz/wp-content/uploads/IRENA_Green_hydrogen_cost_2020.pdf
27. Y. Manoharan, S. E. Hosseini, B. Butler, H. Alzhahrani, B. T. F. Senior, T. Ashuri, and J. Krohn, 'Hydrogen fuel cell vehicles; current status and future prospect', *Applied Sciences*, vol. 9, no. 11, p. 2296, 2019.
28. J. Shin, W. S. Hwang, and H. Choi, 'Can hydrogen fuel vehicles be a sustainable alternative on vehicle market?: Comparison of electric and hydrogen fuel cell vehicles', *Technological Forecasting and Social Change*, vol. 143, pp. 239–248, 2019.
29. P. Fragkos, 'Decarbonizing the international shipping and aviation sectors', *Energies*, vol. 15, no. 24, p. 9650, 2022.
30. M. Sharmina, O. Y. Edelenbosch, C. Wilson, R. Freeman, D. E. H. J. Gernaat, P. Gilbert, A. Larkin, E. W. Littleton, M. Traut, D. P. Van Vuuren, and N. E. Vaughan, 'Decarbonising the critical sectors of aviation, shipping, road freight and industry to limit warming to 1.5–2', *Climate Policy*, vol. 21, no. 4, pp. 455–474, 2021.
31. O. Bethoux, 'Hydrogen fuel cell road vehicles and their infrastructure: An option towards an environmentally friendly energy transition', *Energies*, vol. 13, no. 22, p. 6132, 2020.

32. P. Lubello, M. Pasqui, A. Mati, and C. Carcasci, 2022. Assessment of Hydrogen Based Long Term Electrical Storage in Residential Energy Systems. *Smart Energy*, vol. 8, p.100088. https://www.sciencedirect.com/science/article/pii/S2666955222000260
33. W. Chu, M. Vicidomini, F. Calise, N. Duić, P. A. Østergaard, Q. Wang, and M. da Graça Carvalho, 'Recent advances in low-carbon and sustainable, efficient technology: Strategies and applications', *Energies*, vol. 15, no. 8, p. 2954, 2022.
34. G. A. Reigstad, S. Roussanaly, J. Straus, R. Anantharaman, R. de Kler, M. Akhurst, N. Sunny, W. Goldthorpe, L. Avignon, J. Pearce, and S. Flamme, 'Moving toward the low-carbon hydrogen economy: Experiences and key learnings from national case studies', *Advances in Applied Energy*, vol. 8, p. 100108, 2022.
35. K. E. Jeffrey, P. Hooker, C. Manson-Whitton, C. Spriggs, and S. McEwen, 2019. Safety and Regulatory Challenges of Using Hydrogen/Natural Gas Blends in the UK. https://hysafe.info/uploads/papers/2019/139.pdf
36. F. Salehi, R. Abbassi, M. Asadnia, B. Chan, and L. Chen, 'Overview of safety practices in sustainable hydrogen economy – An Australian perspective', *International Journal of Hydrogen Energy*, vol. 47, no. 81, pp. 34689–34703, 2022.
37. N. V. Emodi, H. Lovell, C. Levitt, and E. Franklin, 'A systematic literature review of societal acceptance and stakeholders' perception of hydrogen technologies', *International Journal of Hydrogen Energy*, vol. 46, no. 60, pp. 30669–30697, 2021.
38. J. A. Gordon, N. Balta-Ozkan, and S. A. Nabavi, 'Homes of the future: Unpacking public perceptions to power the domestic hydrogen transition', *Renewable and Sustainable Energy Reviews*, vol. 164, p. 112481, 2022.
39. P. S. Onen, G. Mokryani, and R. H. Zubo, 'Planning of multi-vector energy systems with high penetration of renewable energy source: A comprehensive review', *Energies*, vol. 15, no. 15, p. 5717, 2022.
40. A. Z. Arsad, M. A. Hannan, A. Q. Al-Shetwi, M. Mansur, K. M. Muttaqi, Z. Y. Dong, and F. Blaabjerg, 'Hydrogen energy storage integrated hybrid renewable energy systems: A review analysis for future research directions', *International Journal of Hydrogen Energy*, vol. 47, no. 39, pp. 17285–17312, 2022.
41. S. Cloete, O. Ruhnau, and L. Hirth, 'On capital utilization in the hydrogen economy: The quest to minimize idle capacity in renewables-rich energy systems', *International Journal of Hydrogen Energy*, vol. 46, no. 1, pp. 169–188, 2021.
42. J. S. Homer, J. S. Rice, K. Mongird, and K. Oikonomou, 2021. *Coordination and Planning for Water and Power System Resilience (Workshop Report)* (No. PNNL-31880). Pacific Northwest National Lab. (PNNL), Richland, WA (United States).
43. S. G. Mungodla, L. Z. Linganiso, S. Mlambo, and T. Motaung, 'Economic and technical feasibility studies: Technologies for second generation biofuels', *Journal of Engineering, Design and Technology*, vol. 17, no. 4, pp. 670–704, 2019.
44. J. Oyekale, M. Petrollese, V. Tola, and G. Cau, 'Impacts of renewable energy resources on effectiveness of grid-integrated systems: Succinct review of current challenges and potential solution strategies', *Energies*, vol. 13, no. 18, p. 4856, 2020.
45. J. Y. Kim, D. Koide, F. Ishihama, T. Kadoya, and J. Nishihiro, 'Current site planning of medium to large solar power systems accelerates the loss of the remaining semi-natural and agricultural habitats', *Science of the Total Environment*, vol. 779, p. 146475, 2021.
46. J. Kurtz, S. Sprik, and T. H. Bradley, 'Review of transportation hydrogen infrastructure performance and reliability', *International Journal of Hydrogen Energy*, vol. 44, no. 23, pp. 12010–12023, 2019.
47. E. Gobina, A. Giwa, and A. Ben-Aron, 'Use of membrane technology to enhance the integration of hydrogen into processing and manufacturing', *Membrane Technology*, vol. 2022, no. 8, 2022.

48. J. T. Perryman, and J. M. Velazquez, 'Design principles for multinary metal chalcogenides: Toward programmable reactivity in energy conversion', *Chemistry of Materials*, vol. 33, no. 18, pp. 7133–7147, 2021.
49. M. G. Rasul, M. A. Hazrat, M. A. Sattar, M. I. Jahirul, and M. J. Shearer, 'The future of hydrogen: Challenges on production, storage and applications', *Energy Conversion and Management*, vol. 272, p. 116326, 2022.
50. M. Yue, H. Lambert, E. Pahon, R. Roche, S. Jemei, and D. Hissel, 'Hydrogen energy systems: A critical review of technologies, applications, trends and challenges', *Renewable and Sustainable Energy Reviews*, vol. 146, p. 111180, 2021.

5 Basics of Computational Fluid Dynamics for Energy Systems

5.1 COMPUTATIONAL FLUID DYNAMICS FUNDAMENTALS

The field of Computational Fluid Dynamics (CFD) utilizes numerical approaches and algorithms implemented in software and executed on computers to simulate mass transfer, heat transfer, and fluid flow (including liquids and gases), thereby predicting the behavior of systems under different physical conditions. According to ANSYS [1], CFD is a science implemented in computer code to predict the flow of liquids and gases, utilizing the governing equations (conservation of mass, energy, and momentum).

Historically, analytical solutions for simple geometries or experimental approaches were used to solve fluid dynamics problems. Over the past several decades, as computer hardware and numerical algorithms have improved (especially in the early 21st century), CFD has become a key tool that supports these methods, allowing for the study of complex geometries, turbulent processes, transient flows, coupled mass and heat transfer, and so on. Recent techniques have expanded to handle complex real-world domains [2]. The essential milestones in CFD evolution are shown in Figure 5.1.

CFD has become practically essential and a much-needed tool in the field of energy systems engineering for the design, optimization, and diagnosis of components such as heat exchangers, combustion chambers, pipes, ducts, etc. The applications are also essential in renewable energy systems, such as wind turbines, fuel

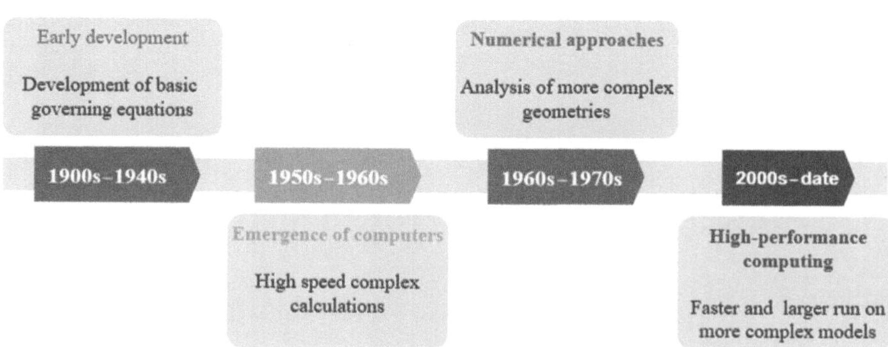

FIGURE 5.1 The milestone of CFD evolution.

DOI: 10.1201/9781003727194-5

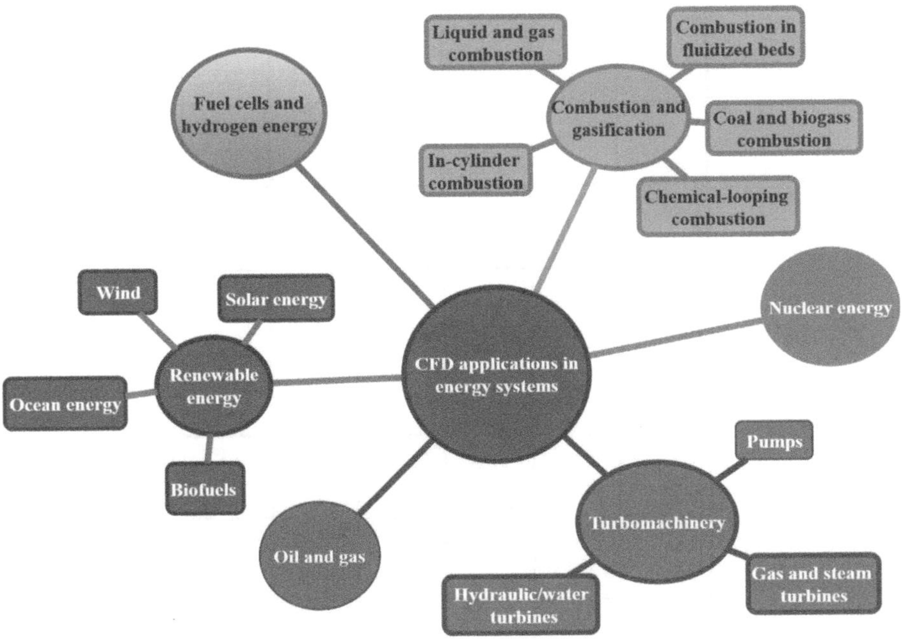

FIGURE 5.2 CFD applications in energy systems [3].

cells, and solar collectors. Figure 5.2 depicts some of the applications of CFD in energy systems engineering.

Any CFD simulation typically consists of three steps or components: pre-processing, solution (or solving), and post-processing. These steps are discussed next.

5.1.1 Pre-processing

The pre-processing primarily involves geometry creation, meshing or discretization, and specification of mathematical or physical models. The creation or acquisition of geometry is typically a CAD model that represents a physical domain, such as the shape of objects.

The domain is discretized or divided into finite or small control volumes, cells, or elements, which could be majorly grouped into structured and unstructured elements [4]. The elements can be one-dimensional (1D), two-dimensional (2D), or three-dimensional (3D). Structured meshing involves dividing the domain into regularly arranged elements, where each cell follows a predictable, grid-like pattern.

These structures are shown in Figure 5.3. The standard features of the structured meshes are in their cell arrangements (typically quadrilateral for 2D or hexahedral for 3D), grid topology (each element has a fixed number of neighbors, which can often be identified using a i, j, and k indexing system), alignment (elements aligning geometry, especially near boundaries or flow direction), and orderly connectivity.

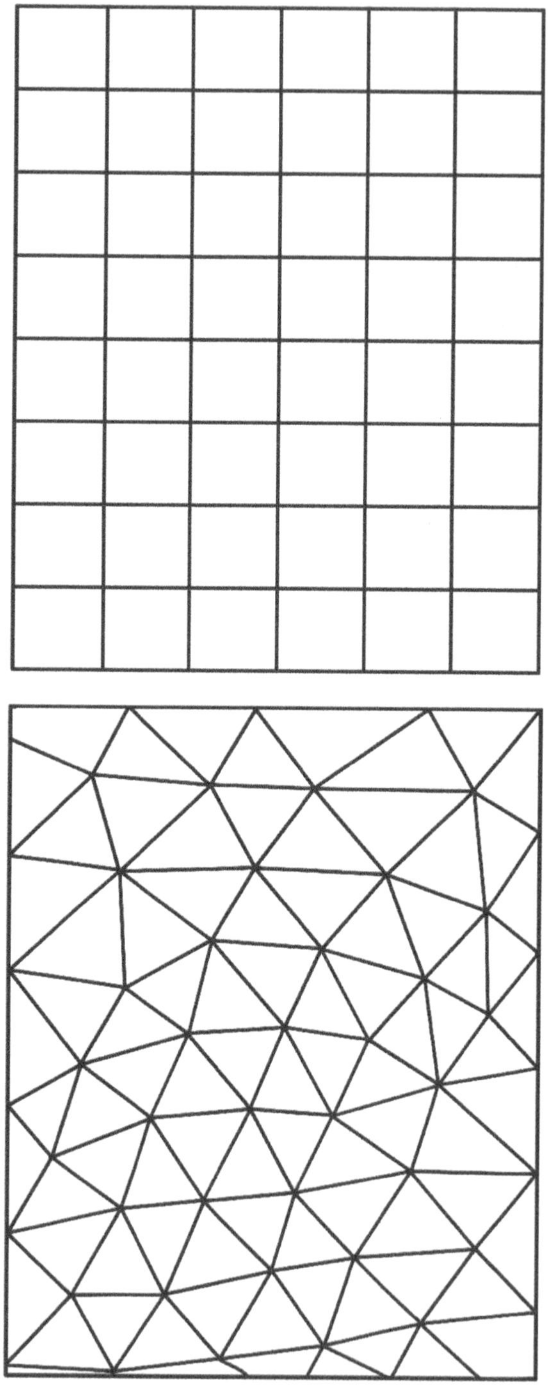

FIGURE 5.3 Two-dimensional (2D) (a) structured and (b) unstructured meshes in CFD.

Structured meshes are suitable for simple geometries. They have high accuracy per cell due to their regular topology and effective solver performance. This results from straightforward data storage and connectivity.

The structured mesh is effective for resolving boundary layers where controlled grid refinement is required. Despite these benefits, their drawbacks include the need for a large number of elements to capture complex shapes, and they have limited flexibility for these shapes [5].

Unstructured mesh, on the other hand, utilizes irregularly arranged elements that vary in shape, size, and connectivity, offering high flexibility for complex geometries, as depicted in Figure 5.3b. They are typically triangular and quadrilateral for 2D geometries and tetrahedra, pyramids, prisms, and polyhedral for 3D geometries and are arranged randomly and irregularly [6]. However, the accuracy of these unstructured meshes is low per element due to their irregular shapes. There is also a high computational cost associated with this type of mesh, which poses significant challenges in accurately modeling boundary layers. In many energy system applications, hybrid meshes are employed due to the combined features of both structures [4, 6, 7].

Specification of the physical models to solve the physical problem, which follows discretization, is the final stage in the pre-processing stage. The choice of these models depends on the specific characteristics of the physical problem.

5.1.2 Solving and Numerical Solution

This solving stage involves converting the governing equations of the fluid flow and other thermodynamic equations into a computerized form that can be handled numerically. Then, this form of equations is systematically solved to derive approximate variable solutions. These variables could include pressure, temperature, velocity, species concentrations, and other relevant factors. This process consists of crucial steps such as discretization, linearization, time stepping, iterative solution, and convergence approaches, each of which is necessary to achieve numerical stability, precision, and convergence.

The discretization of the governing equations (partial differential equations, PDEs) is the first step in the numerical solution process. These may include the continuity equation, the energy equation, and the Navier-Stokes equations. Other transport equations, such as species transport, turbulence models, or reaction kinetics, may also be discretized.

Discretization involves the approximation of the continuous variables, such as temperature, pressure, and velocity at discrete points, or nodes within the computational domain by dividing it up into a finite number of elements. As a result, the PDEs are transformed into an algebraic system of equations that can be solved numerically. Several discretization approaches exist, which are the Finite Difference Method (FDM), Finite Element Method (FEM), and Finite Volume Method (FVM).

The FDM is regarded as the oldest method, and it uses the difference quotient to approximate the differential coefficient that appears in PDEs, resulting in a set of Ordinary Differential Equations (ODEs) that are solved numerically [8,9]. The differential equation is approximated at each grid point by substituting approximations (nodal values of functions) for the partial derivatives [10]. As a result, each grid node has one algebraic equation with the variable value at that node and a specific number of neighbor nodes showing as unknowns. This method has been applied mainly to structured grids.

The FEM uses piecewise continuous functions (shape functions) over finite elements to approximate the elements, which are usually unstructured [10]. The FEM is capable of handling arbitrary geometries and has been demonstrated to have optimality properties for specific kinds of equations and is comparatively simple to analyze mathematically. However, it is more challenging to identify effective solution methods since the matrices of the linearized equations are not as structured as those of the structured grids, thus making it more computationally costly.

The FVM is the most used method for engineering CFD. It ensures both local and global conservation of mass, momentum, and energy by applying the integral form of the conservation laws to every control volume [10]. The method is robust and works well with both structured and unstructured meshes [11, 12]. However, when compared to the FDM, it is more challenging to develop in 3D for approaches of order higher than the second order [10].

After the discretization, the resulting algebraic equations may be nonlinear primarily due to convection terms and turbulence models in the Navier-Stokes equations. Before these equations can be solved effectively, they must be linearized. In linearization, Taylor series expansion and iterative substitution are popular methods that approximate nonlinear terms using values from earlier iterations. This enables the equations to be handled as linear systems at each iteration [10].

The time stepping involves the steady-state and transient simulations. The steady-state simulations assume that the flow variables remain constant over time. Iterations typically continue until the residuals indicate a steady solution, having reached a convergence criterion. Transient simulations are typically required when flow variables change over time, such as in fluid-structured interactions, combustion, or pulsating flows.

The transient simulations involve two schemes: explicit and implicit schemes. The explicit scheme utilizes the current step's known information to compute the solution at the next time step. These approaches require small time steps (i.e., Courant-Friedrichs-Lewy or CFL condition) and are straightforward but conditionally stable. The implicit schemes solve a system of equations involving both present and future time levels. These approaches are appropriate for stiff problems since they allow for larger time steps and are unconditionally stable [12]. The computational resources and stability requirements determine whether to use either the explicit or implicit methods.

The iterative solution involves solving large algebraic systems of equations utilizing the correct solvers. The discretization of the complex physical CFD problem usually leads to an extensive system of coupled algebraic equations. Typical iterative solvers are as follows [10]:

i. Jacobi method: This is the simplest iterative method; however, it is slow to converge, is more expensive, and is rarely used alone in practical CFD problems.

ii. Gauss-Seidel method: This method, although it converges faster (twice) than the Jacobi method, is still insufficient and is often used as a basis for other iterative techniques.
iii. Successive Over-relaxation: This approach improves the Gauss-Seidel convergence by adding an over-relaxation factor. The number of iterations corresponds to the total number of grid points in a single direction when the optimum over-relaxation is used. This represents a considerable improvement over the previous approaches.
iv. Conjugate Gradient (CG): This method is generally effective for symmetric and positive-definite systems, as those found in potential flow problems.
v. Generalized Minimal Residual (GMRES): GMRES is a method that efficiently handles nonsymmetric systems, which are frequently utilized in commercial CFD solvers.

Finally, for convergence, the numerical solution procedure continues until the solution satisfies the specified convergence criteria (which could be the stability of key variables or the residual norms dropping below a threshold). Stability, on the other hand, depends on mesh quality, solver settings, and time step selection. According to Versteeg and Malalasekera [13], divergence typically denotes problems such as poor mesh quality, excessive under-relaxation, or erroneous boundary conditions. Figure 5.4 summarizes the solving and numerical solution stage in CFD.

5.1.3 Post-processing

The final stage of the CFD workflow is the post-processing stage, which follows the solving stage (i.e., numerical solution of governing equations) after the pre-processing stage (i.e., geometry creation and mesh generation). This stage consists of analyzing, visualizing, and interpreting the outputs of the numerical results in order to extract significant physical and engineering insights. This stage serves as

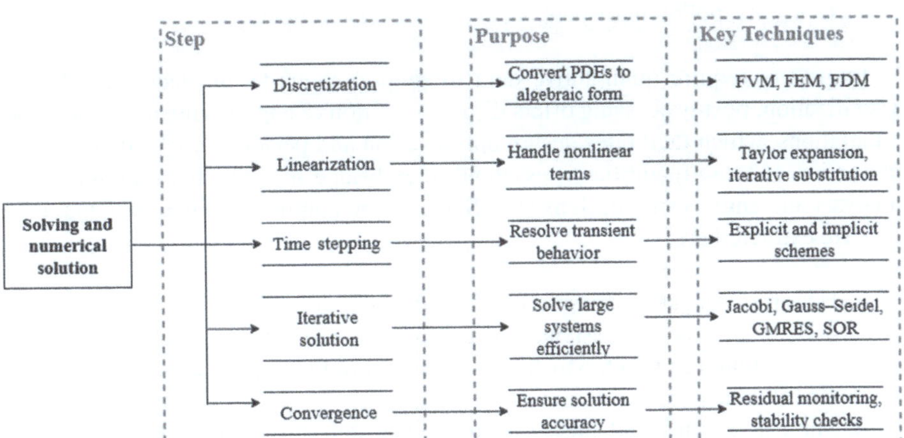

FIGURE 5.4 Summary of the solving and numerical stage.

a diagnostic tool for evaluating the validity, correctness, and stability of the CFD solution [13]. The three main tasks of the post-processing stage involve visualizing flow fields, extracting engineering quantities, and validating/verifying the solution.

One of the most significant aspects of the post-processing stage is visualization. This aspect helps researchers and engineers understand complex 3D flow dynamics by providing simulation output in user-friendly graphical formats. The different approaches in visualization are the velocity vectors and streamlines, contours and color maps, iso-surfaces and volume rendering, particle tracking, and animation.

These visualization tools are discussed as follows:

a. Velocity vectors and streamlines: The velocity vectors display the direction and magnitude of fluid velocity at discrete points, which greatly aid in understanding flow patterns, circulation zones, and recirculation areas. Conversely, streamlines are continuous lines that indicate the trajectory of fluid particles, providing insight into flow topology. These visualization techniques particularly help in aerodynamic or turbomachinery applications in identifying separation points, turbulent mixing, and vortex formation [10].
b. Contours and color maps: The contour plots (for plotting variables, such as pressure, temperature, species concentration, or turbulence intensity) are utilized to find local variation and gradients [11]. Temperature contour plots, for instance, can be utilized to locate hot spots in heat exchangers.
c. Iso-surfaces and particle rendering: Iso-surfaces, such as an iso-surface of velocity magnitude or constant temperature, are regions where a specific variable has a constant value. These visualization techniques are handy for viewing 3D flow structures, such as flame fronts, in combustion simulations [12].
d. Particle tracking and animation: To simulate the trajectory of particles or droplets within a flow, post-processing allows particle tracking and animation. This is particularly crucial for the study of multiphase flows, aerosol transport, pollutant dispersion, and other related fields.

Engineers require quantitative data to support their design decisions. Beyond visualization, post-processing offers direct extraction of engineering quantities from simulations, which facilitates design optimization and performance evaluation. The extraction of these quantities is essential; they bridge the gaps between numerical analysis and engineering designs [10]. Some of the common engineering quantities that are extracted are:

a. Heat transfer rates: These are essential in the analysis of heat exchangers, boilers, and cooling systems. Heat fluxes and heat transfer coefficients can be computed in CFD, which may be compared to experimental and theoretical data.
b. Mass flow rates and pressure drops: The determination of the pressure drops and mass flow rates is essential by analyzing the inlet and outlet flow data. They are needed for system sizing and efficiency assessment.

TABLE 5.1
Post-processing Software Tools

Software	License	Post-processing Capability
OpenFOAM	Open-source	ParaView for visualization, with built-in data extraction tools
COMSOL Multiphysics	Commercial	Integrated visualization and analysis
ANSYS Fluent	Commercial	Velocity vectors, iso-surfaces, animation, and reporting tools
Tecplot 360	Commercial	Advanced 3D visualization and reporting
MATLAB	Commercial	Custom data processing and visualization

c. Forces and moments: These include lift and drag coefficients for aerospace applications, thrust forces in proportion systems, or torque and bending moments in rotary machinery, such as in compressors, turbines, and pumps.
d. Efficiency metrics: System efficiency, such as compressor efficiency, combustor performance, and heat exchanger effectiveness, is computed directly in CFD.

Beyond plotting and interpreting CFD outputs, post-processing plays a crucial role in ensuring the accuracy of the solution. It also entails verifying, validating, and performing error analysis, such as monitoring residuals for convergence and a grid independence study [12, 13].

Comparing CFD outputs against experimental data or recognized benchmark solutions is vital for validation. An example can be taken in pipe flow simulations, where velocity profiles are compared with analytical solutions of fully developed turbulent or laminar flow. Inconsistencies may indicate incorrect boundary conditions or inadequate turbulence modeling, or other modeling errors.

Errors may arise from numerical approximation, computer precision, or assumptions made. Confidence in the simulation outputs is ensured by identifying and reducing these errors.

Some of the post-processing software tools are listed in Table 5.1; they are both commercial and open-source, and they feature powerful post-processing modules. Figure 5.5 further summarizes the core components or core workflow in CFD simulation.

5.2 CFD AND ITS SIGNIFICANCE IN ENERGY SYSTEMS

CFD's ability to precisely simulate fluid dynamics and thermodynamic variables has established it as an essential tool in engineering analysis. Unlike experimental methods, which can be costly and time-consuming, or purely theoretical models that rely on simplifying assumptions, CFD provides a detailed insight into complex physical phenomena [10, 13]. This makes CFD especially useful for energy systems, where performance and efficiency depend heavily on multiphase interactions, such as multiphase flow, combustion, and turbulent mixing.

CFD enables the in-depth examination of thermal processes, as well as fluid and chemical processes in energy systems, leading to improved efficiencies of these

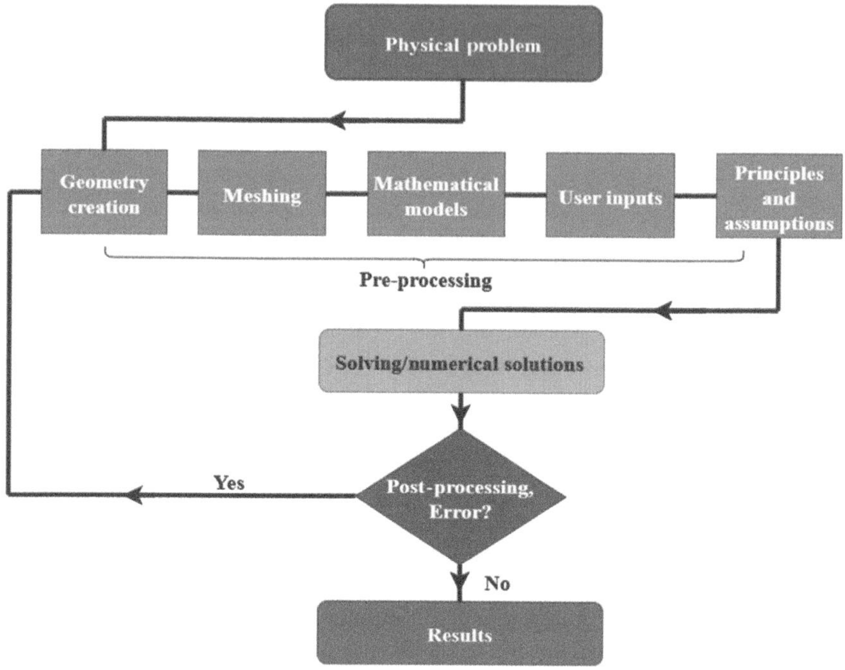

FIGURE 5.5 CFD simulation core workflow.

systems and a reduction of environmental impact. Its benefits encompass renewable energy sources and conventional approaches (such as combustion), as well as emerging technologies such as H_2 electrolysis and fuel cells.

CFD is essential in wind turbine design, providing precise predictions of aerodynamic forces and power outputs. Research indicates that CFD can simulate complex atmospheric turbulence and blade aerodynamics, enhancing turbine placement in wind farms and minimizing wake losses [14].

In thermal systems, CFD is used to model fluid circulation, thermal losses, and heat absorption in flat-plate collectors and concentrating solar power systems. In a study conducted by Ghasemi and Ranjbar [15], CFD analysis was adopted to demonstrate how the heat transfer coefficient was enhanced by 28% to 35% in absorber tubes in a parabolic trough with the use of nanofluids.

In combustors of power systems, CFD predicts flame structure, temperature distribution, and pollutant formation, such as NO_x, SO_x, and CO, in the combustors. As a result, combustion chamber designs are more efficient and cleaner [16]. CFD simulations have been extensively relied on for the development of a low-NO_x burner for the visualization of flame stabilization and minimizing emission hotspots.

CFD is highly useful in analyzing the performance of gas and steam turbines. The operation of a gas turbine depends on the flow distributions within air compressors, combustion chambers, and turbine blades. CFD provides a detailed analysis of the flows and heat transfer within the blade [17].

CFD is employed in H_2 combustion systems. It offers insights into H_2 flame stability and detonation management, thereby enabling safe integration of H_2 into combustion system engines and turbines [18]. In fuel cells, CFD models are used to study thermal management of the systems. These simulations aid in the design of improving the performance of the systems, as well as extending the lifetime of the systems [19].

5.2.1 Advantages and Disadvantages of CFD Over Analytical and Experimental Methods

The benefits of CFD over analytical and experimental approaches include reduced cost and time efficiency, improved visualization of complex phenomena, design optimization, and improved risk reduction.

a. Cost and time requirements: Undertaking full-scale experimental setups in energy systems is frequently too expensive. CFD offers a more affordable option for this by enabling virtual prototypes.
b. Visualization of complex phenomena: Flow fields, heat transfer processes, and turbulence structures that are challenging to measure experimentally are provided by CFD in a more detailed visualization [13].
c. Design optimization: Quick assessment of design variations is made possible by parametric CFD simulations, which help in determining optimal operating conditions.
d. Risks: Before physical testing is undertaken, CFD reduces design risks by simulating extreme operating conditions, such as varying fuel speeds, high temperatures, varying wind speeds, high pressures, and so on.

Despite its significance, CFD faces limitations and challenges that need to be considered. The precision of CFD results heavily relies on the quality of the mesh as well as the accuracy of specifying boundary conditions of the phenomena. Additionally, the high computational power needed restricts its use in industrial settings for high-fidelity CFD [20].

5.3 FUNDAMENTAL GOVERNING EQUATIONS OF FLUID FLOW

The principles of mass conservation (continuity), energy, and momentum form the basis of CFD and describe how fluids behave under various flow conditions. These are mathematical expressions applied to fluid dynamics. These equations in the form of PDEs offer the foundation for accurate modeling and performance energy system analysis [10, 13].

5.3.1 The Continuity Equation

The continuity equation is expressed by the law of conservation of mass. The rate of mass accumulation for a control volume is equal to the net mass flux across, i.e.,

$$\text{Rate of mass accumulation} = \text{Net mass flux across a control volume} \quad (5.1)$$

The detailed formulation of Eq. (5.1) can be found in Ref. [13]. The general continuity equation for unsteady and compressible flow is expressed as:

$$\frac{\partial \rho}{\partial t}+\frac{\partial \rho u}{\partial x}+\frac{\partial \rho v}{\partial x}+\frac{\partial \rho w}{\partial x}=0 \tag{5.2}$$

Equation (5.2) may be written in a compact form as:

$$\frac{\partial \rho}{\partial t}+\nabla\cdot(\rho \boldsymbol{u})=0 \tag{5.3}$$

where ρ is the density of the fluid, and $\boldsymbol{u}$ is the velocity vector. Density variations must be considered for compressible flow for gases and high-speed flows. Applications are found in internal combustion engines and gas turbines, where significant temperature and pressure changes occur.

For incompressible flow with constant ρ, Eq. (5.2) becomes:

$$\frac{\partial u}{\partial x}+\frac{\partial v}{\partial x}+\frac{\partial w}{\partial x}=0 \tag{5.4}$$

or in compact form,

$$\nabla\cdot\boldsymbol{u}=0 \tag{5.5}$$

Equation (5.5) implies that the divergence of the velocity field must vanish. It is widely applied in liquid flows such as water in cooling channels and heat exchangers. The continuity equation is essential because it ensures that no sinks of mass or artificial sources are introduced in CFD simulations, and it is directly coupled with the momentum and energy equations.

5.3.2 Momentum Equation

The momentum equations are the Navier-Stokes equations, derived from Newton's second law. This law states that the rate of change of momentum in a fluid is equal to the sum of forces acting on it, i.e.,

$$\text{Rate of change of momentum in a fluid} = \text{Sum of forces acting on it} \tag{5.6}$$

These forces acting on the fluid are of two types: surface and body forces. The surface forces include viscous stresses and pressure, while the body forces include gravity, electromagnetic forces, centrifugal forces, and Coriolis forces.

The derivation of Eq. (5.6) can be found in Ref. [13]. Figure 5.6 illustrates viscous stress (τ) on the fluid in the three coordinate planes. The momentum equation is expressed in terms of x, y, and z components.

For the x component, the equation is given by:

$$\rho\left(\frac{\partial u}{\partial t}+u.\nabla\boldsymbol{u}\right)=\frac{\partial(-p+\tau_{xx})}{\partial x}+\frac{\partial \tau_{yx}}{\partial y}+\frac{\partial \tau_{zx}}{\partial z}+S_x \tag{5.7}$$

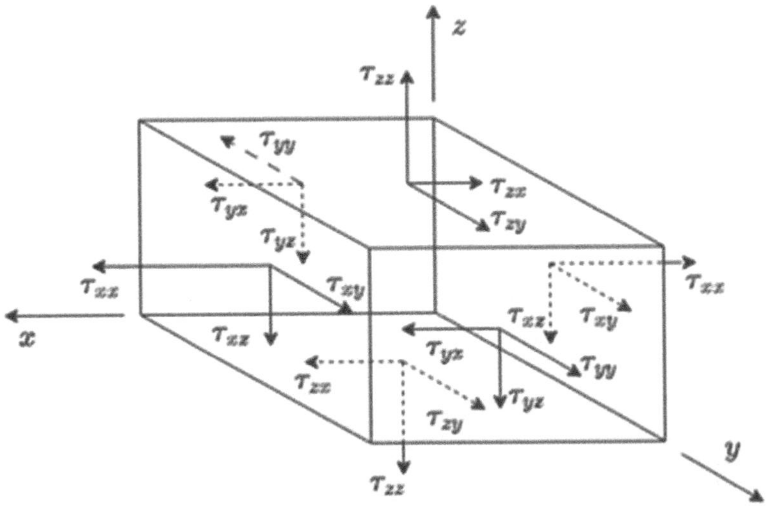

FIGURE 5.6 Viscous stress components on the fluid element in the x, y, and z coordinates.

where S is the source term. For the y component,

$$\rho\left(\frac{\partial v}{\partial t}+v.\nabla \boldsymbol{u}\right)=\frac{\partial \tau_{xy}}{\partial x}+\frac{\partial\left(-p+\tau_{yy}\right)}{\partial y}+\frac{\partial \tau_{zy}}{\partial z}+S_y \tag{5.8}$$

The z component is given by:

$$\rho\left(\frac{\partial w}{\partial t}+w.\nabla \boldsymbol{u}\right)=\frac{\partial \tau_{xz}}{\partial x}+\frac{\partial \tau_{yz}}{\partial y}+\frac{\partial\left(-p+\tau_{zz}\right)}{\partial z}+S_z \tag{5.9}$$

The Left-Hand-Side (LHS) of Eqs. (5.7) to (5.9) represents the LHS of Eq. (5.6), and the Right-Hand-Side (RHS) of Eqs. (5.7) to (5.9) represent the RHS of Eq. (5.6). The viscous stress components in Eqs. (5.7) to (5.9) are obtained from the following equations.

The normal stress components are given by the following:

$$\tau_{xx}=2\mu\frac{\partial u}{\partial x}-\frac{2}{3}\mu\left(\frac{\partial u}{\partial x}+\frac{\partial v}{\partial y}+\frac{\partial w}{\partial z}\right) \tag{5.10}$$

$$\tau_{yy}=2\mu\frac{\partial v}{\partial y}-\frac{2}{3}\mu\left(\frac{\partial u}{\partial x}+\frac{\partial v}{\partial y}+\frac{\partial w}{\partial z}\right) \tag{5.11}$$

$$\tau_{zz}=2\mu\frac{\partial w}{\partial z}-\frac{2}{3}\mu\left(\frac{\partial u}{\partial x}+\frac{\partial v}{\partial y}+\frac{\partial w}{\partial z}\right) \tag{5.12}$$

For the shear stress,

$$\tau_{xy} = \tau_{yx} = \mu\left(\frac{\partial u}{\partial y} + \frac{\partial v}{\partial x}\right) \tag{5.13}$$

$$\tau_{xz} = \tau_{zx} = \mu\left(\frac{\partial u}{\partial z} + \frac{\partial w}{\partial x}\right) \tag{5.14}$$

$$\tau_{yz} = \tau_{zy} = \mu\left(\frac{\partial v}{\partial z} + \frac{\partial w}{\partial y}\right) \tag{5.15}$$

Substituting Eqs. (5.10) to (5.15) into Eqs. (5.7) to (5.9), for the x component, we will have:

$$\begin{aligned}\rho\left(\frac{\partial u}{\partial t} + u.\nabla\boldsymbol{u}\right) &= -\frac{\partial p}{\partial x} + \frac{\partial}{\partial x}\left[2\mu\frac{\partial u}{\partial x} - \frac{2}{3}\mu\left(\frac{\partial u}{\partial x} + \frac{\partial v}{\partial y} + \frac{\partial w}{\partial z}\right)\right] \\ &+ \frac{\partial}{\partial y}\left[\mu\left(\frac{\partial u}{\partial y} + \frac{\partial v}{\partial x}\right)\right] + \frac{\partial}{\partial z}\left[\mu\left(\frac{\partial u}{\partial z} + \frac{\partial w}{\partial x}\right)\right] + S_x\end{aligned} \tag{5.16}$$

Rearranging Eq. (5.16),

$$\begin{aligned}\rho\left(\frac{\partial u}{\partial t} + u.\nabla\boldsymbol{u}\right) &= -\frac{\partial p}{\partial x} + \left[\frac{\partial}{\partial x}\left(\mu\frac{\partial u}{\partial x}\right) + \frac{\partial}{\partial y}\left(\mu\frac{\partial u}{\partial y}\right) + \frac{\partial}{\partial z}\left(\mu\frac{\partial u}{\partial z}\right)\right] \\ &+ \left[\frac{\partial}{\partial x}\left(\mu\frac{\partial u}{\partial x}\right) + \frac{\partial}{\partial y}\left(\mu\frac{\partial v}{\partial x}\right) + \frac{\partial}{\partial z}\left(\mu\frac{\partial w}{\partial x}\right)\right] \\ &+ \frac{\partial}{\partial x}\left[-\frac{2}{3}\mu\left(\frac{\partial u}{\partial x} + \frac{\partial v}{\partial y} + \frac{\partial w}{\partial z}\right)\right] + S_x\end{aligned} \tag{5.17}$$

For incompressible fluid, i.e., $\frac{\partial u}{\partial x} + \frac{\partial v}{\partial y} + \frac{\partial w}{\partial z} = 0$, the RHS of Eq. (5.17) reduces to

$$\rho\left(\frac{\partial u}{\partial t} + u.\nabla\boldsymbol{u}\right) = -\frac{\partial p}{\partial x} + \left[\frac{\partial}{\partial x}\left(\mu\frac{\partial u}{\partial x}\right) + \frac{\partial}{\partial y}\left(\mu\frac{\partial u}{\partial y}\right) + \frac{\partial}{\partial z}\left(\mu\frac{\partial u}{\partial z}\right)\right] + S_x \tag{5.18}$$

$$\rho\left(\frac{\partial u}{\partial t} + u.\nabla\boldsymbol{u}\right) = -\frac{\partial p}{\partial x} + \mu\left[\frac{\partial^2 u}{\partial x^2} + \frac{\partial^2 u}{\partial y^2} + \frac{\partial^2 u}{\partial z^2}\right] + S_x \tag{5.19}$$

Equation (5.19) may be written in a more compact form,

$$\rho\left(\frac{\partial u}{\partial t} + u.\nabla\boldsymbol{u}\right) = -\frac{\partial p}{\partial x} + \mu\nabla^2 u + S_x \tag{5.20}$$

Similarly, y and z components are formulated from Eq. (5.20). For the y component,

$$\rho\left(\frac{\partial v}{\partial t}+v.\nabla \boldsymbol{u}\right)=-\frac{\partial p}{\partial y}+\mu\nabla^2 u+S_y \tag{5.21}$$

while for the z component,

$$\rho\left(\frac{\partial w}{\partial t}+w.\nabla \boldsymbol{u}\right)=-\frac{\partial p}{\partial z}+\mu\nabla^2 u+S_z \tag{5.22}$$

The source terms in Eqs. (5.20) to (5.22) are contributions of body force terms, which may include gravitational or electromagnetic forces. For source terms that include gravitational force, $S_x = 0$, $S_y = 0$, and $S_z = -\rho g$ [13]. μ is the fluid viscosity. Equations (5.20) to (5.22) represent the general form of the Navier-Stokes equations for x, y, and z coordinates on a fluid used in CFD.

5.3.3 The Energy Equation

The first law of thermodynamics is the basis for the energy equation and states that the rate of energy change in a system equals the sum of heat transfer and work done, i.e.,

$$\text{Rate of energy change in a system} = \text{Net rate of heat transfer} + \text{net rate of work done} \tag{5.23}$$

A more detailed derivation of Eq. (5.23) is given in Ref. [13]. The rate of energy change in a system (LHS of Eq. (5.23)) is given by:

$$\frac{\partial(\rho E)}{\partial t}+\nabla(\rho E\boldsymbol{u}) \tag{5.24}$$

where E is the energy, which consists of internal energy (e) and kinetic energy (1/2 $(u^2 + v^2 + w^2)$). The rate of heat transfer (first term in the RHS of Eq. (5.23)) for the three coordinates is given by:

$$-\frac{\partial q_x}{\partial x}-\frac{\partial q_y}{\partial y}-\frac{\partial q_z}{\partial z} \tag{5.25}$$

where q is the heat flux, but from Fourier's law,

$$q_x=-k\frac{\partial T}{\partial x},\ q_y=-k\frac{\partial T}{\partial y},\ \text{and}\ q_z=-k\frac{\partial T}{\partial z} \tag{5.26}$$

where k is the thermal conductivity. Substituting Eq. (5.26) into Eq. (5.25), we will have

$$-\frac{\partial}{\partial x}\left(-k\frac{\partial T}{\partial x}\right)-\frac{\partial}{\partial y}\left(-k\frac{\partial T}{\partial y}\right)-\frac{\partial}{\partial z}\left(-k\frac{\partial T}{\partial z}\right) \tag{5.27a}$$

$$k\left[\frac{\partial}{\partial x}\left(\frac{\partial T}{\partial x}\right)+\frac{\partial}{\partial y}\left(\frac{\partial T}{\partial y}\right)+\frac{\partial}{\partial z}\left(\frac{\partial T}{\partial z}\right)\right] \tag{5.27b}$$

$$k\left[\frac{\partial^2 T}{\partial x^2}+\frac{\partial^2 T}{\partial y^2}+\frac{\partial^2 T}{\partial z^2}\right] \tag{5.27c}$$

$$k\ \nabla^2 T \tag{5.27d}$$

The net rate of work done (i.e., second term of Eq. (2.23)) is given in terms of the shear stress. For the x coordinates, the rate of work done is given by:

$$\frac{\partial\left[u(-p+\tau_{xx})\right]}{\partial x}+\frac{\partial(u\tau_{yx})}{\partial y}+\frac{\partial(u\tau_{zx})}{\partial z} \tag{5.28a}$$

$$-\frac{\partial(up)}{\partial x}+\frac{\partial(u\tau_{xx})}{\partial x}+\frac{\partial(u\tau_{yx})}{\partial y}+\frac{\partial(u\tau_{zx})}{\partial z} \tag{5.28b}$$

Similarly, for the y coordinate,

$$\frac{\partial(v\tau_{xy})}{\partial x}+\frac{\partial\left[v(-p+\tau_{yy})\right]}{\partial y}+\frac{\partial(v\tau_{zy})}{\partial z} \tag{5.29a}$$

$$\frac{\partial(v\tau_{xy})}{\partial x}-\frac{\partial(vp)}{\partial y}+\frac{\partial(v\tau_{yy})}{\partial y}+\frac{\partial(v\tau_{zy})}{\partial z} \tag{5.29b}$$

Similarly, for the z coordinate,

$$\frac{\partial(w\tau_{xz})}{\partial x}+\frac{\partial(w\tau_{yz})}{\partial y}+\frac{\partial\left[w(-p+\tau_{zz})\right]}{\partial z} \tag{5.30a}$$

$$\frac{\partial(w\tau_{xz})}{\partial x}+\frac{\partial(w\tau_{zy})}{\partial y}-\frac{\partial(wp)}{\partial z}+\frac{\partial(w\tau_{zz})}{\partial z} \tag{5.30b}$$

Combining Eqs. (5.28b), (5.29b), and (5.30b), and rearranging, we will have

$$\left[-\frac{\partial(up)}{\partial x}-\frac{\partial(vp)}{\partial y}-\frac{\partial(wp)}{\partial z}\right]+\left[\frac{\partial(u\tau_{xx})}{\partial x}+\frac{\partial(u\tau_{yx})}{\partial y}+\frac{\partial(u\tau_{zx})}{\partial z}+\frac{\partial(v\tau_{xy})}{\partial x}\right.$$
$$\left.+\frac{\partial(v\tau_{yy})}{\partial y}+\frac{\partial(v\tau_{zy})}{\partial z}+\frac{\partial(w\tau_{xz})}{\partial x}+\frac{\partial(w\tau_{zy})}{\partial y}+\frac{\partial(w\tau_{zz})}{\partial z}\right] \tag{5.31a}$$

$$-p\nabla\mathbf{u}+\left[\frac{\partial(u\tau_{xx})}{\partial x}+\frac{\partial(u\tau_{yx})}{\partial y}+\frac{\partial(u\tau_{zx})}{\partial z}+\frac{\partial(v\tau_{xy})}{\partial x}+\frac{\partial(v\tau_{yy})}{\partial y}\right.$$
$$\left.+\frac{\partial(v\tau_{zy})}{\partial z}+\frac{\partial(w\tau_{xz})}{\partial x}+\frac{\partial(w\tau_{zy})}{\partial y}+\frac{\partial(w\tau_{zz})}{\partial z}\right] \tag{5.31b}$$

Combining Eqs. (5.24), (5.27d), and (5.31b), added with energy source term (S_E), the general energy equation is given by:

$$\frac{\partial(\rho E)}{\partial t}+\nabla(\rho E\mathbf{u})=k\ \nabla^2 T+\left[-p\nabla\mathbf{u}+\left[\frac{\partial(u\tau_{xx})}{\partial x}+\frac{\partial(u\tau_{yx})}{\partial y}+\frac{\partial(u\tau_{zx})}{\partial z}+\frac{\partial(v\tau_{xy})}{\partial x}\right.\right.$$
$$\left.\left.+\frac{\partial(v\tau_{yy})}{\partial y}+\frac{\partial(v\tau_{zy})}{\partial z}+\frac{\partial(w\tau_{xz})}{\partial x}+\frac{\partial(w\tau_{zy})}{\partial y}+\frac{\partial(w\tau_{zz})}{\partial z}\right]\right]+S_E \tag{5.32}$$

In more compact form, Eq. (5.32) may be represented as

$$\frac{\partial(\rho E)}{\partial t}+\nabla(\rho E\mathbf{u})=k\ \nabla^2 T+\Phi+S_E \tag{5.33}$$

where Φ represents the viscous dissipation. Heat transfer modes such as conduction, convection, and radiation are incorporated through Fourier's law and boundary conditions. The energy equation thus connects thermodynamics and fluid dynamics, making it essential for analyzing the thermal efficiency of energy systems.

5.3.4 General Scalar Transport Equation

In addition to momentum and energy, many CFD applications include transport of additional scalar quantities, such as turbulence variables, species concentration, and pollutants. The general transport equation is expressed as [13]:

$$\frac{\partial(\rho\phi)}{\partial t}+\nabla(\rho\phi\mathbf{u})=\nabla(\rho\Gamma\nabla\phi)+S_\phi \tag{5.34}$$

where ϕ is the scalar variable, Γ is the diffusion coefficient, and S_ϕ is the source term. Applications of Eq. (5.34) are found in: (a) species transport in fuel cells or hydrogen combustion and (b) turbulence modeling, where ϕ represent the dissipation rate or kinetic energy. The general form demonstrates how versatile CFD is in modeling diverse phenomena within energy systems.

Solving the fundamental governing equations analytically is frequently difficult or impossible; CFD uses boundary conditions and assumptions tailored to specific problems. Boundary conditions are crucial to enable CFD to arrive at realistic solutions. This could be inlet, outlet, wall, symmetrical or cyclic boundary conditions. Mass flow, velocity, temperature, or turbulence intensity may be specified for the inlet boundary conditions. Pressure or outflow conditions may be specified for the outlet, while thermal conditions for heat transfer, or no-slip conditions for viscous flow, may be specified for the wall boundary conditions [13].

5.4 TURBULENCE MODELING

Turbulence, in contrast to laminar flow, is a state of fluid motion marked by chaotic, 3D fluctuations that greatly enhance momentum, mass, and heat transfer [21]. It exhibits irregular eddy formations over a wide range of length and timescales, which makes it extremely difficult to predict compared to laminar flow, which is smooth and predictable. According to Richardson [22], turbulence is an energy cascade in which large-scale eddies gradually transfer energy to smaller eddies until, at a Kolmogorov scale, viscous dissipation is converted from kinetic to heat.

In an energy system, turbulence exhibits a defining role. Turbulence improves air-fuel mixing and reaction rates in combustion chambers of gas turbines and internal combustion engines, which affects emissions as well as efficiencies [23]. CFD relies heavily on turbulence modeling since it is expensive to experimentally resolve turbulence in energy systems, and analytical methods are unable to fully capture its complexity [20].

5.4.1 Reynolds Decomposition and the RANS Equations

The Navier-Stokes equations, as discussed earlier, are the governing equations for fluid flow. The equations are exact; however, solving them directly for turbulent flows at engineering Reynolds numbers is not possible, primarily due to computational demands. To solve this issue, Reynolds developed the concept of Reynolds decomposition, in which the sum of a mean and a fluctuating component is utilized to express each flow variable, such as velocity [24], i.e.,

$$u_i(x,t) = \overline{u_i}(x) + u_i'(x,t) \tag{5.35}$$

where $u_i(x,t)$ is the instantaneous velocity component in the ith direction, i.e., u in the x-direction, v in the y-direction, or w in the z-direction, which changes with space x and time t. $\overline{u_i}(x)$ is the time average velocity component in the ith direction at position x. It eliminates short-term fluctuations and represents the average flow over a long time, i.e.,

$$\overline{u_i}(x) = \frac{1}{t}\int_0^t u_i(x,t)\,dt \tag{5.36}$$

t is a significant time interval. The $u_i'(x,t)$ in Eq. (5.35) is the fluctuating velocity component in the ith direction, which represents turbulent eddies and deviations from the average.

When Eq. (5.35) is substituted into the Navier-Stokes equations and a time-averaging operation is performed, the Reynolds-average Navier-Stokes (RANS) equations are obtained. Although all scales of turbulence are resolved, it adds unknowns in the form of Reynolds stresses [13], i.e.,

$$-\rho\overline{u_i'u_j'} \tag{5.37}$$

Equation (5.37) represents the effect of turbulent fluctuations on the average flow. This results in a closure problem of turbulence. The turbulence models are required to approximate the Reynolds stresses because the RANS equations have more unknowns than equations.

Turbulence models establish constitutive relations for Reynolds stresses and average velocity gradients, commonly through the Boussinesq hypothesis, which assumes isotropic eddy viscosity [13]. The majority of turbulence models used in CFD today are based on this framework, despite the drawbacks in anisotropic flows.

5.4.2 Turbulence Models

The standard turbulence models are the conventional and advanced models. The conventional models include the zero-equation, one-equation, two-equation, and k-ω models. The advanced turbulence models include the Large Eddy Simulation (LES) and Direct Numerical Simulation (DNS).

The zero-equation model is the oldest model for turbulence modeling, which was proposed by Prandtl in 1925. It is predicated on the concept that turbulent eddies behave similarly to molecules in laminar flow, with a distinctive mixing length that defines how much momentum a fluid parcel holds before mixing [25]. Despite the model's simplicity and computational efficiency, mixing length models are unable to handle complex 3D geometries in energy systems and are only applicable to canonical flows, such as boundary layers [13].

A single transport equation (one-equation) for a modified eddy viscosity was introduced by the Spalart-Allmaras model in 1922 [26, 27]. The model balances computational efficiency with increased accuracy compared to the zero-equation model, which makes it widely employed in aerospace and wind turbine simulations [28]. Despite its advantages, it remains less robust in highly swirling regions or strongly separated flows, like in combustor recirculation zones.

The two-equation models offer greater generality by solving transport equations for turbulence quantities. The most widely used model in industrial CFD is the k-ε [29]. It solves turbulent kinetic energy (k) and its dissipation rate (ε). It is less accurate in flows with strong adverse pressure gradients or close to walls [13]. Near-wall performance may be enhanced by variants such as the realizable and RNG k-ε. However, the k-ε model is computationally cheap and efficient for free-shear flows, such as mixing layers and jets.

In contrast, the k-ω model solves for turbulence kinetic energy (k) and specific dissipation rate (ω) [30]. This works better in regions near the wall and in boundary-layer-dominated flows. The strengths of the k-ε in the free stream and the k-ω near wall are being combined in Menter's Shear Stress Transport (SST) k-ω model [31], which makes its applicability in complex energy systems (such as heat exchangers and gas turbine cooling channels) suitable.

These turbulence models are the LES and DNS models. The LES uses the subgrid-scale models to model more minor or dissipative scales while resolving large-scale dynamics [20]. It is increasingly used in applications such as combustion instability studies, wind turbine wake, and urban flow modeling [32]. Despite these, its computational cost is still considerably high, particularly for flows that are wall-bounded [33].

All turbulence scales are fully resolved by DNS without requiring any modeling assumptions. DNS offers fundamental insights; however, because of its extreme computational requirements, it is limited to academic problems and very low Reynolds numbers [10].

5.4.3 Choice of Turbulence Model Selection

The application and computational resources determine the choice of the turbulence model. Although LES is being utilized more in research, the standard or realizable k-ε model is frequently sufficient for combustion chambers of gas turbines where mixing and recirculation predominate, due to low cost and robustness. LES or hybrid RANS-LES models are recommended for wind turbine blade aerodynamics to account for transient wake effects and inflow turbulence.

The SST k-ω is the most reliable source for heat exchanger cooling systems due to its precise boundary-layer predictions [27, 31]. Finally, engineers must select a turbulent model based on the physical problem, computational feasibility, and the required accuracy [10, 20]. Table 5.2 compares the advantages and disadvantages of both conventional and advanced turbulence models, as well as their applicability to energy systems.

TABLE 5.2
Advantages and Disadvantages of the Conventional and Advanced Turbulence Models

Class	Models	Applications	Advantages	Disadvantages
Conventional models	Zero-equation (mixing length)	Educational use	Simple and low cost	Limited to simple flows
	One-equation (Spalart-Allmaras)	Wind turbines	Good for external aerodynamics	Poor in separated flows
	k-ε	Gas turbine combustors	Robust, inexpensive	Weak near walls
	k-ω	Heat exchangers	Accurate near walls	Sensitive to free-stream ω
	SST k-ω	Gas turbines, cooling	Combines k-ε and k-ω	More costly than k-ε
Advanced models	LES	Combustion, wind farms	Captures unsteady features	Very expensive
	DNS	Academic research	Most accurate	Unfeasible for engineering

5.5 MECHANISMS OF HEAT AND MASS

The efficiency, reliability, and safety of modern energy systems are governed by heat and mass transfer. Accurate prediction of mechanisms (for example, from the cooling of gas turbine blades to hydrogen transport in fuels and pollutant dispersion in combustion chambers) is crucial for system optimization. Engineers can now capture intricate transport phenomena that are frequently difficult to obtain experimentally with the aid of CFD. This offers a framework for solving coupled energy and species conservation equations [13, 34].

In energy systems, heat transfer typically occurs through conduction, convection, and radiation modes, and it frequently occurs in a coupled manner [35, 36]. According to Incropera and DeWitt [35], thermal energy is in transit as a result of a spatial temperature difference. They further stated that when there is a temperature difference in a medium or between media, heat transfer must occur.

Conduction is the energy transfer as a result of interactions of molecules in a stationary fluid and solids. Fourier's law is used in CFD, where the temperature gradient is the determinant of conductive heat flux [35], i.e.,

$$q = -k\nabla T \Rightarrow q_x = -k\frac{\partial T}{\partial x},\; q_y = -k\frac{\partial T}{\partial y},\; \text{and } q_z = -k\frac{\partial T}{\partial z} \tag{5.38}$$

where q_x, q_y, and q_z are the heat flux along the three coordinate planes.

Convection is a dominant phenomenon in many energy systems, including condensers, heat exchangers, and boilers. It is the combination of conduction and fluid motion. Convection is heat transfer occurring between a moving fluid and a surface when both have distinct temperatures [35]. Energy and Navier-Stokes equations are employed in CFD to resolve convective heat transfer for laminar and turbulent flows [20].

Radiation, as the third mode of heat transfer, occurs when a surface emits energy in the form of electromagnetic waves, even at finite temperatures [20]. It becomes significant at high temperatures, which is often experienced in combustion chambers and furnaces. Radiative heat flux is approximated utilizing CFD models that capture interactions with solid walls and gases [37].

Generally, in practical energy systems, conjugate heat transfer (CHT) describes the simultaneous and strongly coupled interaction between heat conduction in solid components and convective heat transfer in adjacent fluid flows [10]. This CHT predicts the spatial distribution of temperature, heat transfer, and other essential thermal parameters, which are critical for the optimization of thermal systems [38].

In thermal systems such as fuel cells, combustion, and pollution transport, where species concentrations change, mass transfer plays a crucial role. The species transport equation is utilized in CFD to model mass transfer of chemical species in reacting or non-reacting flows, i.e.,

$$\frac{\partial(\rho Y_i)}{\partial t} + \nabla(\rho \boldsymbol{u} Y_i) = -\nabla J_i + S_i \tag{5.39}$$

where Y_i is the mass fraction of species i, ∇J_i is the diffusion term, and S_i is the source term (net production due to chemical reactions) [39].

5.5.1 Coupling Heat and Mass Transfer

Heat and mass transfer are closely linked in many energy system applications, such as in fuel cells, phase change processes, and combustion systems. In the porous electrodes of fuel cells, the mass transfer of water and gases interacts with heat generation from electrochemical reactions, producing a conjugate heat and mass transfer [40]. Water and thermal management are essential in fuel cells. With this, flooding is prevented and a uniform temperature distribution is ensured.

In phase change processes, including evaporation, condensation, and boiling, mass and heat transfer occur simultaneously. These are usually modeled using CFD approaches such as Eulerian multiphase solvers or the Volume of Fluid (VOF) approach [41].

Strong coupling is demonstrated in combustion systems, where species distribution is governed by turbulent mixing and heat is released through exothermic chemical reactions, which in turn feed back into the flow structures. These interactions are captured by CFD models utilizing turbulence-chemistry interaction closures (such as flamelet models or eddy dissipation) [13, 20].

Several other applications in energy systems where CFD is useful for understanding and optimizing coupled heat and mass transfer are in gas turbine blade cooling, solar thermal systems, and heat exchangers. In gas turbine blade cooling, convective and film cooling are simulated in CFD, making it possible to design channels and cooling holes to keep the blade temperatures below critical thresholds [42–44].

5.6 STRATEGIES IN MESHING AND GRID INDEPENDENCE

One of the essential processes in CFD analysis is the meshing process. It establishes how the computational domain is discretized into finite elements where the fundamental equations are solved. Meshing has a direct and significant impact on the accuracy, convergence, and stability of numerical solutions [13]. An effective grid generation must reflect the fundamental physical features of the flow. The difficulties in numerical solution convergence, numerical diffusion, and false gradients may be attributed to the quality of meshing. However, high-quality meshing is always prioritized in CFD analysis.

As mentioned in Section 5.1, mesh types are typically categorized into three: structured, unstructured, and hybrid. The structured meshes are usually quadrilateral in 2D and hexahedral in 3D. They are regular and grid-like elements organized systematically, as depicted in Figure 5.3. According to Ferziger et al. [10], structured grids are suitable for basic geometries where coordinate lines align with the domain boundary. However, they are less adaptable for intricate geometries.

Unstructured grids, on the other hand, may adapt to complicated geometrical domains. They are made up of irregularly connected elements, including triangles in 2D and tetrahedra in 3D (Figure 5.3). Although they offer greater flexibility and automatic grid generation, their uneven grid spacing makes them computationally expensive and less accurate per element [12].

By combining the features of structured and unstructured meshing, hybrid meshes enable unstructured flexibility in complex regions and structured flexibility in crucial areas, such as boundary layers. In contrast to tetrahedral meshes, polyhedral elements have recently drawn attention for their capacity to improve convergence and decrease numerical diffusion with fewer elements [45].

The essential feature of the mesh is the mesh quality. It is assessed based on parameters such as the skewness, orthogonality, and aspect ratio. Skewness measures the degree to which an element deviates from its ideal shape. Skewness values close to zero (0) signify a high-quality mesh, while values around one (1) signify a poor-quality mesh [45]. However, within this skewness range, it is recommended that for a tetrahedral mesh, the targeted skewness should be at least 0.8. This high skewness value applies only to the tetrahedral mesh. Orthogonality quantifies the alignment of grid lines with flow directions, which influences convergence and accuracy. The orthogonal range is within 0 and 1, where 0 represents the worst quality and a value of 1 represents the best.

5.6.1 Mesh Refinement Approaches

Mesh refinement enhances precision by increasing element density in regions with steep gradients. In wall-bounded flows, where temperature and velocity gradients are large near solid surfaces, boundary layer meshing becomes crucial. To accurately capture heat transfer and shear stress, especially close to walls, inflation layers are generated. Such layers consist of hexahedra or thin prismatic cells [13]. This is demonstrated in Figure 5.7, where a mesh is being refined to capture heat transfer in a pipe wall.

Another refinement approach is Adaptive Mesh Refinement, which dynamically adjusts mesh resolution during the simulation by refining regions characterized by steep temperature or velocity gradients. This targeted refinement strategy minimizes computational cost while preserving solution accuracy and numerical stability [12]. Accuracy, stability, and computational costs while maintaining high accuracy. Balancing accuracy, stability, and computational efficiency is essential for effective meshing. This approach can be useful when structured grids are employed in simple geometric regions, while unstructured grids are employed in more complex areas. Additionally, preventing solver divergence involves using pre-simulation mesh diagnostics, such as checks for skewness, cell count, and aspect ratio [45].

Ultimately, the mesh must strike a balance between computational efficiency and the resolution of essential flow characteristics, such as shear layers, vortices, and recirculation regions. The credibility and accuracy of CFD predictions in energy system simulations are ensured when a well-structured mesh approach and an appropriate grid independence test are both undertaken.

5.6.2 Grid Independence Study

This study ensures that the simulation results accurately reflect the physical solution and are not influenced by mesh size. Grid independence study entails performing simulations with increasingly finer meshes and keeping an eye on essential output parameters, such as velocity, pressure drop, or Nusselt number, until successful refinement produces insignificant changes, often <1–2% [10]. This study confirms the reliability of the model and its numerical convergence.

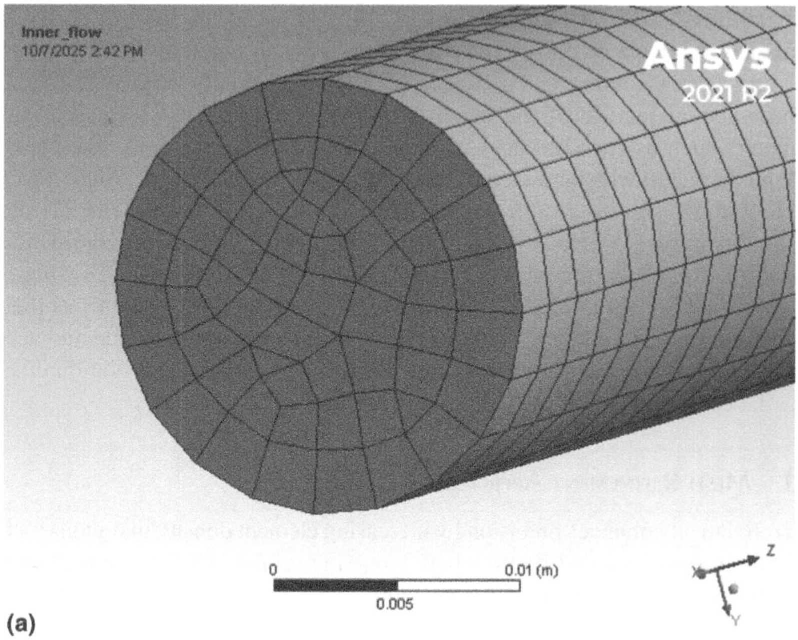

(a)

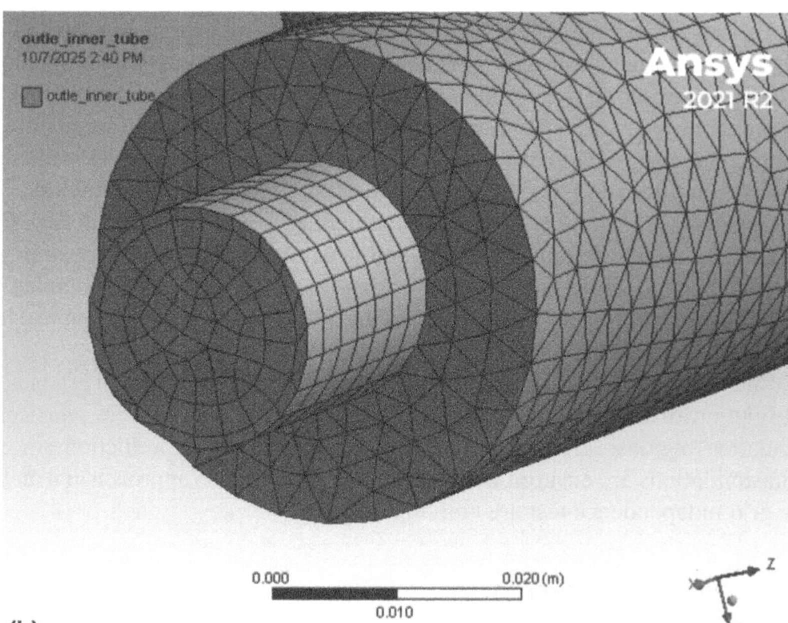

(b)

FIGURE 5.7 Meshing indicating application of inflation layers at the boundary between fluid and solid. (a) Outer mesh (indicating solid mesh) and inner mesh (indicating fluid mesh). (b) Fluid mesh before applying inflation at the boundary. (c) Fluid mesh after applying inflation at the boundary. (d) Fluid and solid domain mesh. *(Continued)*

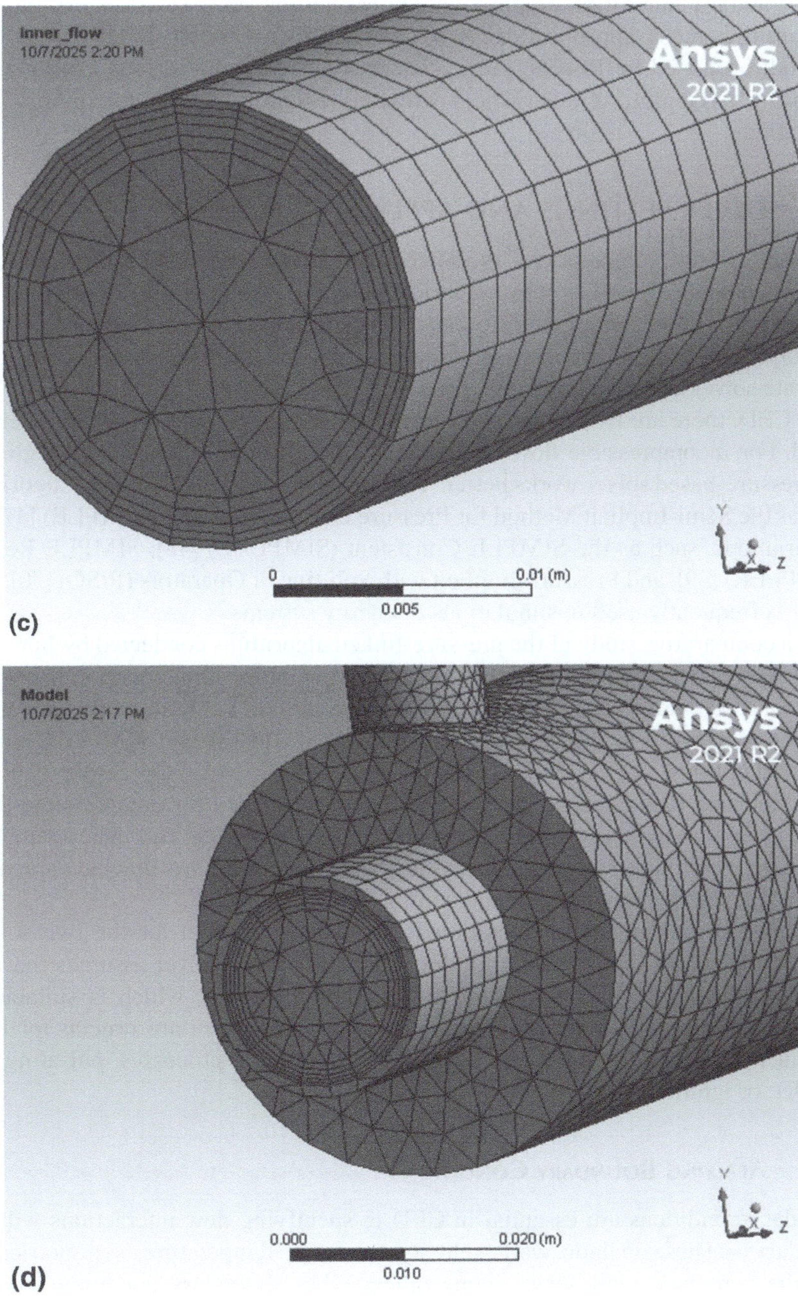

FIGURE 5.7 *(Continued)*

A standard quantitative approach for estimating numerical uncertainty and determining mesh appropriateness is the Grid Convergence Index (GCI). This method, proposed by Roache [46], compares results from various grid densities to ensure asymptotic convergence, which is essentially vital for quality CFD validation.

5.7 SOLVER SETTINGS AND APPLYING BOUNDARY CONDITIONS

Through iterative computations, the solver, being the core of CFD simulation, translates mathematical models into numerical solutions. Appropriate solver setups are influenced by flow, compressibility, and transient properties of a physical problem [12, 13]. Complex flow properties are often involved in energy systems, requiring an accurate solver and carefully configured boundary conditions.

In CFD, there are two basic solver formulations: the pressure-based and density-based. For incompressible flow applications where density changes are negligible, the pressure-based solver works better. The pressure-based solver utilizes algorithms such as the Semi-Implicit Method for Pressure Linked Equations (SIMPLE) [47] and its variations, such as the SIMPLE Consistent (SIMPLEC) [48], SIMPLE Revised (SIMPLER) [49], and Pressure-Implicit with Splitting of Operators (PISO) [50]. This solver is frequently used in simulations of energy systems.

In a comparative study of the pressure-linked algorithms conducted by Jang et al. [51], it was concluded that, for problems where the momentum equation is not coupled to a scalar variable, the PISO algorithm is preferred. For scalar variables, which are closely coupled to the momentum equation, SIMPLER and SIMPLEC are the preferred options.

The density-based solver, on the other hand, is utilized for compressible flows, where density variation is significant, such as in supersonic gas turbines, combustion chambers, or jet engines. For numerical stability, it often employs upwind approaches to solve the energy, momentum, and continuity equations [39].

There are two modes of operation (with respect to time) for the two solvers, namely, the steady and transient modes. The steady-state solver assumes that flow conditions remain constant over time (i.e., time-invariant), which is suitable for systems with steady operating conditions. Transient solvers are crucial for time-dependent flow conditions, such as unsteady combustion processes, pulsating heat transfer, or ignition [10].

5.7.1 Applying Boundary Conditions

Boundary conditions are essential in CFD in specifying flow interactions with the boundaries. These include wall, velocity, pressure, temperature, symmetric, and periodic boundary conditions, among others. These boundary conditions are discussed next.

a. Wall boundary: This imposes thermal conditions and no-slip velocity. These boundary conditions may include constant heat flux, convective heat transfer coefficients, or constant temperature.

b. Velocity inlet: This boundary condition is ideal for compressors, pipe inlets, and turbines. It specifies the velocity magnitude or component with turbulence quantities.
c. Pressure outlet: This boundary condition is typical of diffuser and exhaust exits. It defines the static pressure at the outlet of a geometry, allowing backflow conditions when required.
d. Symmetrical conditions: Used for mirror effects to decrease the computational domain size, particularly.
e. Periodic boundary: Ideal for repeating geometries such as heat exchanger channels or turbine blade passages.

5.8 MODEL VALIDATION IN CFD

The predictive capability of numerical models used in the simulation of energy systems in CFD is heavily relied on validation to ensure their reliability. Validation gives assurance that, within acceptable limits, the numerical outputs accurately depict physical reality in the context of engineering simulations. Without validation, CFD findings may be inaccurate [10, 46, 52].

5.8.1 Verification and Validation

The terms verification and validation are frequently used synonymously; however, in the context of CFD, they refer to different processes [53]. In verification, it aids in answering the question:

- Are we solving the selected governing equations appropriately?

The questions help to concentrate on the mathematical and numerical accuracy of implementation, which ensures that the equations are accurately discretized and solved for the physical problem. Verification includes checking iterative, convergence, grid convergence, and error estimation using GCI [52].

Validation, on the other hand, helps in answering this question:

- Are we solving the correct equations?

To answer the question, the outcomes from CFD are compared with experimental results to determine how accurately the selected models and assumptions represent the real phenomena.

Simply put, verification assesses the accuracy of the CFD computation, while validation assesses the accuracy of the model. The two procedures work in tandem: verification guarantees internal consistency, while validation offers outward credibility [53, 54]. Figure 5.8 illustrates the validation process. The process involves defining the physical problem and defining the objectives [13, 39], selection of benchmark or experimental data [55], model development and setup [10], simulation and post-processing [12], comparison with experimental results, uncertainty quantification [52], model refinement [53], and finally the validated model.

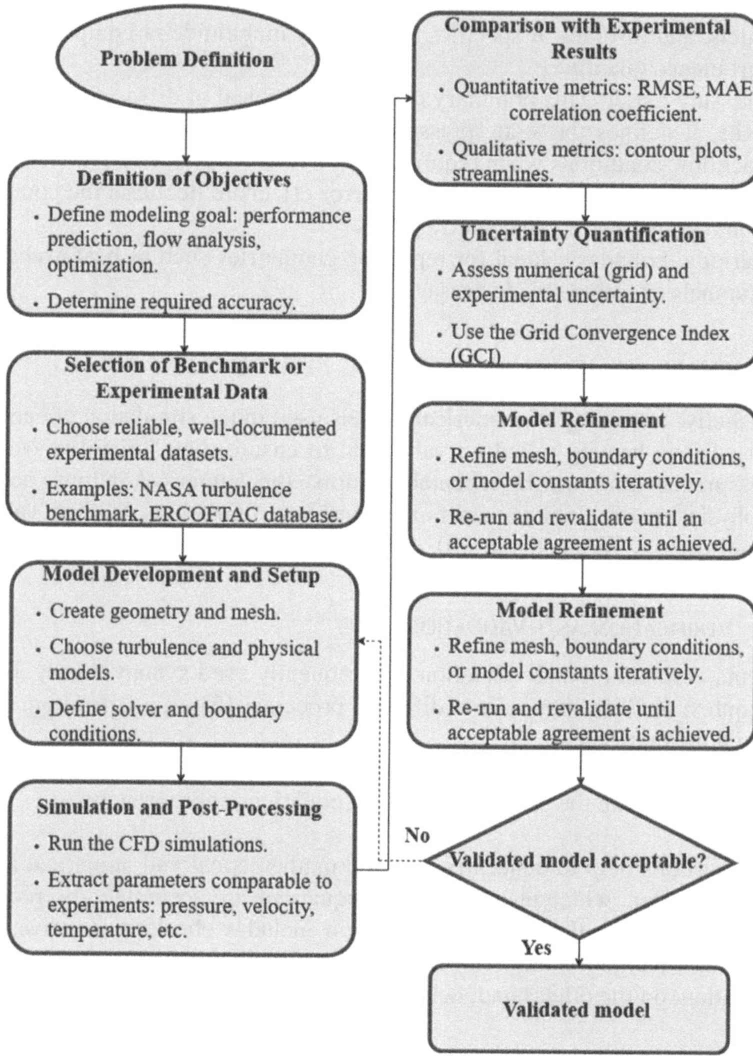

FIGURE 5.8 Validation processes in CFD.

REFERENCES

1. ANSYS, 'What is Computational Fluid Dynamics (CFD)?', 2025. [Online]. Available: https://www.ansys.com/simulation-topics/what-is-computational-fluid-dynamics
2. C. R. Maliska, *Fundamentals of Computational Fluid Dynamics: The Finite Volume Method*, in Fluid Mechanics and Its Applications, vol. 135. Springer International Publishing, 2023. doi: 10.1007/978-3-031-18235-8.
3. A. Iranzo, 'CFD applications in energy engineering research and simulation: An introduction to published Reviews', *Processes*, vol. 7, no. 12, p. 883, 2019. doi: 10.3390/pr7120883.

4. I. Sadrehaghighi, 'Mesh generation in CFD', 2020. [Online]. Available: https://www.researchgate.net/publication/318456955_Mesh_Generation_in_CFD#fullTextFileContent
5. D.da S. Tonon, J. T. Tomita, E. C. Garcia, C. Bringhenti, R. B. Díaz, & L. H. L. Whitacker, "Comparative study between structured and unstructured meshes applied in Turbopump's Hydraulic Turbine." *Proceedings of the ASME Turbo Expo 2020: Turbomachinery Technical Conference and Exposition. Volume 6: Education; Electric Power.* Virtual, Online. September 21–25, 2020. V006T08A006. ASME. doi: 10.1115/GT2020-15004.
6. I. Sadrehadighighi, 'Unstructured meshing for CFD', 2022. [Online]. Available: https://www.researchgate.net/publication/339285304_Unstructured_Meshing_for_CFD#fullTextFileContent
7. M. Tomac, and D. Eller, 'Towards automated hybrid-prismatic mesh generation', *Procedia Engineering*, vol. 82, pp. 377–389, 2014. doi: 10.1016/j.proeng.2014.10.398.
8. Y. Nakayama, 'Computational fluid dynamics', in *Introduction to Fluid Mechanics*, Elsevier, 2018, pp. 293–327. doi: 10.1016/B978-0-08-102437-9.00015-2.
9. T. W. Kerlin, and B. R. Upadhyaya, 'Space-time kinetics', in *Dynamics and Control of Nuclear Reactors*, Elsevier, 2019, pp. 105–109. doi: 10.1016/B978-0-12-815261-4.00009-3.
10. J. H. Ferziger, M. Perić, and R. L. Street, *Computational Methods for Fluid Dynamics*, Springer International Publishing, 2020. doi: 10.1007/978-3-319-99693-6.
11. S. V. Patankar, *Numerical Heat Transfer and Fluid Flow*, Hemisphere Pub. Corp.; McGraw-Hill, 1980.
12. F. Moukalled, L. Mangani, and M. Darwish, *The Finite Volume Method in Computational Fluid Dynamics: An Advanced Introduction with OpenFOAM and Matlab.* Springer, 2016.
13. H. K. Versteeg, and W. Malalasekera, *An Introduction to Computational Fluid Dynamics: The Finite Volume Method*, 2nd ed., Pearson Education Ltd, 2007.
14. R. J. Barthelmie, and L. E. Jensen, 'Evaluation of wind farm efficiency and wind turbine wakes at the Nysted offshore wind farm', *Wind Energy*, vol. 13, no. 6, pp. 573–586, 2010. doi: 10.1002/we.408.
15. S. E. Ghasemi, and A. A. Ranjbar, 'Thermal performance analysis of solar parabolic trough collector using nanofluid as working fluid: A CFD modelling study', *Journal of Molecular Liquids*, vol. 222, pp. 159–166, 2016. doi: 10.1016/j.molliq.2016.06.091.
16. S. R. Turns, *An Introduction to Combustion: Concepts and Applications*, 3rd ed., McGraw-Hill, 2012.
17. A. H. Lefebvre, and D. R. Ballal, *Gas Turbine Combustion*, CRC Press, 2010. doi: 10.1201/9781420086058.
18. H. Bundele, C. Kurien, P. S. Varma, and M. Mittal, 'Experimental and computational study on the enhancement of engine characteristics by hydrogen enrichment in a biogas fuelled spark ignition engine', *International Journal of Hydrogen Energy*, vol. 47, no. 71, pp. 30671–30686, 2022. doi: 10.1016/j.ijhydene.2022.07.029.
19. H. Pourrahmani, and J. Van Herle, 'Water management of the proton exchange membrane fuel cells: Optimizing the effect of microstructural properties on the gas diffusion layer liquid removal', *Energy*, vol. 256, p. 124712, 2022. doi: 10.1016/j.energy.2022.124712.
20. S. B. Pope, *Turbulent Flows*, 1st ed., Cambridge University Press, 2000. doi: 10.1017/CBO9780511840531.
21. H. Tennekes, and J. L. Lumley, *A First Course in Turbulence*, The MIT Press, 1972.

22. L. F. Richardson, *Weather Prediction by Numerical Process*, Cambridge University Press, 1922. [Online]. Available: https://dn721509.ca.archive.org/0/items/weatherpredictio00richrich/weatherpredictio00richrich.pdf
23. A. R. Uyanwaththa, 'CFD Modelling of Gas Turbine Combustion Processes', Wolfson School of Mechanical Electrical and Manufacturing Engineering Loughborough University, UK, 2018. [Online]. Available: https://core.ac.uk/download/288359572.pdf
24. D. Jackson, and B. Launder, 'Osborne Reynolds and the publication of his papers on turbulent flow', *Annual Review of Fluid Mechanics*, vol. 39, no. 1, pp. 19–35, 2007. doi: 10.1146/annurev.fluid.39.050905.110241.
25. H. Schlichting, and K. Gersten, *Boundary-Layer Theory*, Springer, 2017. doi: 10.1007/978-3-662-52919-5.
26. R. Spalart and S. R. Allmaras, "A one-equation turbulence model for aerodynamic flows," in *Proc. 30th Aerospace Sciences Meeting and Exhibit*, AIAA Paper 92-0439, 1992.
27. P. R. Spalart, and S. R. Allmaras, 'One-equation turbulence model for aerodynamic flows', *Recherches. Aérospatiales*, vol. 1, pp. 5–21, 1994.
28. D. C. Wilcox, *Turbulence Modeling for CFD*, 3rd ed., DCW Industries, 2006.
29. B. E. Launder, and D. B. Spalding, 'The numerical computation of turbulent flows', *Computer Methods in Applied Mechanics and Engineering*, vol. 3, no. 2, pp. 269–289, 1974. doi: 10.1016/0045-7825(74)90029-2.
30. D. C. Wilcox, 'Reassessment of the scale-determining equation for advanced turbulence models', *AIAA Journal*, vol. 26, no. 11, pp. 1299–1310, 1988. doi: 10.2514/3.10041.
31. F. R. Menter, 'Two-equation eddy-viscosity turbulence models for engineering applications', *AIAA Journal*, vol. 32, no. 8, pp. 1598–1605, 1994. doi: 10.2514/3.12149.
32. P. Sagaut, *Large Eddy Simulation for Incompressible Flows*, in Scientific Computation, Springer-Verlag, 2006. doi: 10.1007/b137536.
33. C. Meneveau, and J. Katz, 'Scale-invariance and turbulence models for large-eddy simulation', *Annual Review of Fluid Mechanics*, vol. 32, no. 1, pp. 1–32, 2000. doi: 10.1146/annurev.fluid.32.1.1.
34. A. Bejan, *Convection Heat Transfer*, John Wiley & Sons, 2013.
35. Incropera and DeWitt, *Fundamentals of Heat and Mass Transfer*, John Wiley & Sons, 2006.
36. Y. A. Çengel, and A. J. Ghajar, *Heat and Mass Transfer: Fundamentals & Applications*, 5th ed., McGraw Hill Education, 2015.
37. M. F. Modest, *Radiative Heat Transfer*, 3rd ed., Academic Press, 2013.
38. 'Conjugate Heat Transfer: Basics, Best Practices & Applications', Neural Concept. [Online], 2025. Available: https://www.neuralconcept.com/post/conjugated-heat-transfer-best-practices-applications#:~:text=Conjugate%20Heat%20Transfer%2C%20or%20CHT,representing%20thermal%20interactions%20between%20media.
39. J. D. Anderson, *Computational Fluid Dynamics: The Basics with Applications*, McGraw-Hill, 2006.
40. R. P. O'Hayre, S.-W. Cha, W. G. Colella, and F. B. Prinz, *Fuel Cell Fundamentals*, 3rd ed., John Wiley & Sons Inc, 2016.
41. S. G. Kandlikar, 'History, advances, and challenges in liquid flow and flow boiling heat transfer in microchannels: A critical review', *Journal of Heat Transfer*, vol. 134, no. 3, p. 034001, 2012. doi: 10.1115/1.4005126.
42. M. M. Aslam Bhutta, N. Hayat, M. H. Bashir, A. R. Khan, K. N. Ahmad, and S. Khan, 'CFD applications in various heat exchangers design: A review', *Applied Thermal Engineering*, vol. 32, pp. 1–12, 2012. doi: 10.1016/j.applthermaleng.2011.09.001.
43. J.-C. Han, S. Dutta, and S. Ekkad, *Gas Turbine Heat Transfer and Cooling Technology*, CRC Press, 2012. doi: 10.1201/b13616.

44. S. Kalogirou, *Solar Energy Engineering: Processes and Systems*, Elsevier/Academic Press, 2009.
45. ANSYS, 'ANSYS Meshing User's Guide', 2025. [Online]. Available: https://ansyshelp.ansys.com/public/Views/Secured/corp/v251/en/pdf/ANSYS_Meshing_Users_Guide.pdf
46. P. J. Roache, *Fundamentals of Verification and Validation*, Hermosa publishers, 2009.
47. S. V. Patankar, and D. B. Spalding, 'A calculation procedure for heat, mass and momentum transfer in three-dimensional parabolic flows', *International Journal of Heat and Mass Transfer*, vol. 15, no. 10, pp. 1787–1806, 1972. doi: 10.1016/0017-9310(72)90054-3.
48. J. P. Van Doormaal, and G. D. Raithby, 'Enhancements of the simple method for predicting incompressible fluid flows', *Numerical Heat Transfer*, vol. 7, no. 2, pp. 147–163, 1984. doi: 10.1080/01495728408961817.
49. S. V. Patankar, 'A calculation procedure for two-dimensional elliptic situations', *Numerical Heat Transfer*, vol. 4, no. 4, pp. 409–425, 1981. doi: 10.1080/01495728108961801.
50. R. I. Issa, A. D. Gosman, and A. P. Watkins, 'The computation of compressible and incompressible recirculating flows by a non-iterative implicit scheme', *Journal of Computational Physics*, vol. 62, no. 1, pp. 66–82, 1986. doi: 10.1016/0021-9991(86)90100-2.
51. D. S. Jang, R. Jetli, and S. Acharya, 'Comparison of the PISO, SIMPLER, and SIMPLEC algorithms for the treatment of the pressure-velocity coupling in steady flow problems', *Numerical Heat Transfer*, vol. 10, no. 3, pp. 209–228, 1986. doi: 10.1080/10407788608913517.
52. P. J. Roache, *Verification and Validation in Computational Science and Engineering*, Hermosa Publishers, 1998.
53. W. L. Oberkampf, and C. J. Roy, *Verification and Validation in Scientific Computing*, 1st ed., Cambridge University Press, 2010. doi: 10.1017/CBO9780511760396.
54. American Institute of Aeronautics and Astronautics, Ed., *AIAA Guide for the Verification and Validation of Computational Fluid Dynamics Simulations*, American Institute of Aeronautics and Astronautics, 1998.
55. L. Vervisch, D. Roekaerts, and E. Kommission, Eds., *Computational Fluid Dynamics of Turbulent Combustion*, in ERCOFTAC best practice guidelines, ERCOFTAC, 2016.

6 Applications of CFD in Hydrogen Generation and Combustion

6.1 CFD ROLES IN GREEN HYDROGEN GENERATION AND STORAGE

Hydrogen (H_2) generated by water electrolysis using renewable energy is now at the core of long-term decarbonization strategies, driven by the global shift to low-carbon energy systems [1, 2]. Understanding the various coupled transport and chemical phenomena that occur across multiple spatial and temporal scales is crucial for designing, optimizing, and safely integrating H_2 generation and storage systems.

With the ability of CFD to predict fluid flow, heat transfer, multiphase behavior, species transport, chemical reactions, and their mutual couplings both spatially and temporally, it has become a fundamental tool for predicting these multiphysics interactions [3]. According to recent systematic reviews, CFD-based modeling is already a well-established pillar for H_2 combustion, alkaline electrolyzer design, and storage-safety studies, and its use is only growing as systems scale and get more diverse [4].

6.1.1 Typical CFD Modeling Approaches for Electrochemical Energy Systems

Two primary CFD strategies exist for electrochemical energy system modeling: resolved multiphase CFD and porous/homogenized electrochemical CFD. The resolved multiphase CFD consists of Volume-of-Fluid (VOF) and Euler-Euler approaches. These approaches are utilized to resolve macroscopic bubble plumes and gas hold-up patterns in a cell and small stacks. In 2D and 3D cell-level analysis, VOF is often utilized to view bubble plumes and detachment behavior [5]. Euler-Euler methods are employed when the bubble population is dense and a continuum description is more effective [6, 7]. The two approaches are frequently combined with species transport and energy equations to measure local conductivity, temperature, and dissolved-gas fields [8].

In porous/homogenized electrochemical CFD, porous-media formulations or macro-homogenized electrochemical models are employed, where resolving porous electrodes is not feasible or on larger scales [9, 10]. While employing CFD for flow and heat transfer in free-flow channels, they embed bulk resistances and electrochemical source terms, such as Butler-Volmer or simplified Faradaic relations, within porous regions.

Some of the possible outcomes CFD reveals are bubble plume dynamics and removal, local current density non-uniformity, thermal hotspots and cooling

DOI: 10.1201/9781003727194-6

requirements, and design trade-offs [10–12]. For instance, in bubble plume dynamics and removal, CFD demonstrates zones of bubble coalescence and preferred detachment paths that raise the local gas fraction and effective ohmic resistance [11].

6.1.2 Hydrogen Storage Using CFD: Tanks, Transient Flows, and Underground Storage

Metal hydrides, compressed gas tanks, cryo-liquid H_2, and underground storage are forms of H_2 storage. Across these scales, CFD is utilized for a variety of purposes, such as thermo-structural interaction, temperature and pressure transients in tanks, porous media flow for underground H_2 storage, and plume and dispersion modeling for safety evaluations [13]. According to recent reviews, CFD, which is often extended with reactive transport and geomechanical modules, is crucial for risk analysis and injection/withdrawal optimization in underground H_2 storage and salt cavern systems because they pose unique multiphase and geochemical challenges [13, 14].

6.1.3 CFD in Engine and Hydrogen Combustion

The use of H_2, particularly combustion in turbines and engines, defines realistic end use performance and emissions trade-offs, whereas generation and storage deal with supply [15]. CFD for H_2 combustion highlights quick reaction kinetics and high-fidelity turbulence, such as fast reaction kinetics, chemistry interaction, and the unique flame properties (i.e., high laminar flame speed, high diffusivity, or low ignition energy) [16, 17].

In port-injected systems, these properties lead to rapid premixed combustion and an increased risk of pre-ignition or backfire; however, they also offer opportunities for improved thermal efficiency if properly controlled. The preferred approach for visualizing flame propagation, quantifying local temperature fields (which are directly connected to NO_x formation), and testing the strategies of injection and ignition in silico is CFD's Reynolds-Average Navier-Stokes (RANS) and Large-Eddy Simulation (LES) with suitable combustion models, or even Direct Numerical Simulation (DNS) for fundamental investigations [18]. Recent numerical studies of pure H_2 and dual-fuel (H_2-diesel) engines show how CFD quantifies attainable soot and CO reductions using concurrent NO_x management techniques [19, 20].

The methods in CFD for engine analysis include turbulent-chemistry coupling, spray and ignition modeling, and LES/RANS [21–23]. In turbulence-chemistry coupling, for instance, depending on the level of fidelity required, flamelet models, the Eddy Dissipation Concept (EDC), or Partially Stirred Reactor (PaSR) models are adopted [21, 22]. Reduced but validated kinetics are necessary for H_2 chemistry to strike a balance between computational cost and accuracy.

Due to its cheaper computing cost, RANS remains the standard for engineering parametric sweeps, whereas LES is often used for research where capturing transient flame structure is essential. Recent studies on the comparison between RANS and LES indicate that LES better resolves flame kernels and the abundance of high-temperature pockets that produce NO_x [23].

It is recommended to adopt these practices for providing results and recipes, as the case study in the subsequent section serves as a demonstrative educational example, designed to highlight typical CFD applications rather than provide novel experimental data. These recipes are as follows:

a. Stating model assumptions explicitly: State the assumptions of the model explicitly. For instance, indicate whether bubble nucleation is prescribed, whether wall heat transfer is adiabatic, or whether electrodes are modeled as porous areas or impermeable walls.
b. Use dimensionless trends and metrics: Results are more generalizable across scales and platforms when normalized metrics are presented. For instance, the relative change in cell voltage.
c. Validate against benchmarks in the literature: Rather than using absolute numbers to validate trends in the absence of experimental data, use the outcomes of peer-reviewed papers in the literature, such as pressure trace shape, NO_x sensitivity to temperature, or bubble plume trends, for instance. Recent parametric CFD studies provide relevant benchmarks and review papers [19, 24].
d. Provide repeatable recipes: Provide readers with geometric drawings, mesh specifications, boundary condition values, solver options, and, if relevant, references to UDF pseudo-code or mechanism. This repeatable recipe is used in the demonstration case study in the subsequent section.
e. Talk about uncertainties and limitations: At the very least, discuss the expected sensitivity directions or include short sensitivity studies.

The internal operations of green H_2 technologies can be observed in a uniquely detailed manner thanks to CFD, which can resolve flow fields, multiphase behavior, heat and mass transfer, and reaction zones, all of which impact the performance, durability, and safety of H_2 systems. CFD demonstration recipes for a pedagogical case study, such as that in the subsequent section, accomplish two goals: (a) teach how to use real-world solver choices to translate physics into numerical models and (b) offer interpretable, transferable insights suitable for designers, students, and early-stage researchers.

6.2 ELECTROLYZER SYSTEM CASE STUDY MODELING IN CFD

This section aims to demonstrate an electrolyzer modeling approach using ANSYS Fluent. The case study involves CFD modeling of a typical Hydrox-type alkaline electrolyzer demonstrated in Fluent. Hydrox Holdings Ltd., based in South Africa, patented the Direct Electrode Flow-Through (DEFT) electrolyzer system [25]. The purpose of this demonstration is to show how CFD can be used to study electrolyte flow, gas evolution and detachment, temperature distribution, and current-density non-uniformity in an alkaline electrolysis cell.

One of the most direct and cleanest approaches for generating green H_2 is water electrolysis. The alkaline electrolyzer remains the most developed and cost-effective electrolyzer technology for large-scale applications [26]. An alkaline electrolyzer typically has 25–30 wt% KOH as alkaline electrolyte [27], and nickel-based electrodes are utilized in industrial systems, such as those utilized by Hydrox Holdings, to generate H_2 and O_2 gases.

According to Hydrox, their system can operate at intermediate temperatures between 200°C and 300°C compared to the Proton Exchange Membrane (PEM) electrolyzer. Also, their findings suggested that at 250°C, there is 3.5 times better conductivity when compared to the PEM operating temperature of 80°C [28]. Other detailed operating parameters of Hydrox's commercial units are not publicly disclosed; however, parameters for general alkaline electrolyzers can be found in Ref. [27].

This case study illustrates the use of the VOF approach in conjunction with species transport and electrochemical source terms given by UDF to model a single-cell alkaline electrolyzer in ANSYS Fluent. The simulation shows how CFD makes it possible to analyze fluid flow, heat transfer, gas evolution, and reaction kinetics in tandem. These parameters have a significant impact on the longevity and efficiency of industrial electrolyzers.

Recent studies have validated the potential of CFD in this type of study. For instance, in the study conducted by Rodríguez and Amores [12] in CFD modeling on an alkaline electrolyzer, they utilized transport equations for gas and liquid phases, and electric current equations for fluid dynamic and electrochemical phenomena. Likewise, Khalighi et al. [29] simulated the growth of hydrogen bubbles from a vertical electrode using a 30 wt% KOH electrolyte, investigating the effects of operating pressure and mass flow rate on bubble growth behavior. The case study in this section builds on these developments by using a similar numerical approach tailored to Hydrox-type electrolyzer geometry.

6.2.1 Geometry and Physical Description

The 2D geometry of the case study, indicating dimensions for the simulation of the alkaline-type single cell electrolyzer, is shown in Figure 6.1. The dimensions in the

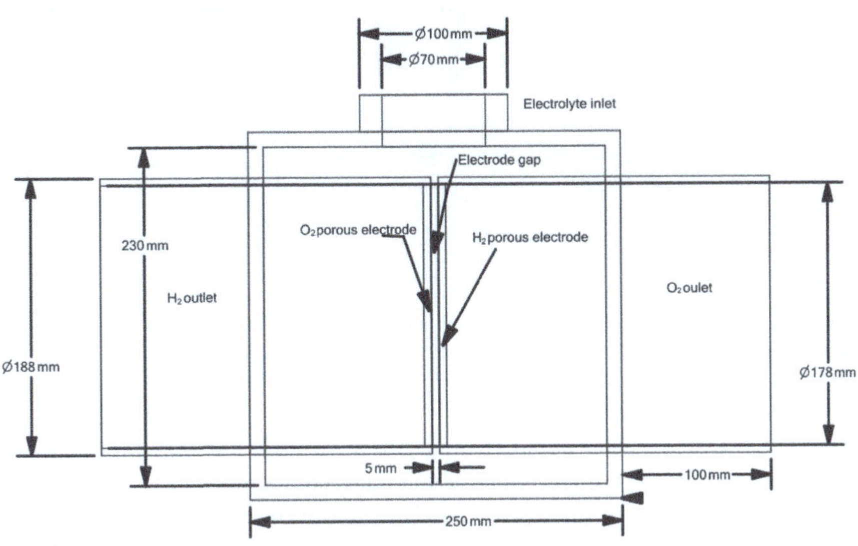

FIGURE 6.1 A single-cell alkaline electrolyzer.

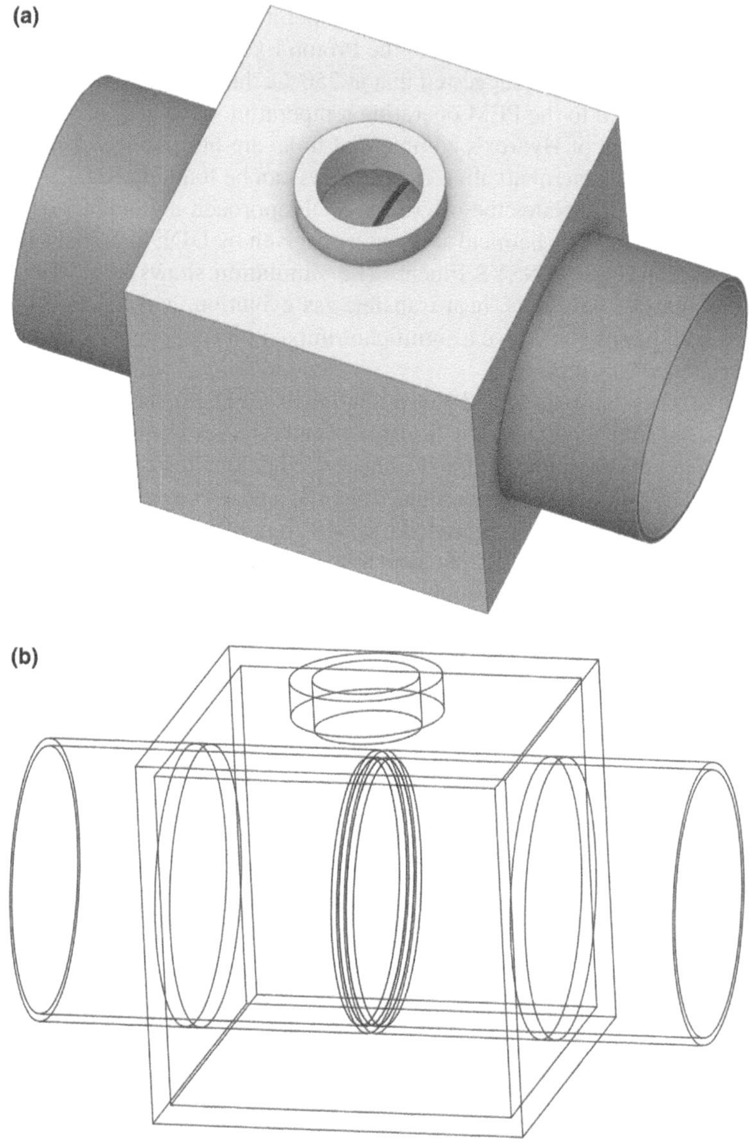

FIGURE 6.2 The full 3D geometry of the electrolyzer: (a) shaded graphics and (b) wireframe graphics.

figure do not represent the actual Hydrox DEFT electrolyzer. The figure represents a single inter-electrode compartment of an industrial alkaline cell. The cathode (for the hydrogen evolution reaction, HER) and anode (for the oxygen evolution reaction, OER) electrodes used are porous. Figure 6.2 illustrates the 3D geometry used in the simulation.

TABLE 6.1
Parameters of the Geometry for the Case Study

Parameter	Value	Description
Electrolyte diameter	89 mm	Liquid region before gas disengagement
Electrode gap	5 mm	Distance between cathode and anode
Electrode thickness	5 mm	The thickness of each electrode
Electrode active area	248.95 cm^2	Equivalent to radius ≈ 8.9 cm

The primary geometric parameters used for this simulation are given in Table 6.1. These values do not represent the actual values for Hydrox-type alkaline electrolyzer or other typical electrolyzers. The values are used for demonstration purpose.

Figure 6.3 illustrates the fluid extracted from the fluid domain for the geometry. The fluid domain represents a single fluid domain comprising the three fluids in the electrolyzer, i.e., the electrolyte fluid (liquid),

Figure 6.3 illustrates the extracted fluid domain for the geometry. This domain represents a multiphase region comprising the electrolyte (liquid), H_2 gas generated at the cathode porous electrode, and O_2 gas generated at the anode porous electrode, as shown in Fig. 6.3a. The domain is subsequently segmented to define the porous electrode regions. The sliced portions corresponding to the gas regions, which represent the porous electrodes, are shown in Fig. 6.3b. These regions indicate the zones where H_2 and O_2 are generated and transported through their respective gas channels.

(a)

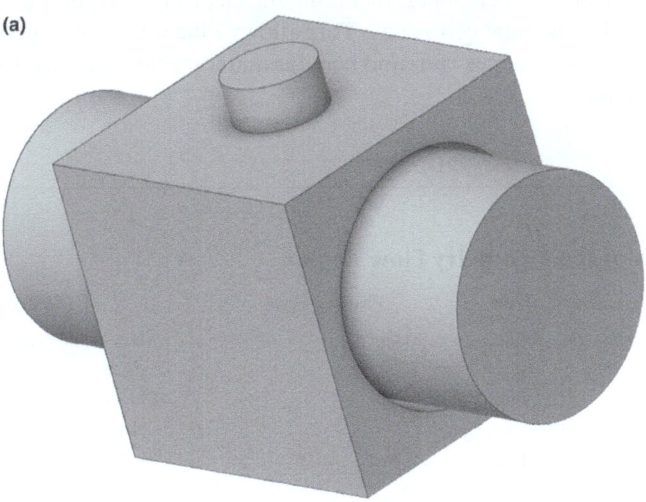

FIGURE 6.3 Fluid domain extracted from the geometry: (a) Entire fluid domain and (b) fluid domain for the species (H_2 and O_2) and porous electrodes. *(Continued)*

FIGURE 6.3 (*Continued*)

Table 6.2 summarizes the descriptions of the configurations of the domains. This setup in the table ensures that Fluent interprets the regions correctly and maintains continuous flow in the electrolyzer.

Boundary naming is also applied to the sliced single-fluid domain (i.e., the electrolyte region, H_2 porous region, and O_2 porous region), each of which is enclosed by its own outer solid boundary. The electrolyzer container encloses the electrolyte region, while solid pipes enclose the porous regions. The fluid zone naming is summarized in Table 6.3.

6.2.2 Meshing Fluid Domain

The meshing strategies for the fluid domain involve meshing the entire fluid domain as a continuous mesh of cell zones, including the electrolyte, H_2 porous electrode, O_2 porous electrode, and pipe cell zones. This allows Fluent to apply the porous media model to the electrode zones. Tet- and hex-dominant meshes are used for the bulk of the fluid domain.

TABLE 6.2
Description of the Geometry Flow Region

Region	Type	Physics State	Purpose
Electrolyte volume	Fluid	Active	Main liquid region where electrolysis occurs
O_2 pipe volume	Fluid	Active	Flow of O_2 gas through porous electrode outlet
H_2 pipe volume	Fluid	Active	Flow of H_2 gas through porous electrode outlet
Solid electrodes	Solid	Suppressed for physics	Represents the physical barrier, not simulated in flow

TABLE 6.3
Fluid Zone Boundary Naming

Boundary	Type	Note
Electrolyte inlet	Mass-flow-inlet or velocity-inlet	Where electrolyte enters
O_2 outlet	Pressure-outlet or mass-flow-outlet	O_2 gas out of the O_2 porous electrode
H_2 outlet	Pressure-outlet or mass-flow-outlet	H_2 gas out of the H_2 porous electrode
Walls (electrolyte, O_2, and H_2)	Wall	Non-slips or specified potential conditions

Inflation layers are added to the pore/electrolyte interface and to the solid wall interface to resolve boundary layers. The porous electrode thickness is resolved by using multiple cells across its thickness to capture the fluid behavior in the electrodes adequately.

The mesh is refined where gradients are expected, such as in the following:

i. Across the porous electrode thickness: Four to ten cells may be targeted across the electrode thickness. The thin electrodes need more cells per thickness to capture both pressure drop and species generation.
ii. At electrolyte–porous interfaces: At the electrolyte-porous interfaces, refine to capture interfacial gradients.
iii. Near inlet and gas pipe exits: Refine to capture developing flow and outlet pressure boundary layers.
iv. Around sharp geometry: Corners or small gaps should be refined to avoid highly distorted cells.

Face sizing may be used on the porous-volume interfaces, and body sizing may be applied to porous volumes with an element size smaller than that of the electrolyte. Figures 6.4–6.6 illustrate the increase in element size across the porous electrode thickness and within the porous zones. Figure 6.4 shows the default mesh size generated for the porous zones before remeshing zones or applying element sizing.

For targeted element resizing along the porous electrode thickness, a quick formulation, as given in Eq. (6.1), may be adopted.

$$\text{Element sizing} = \frac{T}{N} \tag{6.1}$$

where T is the thickness of the porous electrode and N is the number of elements desired for the porous thickness. For this case, the porous thickness is 5 mm (0.005 m); we may say N is 6, i.e., 0.005/6 = 0.00083 m. The larger the N, the smaller the element size and the more accurate the result. Equation (6.1) is only applied for the

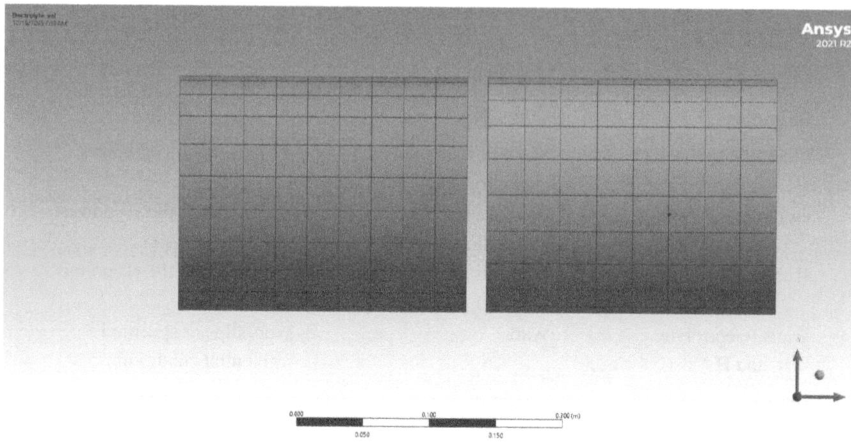

FIGURE 6.4 Default mesh across porous zones.

FIGURE 6.5 The porous zone body sizing meshing is performed using an element size of 0.00083 m with a growth rate of 1.1.

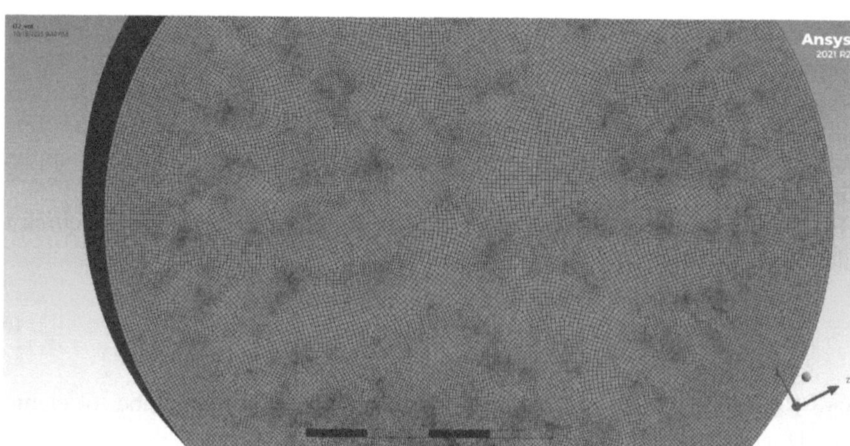

FIGURE 6.6 Applying an inflation layer, electrode, and electrolyte interface.

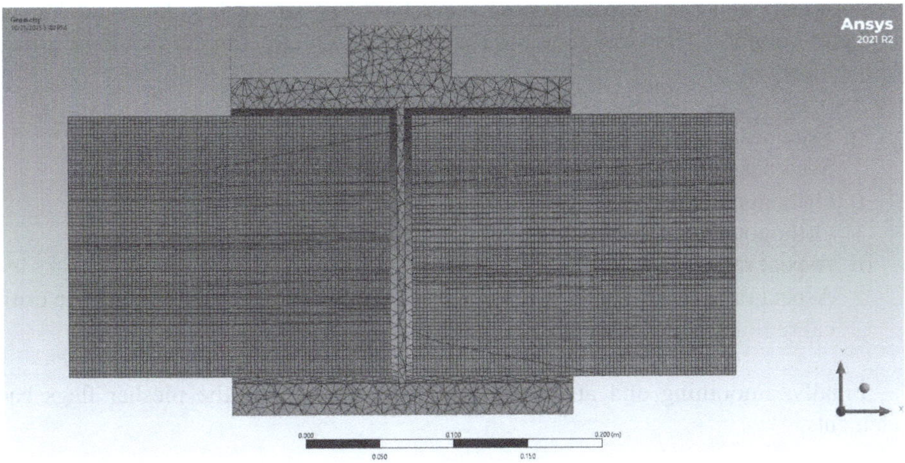

FIGURE 6.7 Cross section of the electrolyte zone body sizing mesh using an element size of 0.005 m and the porous zones.

porous electrode thickness, where a uniform mesh along the electrode thickness is expected. However, for other geometries, Eq. (6.1) is not relevant and should not be adopted for resizing the mesh. Figure 6.5 illustrates the application of body sizing with an element size of 0.00083 m (as determined by Eq. (6.1)) for the porous zones and a growth rate of 1.1.

Figure 6.6 illustrates the application of an inflation layer to the porous electrode and electrolyte interface across the thickness of the porous electrode, capturing interfacial gradients. While Figure 6.7 shows the cross section of the meshed single-fluid domain, comprising the electrolyte, H_2 porous, and O_2 porous zones. Body sizing is applied for the electrolyte zone, where the element size of 0.005 m is used.

6.2.3 Boundary Layer and Inflation Settings

As pointed out earlier, inflation layers may be added to the faces and walls of the fluid domain. Additionally, inflation layers are added to the faces adjacent to porous zones to account for significant near-wall gradients. A growth rate of 1.1 to 1.3 should be chosen for a smoother transition. First layer thickness may be set based on the y+ target [30, 31], such as:

i. If resolving the viscous sublayer is required, i.e., low *Re* and turbulence model with near-wall resolution, $y+ \lesssim 1$.
ii. If wall functions are used, $y+ \approx$ 30–300.

If porous flow dominates and the pore Reynolds number is low, extremely fine wall resolution within the porous zone may not be necessary; however, cells across the thickness may be retained.

As mentioned earlier, the element type should be tet- or hex-dominant for the complex volumes. These are required for efficiency. Quality targets should be aimed at meeting:

i. Skewness:
   Skewness <0.9 is fine, and <0.5 is excellent.
ii. Orthogonal quality:
   Orthogonal quality should be >0.2, and >0.4 should be aimed.
iii. Aspect ratio:
   Aspect ratio should be kept reasonable, i.e., <50 locally and closer to 1 in critical regions.

Finally, smoothing and auto-smoothing may be used if the mesher flags bad elements.

6.2.4 Setting up Physics: General Setup, Multiphase Model, and Species Transport

The meshed fluid domain is imported to Fluent for setting up the physics. Before launching Fluent, double precision is chosen, and the number of computer cores is increased from the default, which is usually 1 (in this case, 4 is chosen). The fluid domain geometry in Fluent, ready for physics, is shown in Figure 6.8. This enables users to confirm that the boundaries (inlets, outlets, walls) and cell zones are correctly presented in Fluent before applying physics.

For the general setup, the pressure-based solver type is selected, and the transient solver is chosen. The multiphase numerical setup involves selecting the model type, specifying phases, and the phase interactions. The models in multiphase modeling consist of the VOF, Mixture, and higher-fidelity Eulerian models. For this analysis, the Mixture model is adopted.

Three phases were selected for the case study, which involve the electrolyte phase as the primary phase and H_2 and O_2 gases as the secondary phases. A material having similar properties to those of a 30 wt% KOH solution is created for the primary electrolyte phase. It is created by specifying density, specific heat, dynamic viscosity, thermal conductivity, and molecular weight. Further setup based on choosing the Mixture model consists of the mixture formulations, slip velocity, surface tension, and interface discretization.

Species transport is enabled for the modeling. The phases (primary and secondary) are assigned to the proper materials. The primary phase is assigned to the KOH solution, the first secondary phase is assigned to the H_2 gas, while the second secondary phase is assigned to the O_2. In this case, the Reaction is switched off to enable UDF to handle gas generation in the porous media.

For volumetric porous electrodes modeling, porous zones are enabled for the H_2 porous zone and the O_2 porous zone to capture the two porous electrodes adequately. Material is assigned to the porous electrodes. In this case, Nickel (Ni) is assigned for the porous electrodes. The porous electrodes have two interfaces, including the electrolyte-nickel and nickel-gas interfaces. With this, the liquid

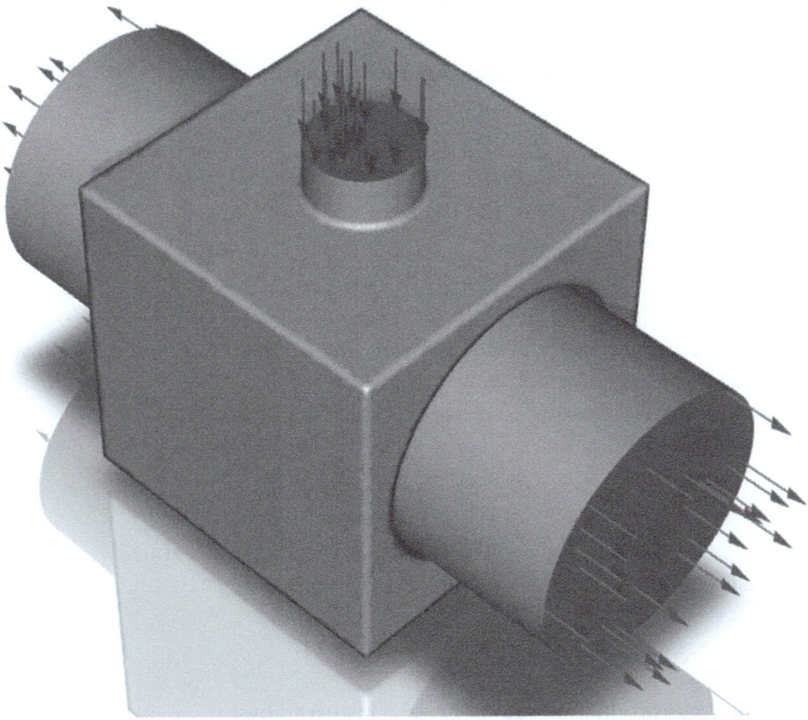

FIGURE 6.8 The extracted fluid domain indicates boundaries in Fluent ready for applying physics.

electrolyte (KOH solution) serves as the primary material and the Ni (solid) as the porous zone material.

Correct permeability and porosity values are assigned to each porous zone, and optional thermal/anisotropic values may also be assigned. The permeability value is used for the viscous resistance in the porous zones, and it is given by:

$$\text{Viscous resistance} = \frac{1}{K} \tag{6.2}$$

where K is the permeability of the porous electrode material in m^{-2}. The viscous resistance is specified for the three coordinates (x, y, and z). When the K is uniform across x, y, and z, an equal value is specified; otherwise, different values may be specified for each coordinate. The energy equation is enabled to include temperature effects in the modeling. It may be kept off for isothermal studies if needed.

6.2.4.1 Cell Zone Conditions

For each cell zone, the correct phase is assigned, and the corresponding source function is hooked to the H_2 porous zone, the O_2 porous zone, and the liquid zone. A UDF is written for this case that computes mass production rates from electrochemical current density using Faraday's law and adds sources to gas phases while removing mass from the electrolyte.

For the cathode, H_2 molar production rate ($\dot{n}_{H_2}$) from Faraday's law is given as:

$$\dot{n}_{H_2} = \frac{jA}{2F} \tag{6.3}$$

In terms of mass flow rate of the H_2, Eq. (6.3) is given by:

$$\dot{m}_{H_2} = \frac{jA}{2F} M_{H_2} \tag{6.4}$$

where j is the current density, A is the porous electrode area, F is the Faraday's constant, and M_{H_2} is the molecular weight of H_2.

A volumetric source term $\dot{S}_{H_2}$ is added, which is obtained as:

$$\dot{S}_{H_2} = \frac{\dot{m}_{H_2}}{V_{H2\_zone}} \tag{6.5}$$

where V_{H2_zone} is the electrode active volume in the H_2 porous zone.

For the O_2 anode, using the stoichiometry:

$$2H_2O \rightarrow 2H_2 + O_2 \tag{6.6}$$

The molar gas production rate of O_2 ($\dot{n}_{H_2}$) is given by:

$$\dot{n}_{O_2} = \frac{\dot{n}_{H_2}}{2} \tag{6.7}$$

Similarly, a volumetric source term is formulated for the O_2 porous zone.

With Eqs. (6.3) to (6.7), the UDF is written as given in Appendix.

6.2.4.2 Solution, Initialization, Monitors/reports, and Run Controls

Solution control includes settings such as pressure–velocity coupling, pressure, momentum, volume fraction, and energy discretization schemes. In this case, the PISO scheme is employed for pressure–velocity coupling due to the transient operating conditions. The PRESTO scheme is used for pressure interpolation, while the Second-Order Upwind scheme is applied to momentum and energy equations. The Compressive scheme is adopted for the volume fraction to ensure sharp interface capturing.

Initializing the domain, two methods are often used, that is, the hybrid or standard initialization. For this case, hybrid initialization is used. Here, all zones have proper initial values. For example, an initial gas fraction of 0.001 is set to help stability.

Monitors and reports may be created to track the simulation's progress. In this case, residual monitors are created for continuity, momentum, volume fraction, and species. A report, such as a mass balance report, may be added to ensure that the generated gas matches stoichiometry.

In the run settings, the simulation may start with a steady state for convergence testing. However, in this case, the transient mode is employed, which facilitates the study of bubble evolution or flow oscillations. For a transient mode, a small number of time steps is recommended, small enough to resolve bubble growth. Iterations of 500 to 1000 are recommended to monitor convergence.

6.2.5 Post-Processing

This process involves what output to plot or emphasize for the simulation. Contour plots and animations of gas volume fraction, velocity streamlines in the electrolyte, displaying recirculation, and local current density are essential (necessary when the UDF calculates spatially varying electrochemical reactions). Other outputs, such as temperature contours of the temperature distribution across the cells, are also essential.

Figures 6.9 and 6.10 display the demonstrative CFD outputs of pressure distribution and volume fractions of the phases. Figure 6.9 depicts the pressure distribution in the electrolyzer. Higher pressure is observed along the H_2 porous zone than the O_2 porous zone.

The volume fraction which is one of the crucial parameters for the electrolyzer indicates that more electrolyte penetrates through the H_2 porous zone than the O_2 porous zone, as evident in Figure 6.10a. Figure 6.10b represents a crucial CFD output, which shows the volume fraction of the H_2 phase in the O_2 porous zone. This indicates that not all H_2 gas leaves through the H_2 porous zone; some amount of the gas leaves through the O_2 porous zone. In Figure 6.10c, O_2 gas entering the H_2 porous zone is minimal when compared to that of the H_2 gas.

The simulation of the typical alkaline DEFT electrolyzer, undergone in this case study, is for demonstrative purposes only. The dimensions of the geometry and

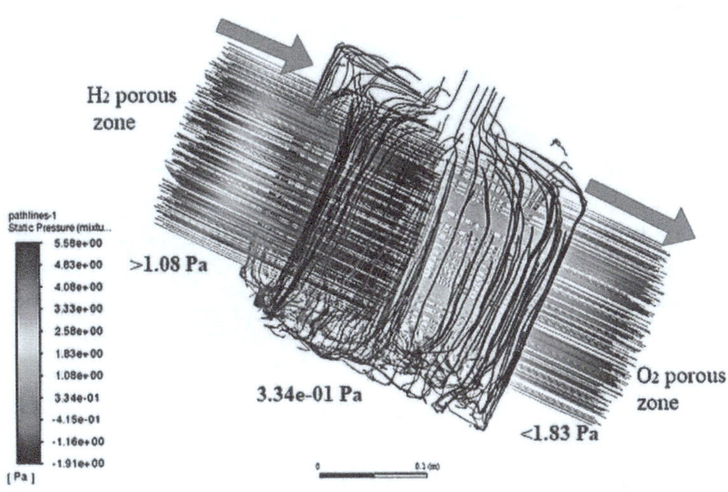

FIGURE 6.9 Pressure distribution along the porous zones.

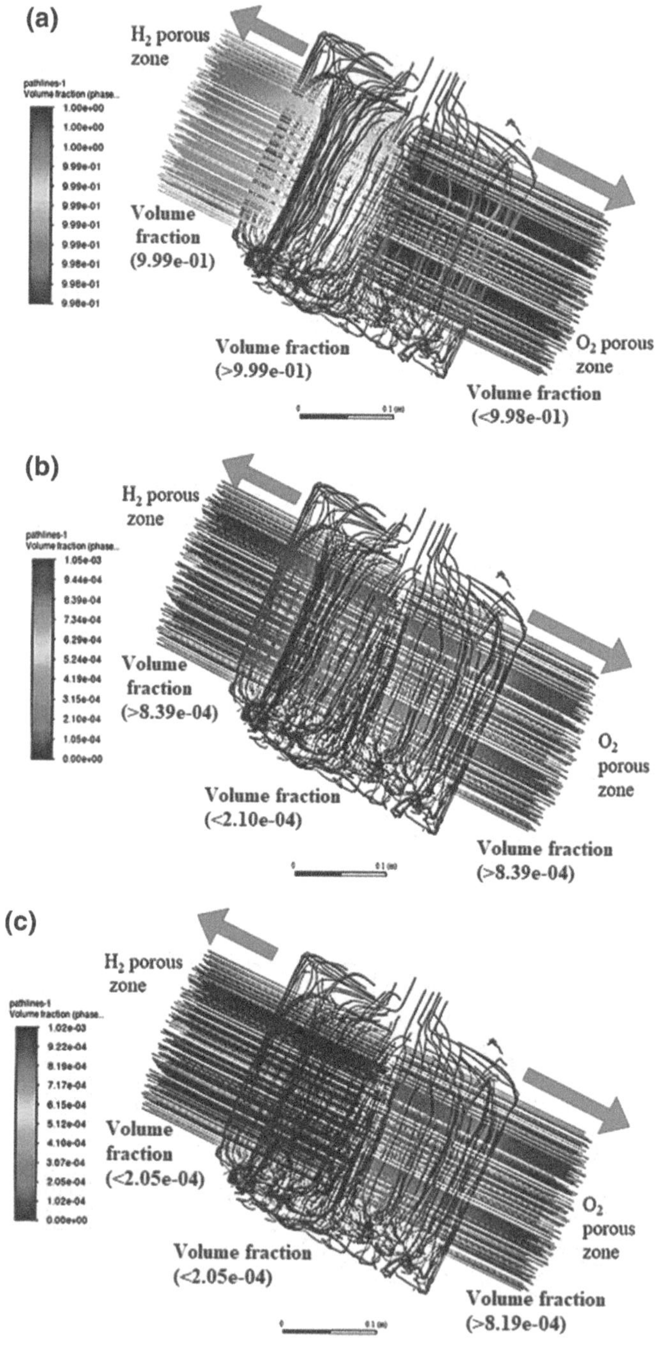

FIGURE 6.10 Volume fraction of (a) electrolyte phase, (b) hydrogen phase, and (c) oxygen phase.

properties are not the same as those of the existing Hydrox electrolyzer. The outputs of the simulation cannot be relied on. The electrochemical kinetics used in the simulation are simplified to a Faradaic source; a complete kinetics model would require coupling detailed Butler-Volmer kinetics and ion transport in porous electrodes.

The mixture or the VOF models of the multiphase modeling used in this case study neglect microscale bubble nucleation physics; tuning of bubble size nucleation is needed. Validation of the outputs is required to compare trends, rather than exact values, with experimental or literature data.

6.3 THERMOCHEMICAL HYDROGEN GENERATION SIMULATION

Thermochemical hydrogen generation refers to processes that utilize heat, typically at high temperatures, to drive chemical reactions or cycles that produce H_2 without relying solely on electrical energy. The primary methods are (i) hydrocarbon reforming, such as SMR, auto-thermal reforming, and partial oxidation, (ii) gasification of biomass or coal followed by shift conversion, and (iii) high-temperature water-splitting cycles that use nuclear, concentrated solar, or waste heat, such as copper-chlorine, sulfur-iodine, or metal-oxide redox cycles. Despite being less developed, the thermochemical H_2 generation category accounts for about 22% of H_2 generation case studies, compared to 74% of electrochemical case studies, according to a recent review [32].

Thermochemical pathways enable high productivity through high-throughput reactor designs and offer the advantage of utilizing thermal energy, which can be abundant or low-cost when sourced from solar or waste heat. However, several challenges arise in their modeling, including high operating temperatures, multiphase flows (such as gas–solid–slag–metal systems), strong coupling between transport phenomena (fluid flow, heat, and mass transfer) and chemical kinetics, as well as complex reactor geometries. These factors necessitate detailed simulations for accurate design, scale-up, and optimization [33].

This section covers the use of CFD in thermochemical hydrogen-generating system modelling. It shows the main physical mechanisms and model selections. It also reveals the way CFD supports system integration and optimization.

6.3.1 Core Physical Phenomena and Modeling Challenges

Usually, endothermic reactions (e.g., $CH_4 + H_2O \rightarrow CO + 3H_2$) or high-temperature redox cycles (such as hydrolysis, metal oxide reduction producing H_2) are used in thermochemical H_2 generation. The mass and heat transfer properties within the catalyst have a significant impact on the reaction kinetics in reforming reactors.

Reaction rate models (which are frequently expressed in Arrhenius or power-law forms) must be incorporated into a CFD solver that concurrently resolves momentum, species, energy, and, sometimes, solid-phase heat conduction. The interaction of these integrated models is complicated, and local hot spots in the bed may cause catalyst sintering or deactivation, whereas local cold spots in the bed lower reaction rates.

Accurate thermal management is essential for thermochemical systems. Reformers require heat to be supplied (which is usually a heat exchanger or combustion) and distributed throughout the reactor. Heat recovery and thermal storage are integral components of the systems. CFD plays a crucial role in identifying temperature gradients, thermal resistances in foams or packed beds, and boundary layer limitations. CFD has been used in a tubular packed-bed reformer with metal form channels to optimize radial heat transfer and raise H_2 yield [34].

A critical review conducted by D'Souza [35] analyzed thermochemical water splitting based on redox and noted that overall performance is dominated by heat management and kinetics. Similar studies have been conducted where heat transfer and reaction coupling in a packed-bed steam reformer for distributed H_2 generation were investigated using CFD [36].

Common thermochemical reactors involve slag flows, membranes, metallic forms, solid catalysts, or moving parts (such as redox cycles). The interactions among thermal conduction, solid structures, fluid (steam and gas), and occasionally particle fluxes or packed beds must all be resolved with the aid of CFD. A study conducted by Zhao et al. [37] developed a CFD model coupling membrane oxygen transport, methane combustion, and hydrogen generation in a membrane reactor.

The impact of model scale and geometry is enormous: heat and mass gradients, as well as local conversion, are influenced by reactor length, diameter, bed shape, and catalyst distribution. By simulating the reactor domains as opposed to only representative slices, CFD helps to provide insights into hotspot formation, flow bypass or channeling, and maldistribution.

6.3.2 CFD Modeling Framework

A systematic framework ensures the reliability and reproducibility of CFD for thermochemical H_2 generation. Depending on the complexity and processing capability, a 2D axisymmetric reactor, a quasi-3D reactor, or a full 3D reactor may be chosen to simulate the process. For packed beds, the domain may be segmented into fluid and solid regions. This has been demonstrated in the study conducted by Han et al. [34] where metal foam channels were used in a tubular shape.

The physical models required are:

i. Momentum/Flow: This involves the Navier-Stokes (RANS or LES) equations for fluid. If the packed bed is treated as a continuum, then porous media models may be applicable.
ii. Energy: When the temperature exceeds 800°C, conjugate heat transfer (which includes fluid and solid) and radiative heat transfer are solved.
iii. Species/Reaction: The species transport model is enabled, and reaction mechanisms are incorporated using heterogeneous catalysis or embedded chemical kinetics (either global or detailed mechanisms).
iv. Mass transfer and solid phases: User-defined models or multi-domain coupling approaches may be incorporated, where applicable, to account for

processes such as membrane permeation and solid–gas reactions User-defined models may be incorporated or multi-domain coupling, if applicable, such as membrane permeation and solid/gas reactions.

Specify the solid catalyst zones, inlet gas/steam flows, temperature, and hydraulic/thermal boundary conditions. Heterogeneous catalysis can be represented by internal source terms in reaction zones (i.e., mass sink for CH_4 and mass source for H_2). UDFs may be used to implement the wall/permeation mass flux of H_2 in a membrane reactor.

The mesh should capture the necessary gradients in species, temperature, and velocity, such as those found at bed surfaces or in boundary layers. The proper time steps for transient phenomena (i.e., startup, heat pulse) should be used, while pseudo-steady-state may be adequate for steady-state design sweeps.

Essential parameters can be compared with data from experiments or literature, such as H_2 conversion yield or temperature profiles. The impacts of heat transfer coefficients, mesh size, and kinetic parameters may be assessed through sensitivity studies. Detailed CFD modeling methods and validation practices for reformers have been provided [33, 36].

Thermochemical H_2 generation CFD outputs often provide valuable design insights, such as helping in: (i) optimizing bed shape, heat-flux distribution, and insulation design, (ii) investigating the impact of scaling, (iii) feeding into system-level models for cycle efficiency, integration, and costs, and (iv) component-level optimization for H_2 generation and deployment.

6.4 MECHANISMS IN HYDROGEN COMBUSTION AND CHARACTERISTICS OF FLAMES

Several features set H_2 combustion apart from hydrocarbon fuels and have direct implications for engine control, engine design, and CFD modeling. Hydrogen has a very wide flammability range, a very high diffusivity, a high laminar flame speed, and a low minimum ignition energy when compared to other common hydrocarbon fuels [38]. These features cause rapid flame propagation and have a strong tendency to pre-ignite or backfire in port injection systems and produce different NO_x formation features that call for specialized control strategies.

To simulate H_2 combustion accurately and design emission mitigation plans for H_2-fueled engines, it is crucial to understand the combustion processes, including flame speed, flame regimes, ignition behavior, and NO_x chemistry. These unique features and their technological implications are summarized in comprehensive reviews [39, 40].

Depending on how H_2 is delivered in relation to the oxidizer, it can participate in all three of the conventional combustion regimes, such as the premixed, partially premixed, and non-premixed [38]. The mixture is primarily premixed in spark-ignition or port-injected H_2 engines; however, partially premixed fields where local equivalence ratio gradients are significant can be produced by direct injection of H_2 (or stratified injection techniques).

TABLE 6.4
Comparison of Hydrogen-Air and Hydrogen-Oxygen Flames

Flame Properties	H_2-air flame	H_2-O_2 flame
Flame temperature and speed	Low adiabatic temperature	Higher adiabatic temperature as a result of eliminating N_2
	Lower lamina flame speed	Higher laminar flame speed
NO_x potential	There is thermal NO_x formation	No thermal NO_x formation
		O_2 enrichment increases NO_x formation in the presence of N_2 (since oxidizers in systems contain air) [40]

The regime affects the stability and structure of the flame. Premixed H_2 flames are characterized by their thinness and speed, which are highly sensitive to the initial pressure, temperature, and equivalence ratio. Non-premixed H_2 flames (less common in engines) exhibit strong diffusion effects and can produce localized hot spots where NO_x may form. As a result, CFD users need to select an appropriate turbulence model (such as the PaSR or flamelet) that is suitable for the regime and the relevant spatial and temporal scales. RANS model captures mean behavior and is suitable for parametric studies, but the LES model is increasingly employed when resolving transient flame kernels and local extinction/re-ignition events are significant [41].

Comparing hydrogen-air and hydrogen-oxygen flames demonstrates the separate roles of oxidizer content and diluent on flame behavior. Table 6.4 summarizes these.

These distinctions (Table 6.4) are crucial when putting up CFD scenarios. While H_2-air CFD must resolve the N_2 chemistry and its temperature sensitivity, pure O_2 oxidizers require different chemistry options and often lack NO formation routes.

The physics discussed above translate into applicable practical requirements from the perspective of CFD modeling. When selecting kinetic mechanisms and computational cost is a concern, validated mechanisms for H_2 combustion (with NO chemistry) can be used, and mechanism reduction may be considered. Validation against ignition delay and laminar flame should always be carried out [38].

The mesh and model choices that affect temperature prediction, such as wall heat transfer, radiation, and turbulence models, have a significant impact on NO prediction because NO_x is exponentially sensitive to peak temperature. The temperature fields should be validated first before NO_x outputs [41, 42]. Figure 6.11 depicts a cross section of a single cylinder internal combustion (IC) engine with extracted cylinder volume that may be used for CFD hydrogen combustion modeling.

6.5 PREDICTION AND CONTROL OF EMISSION

In an H_2-based energy conversion system, accurately predicting and controlling emissions remains a significant challenge. Despite being carbon-free by nature, H_2 combustion can still produce NO_x through high-temperature processes (as highlighted in

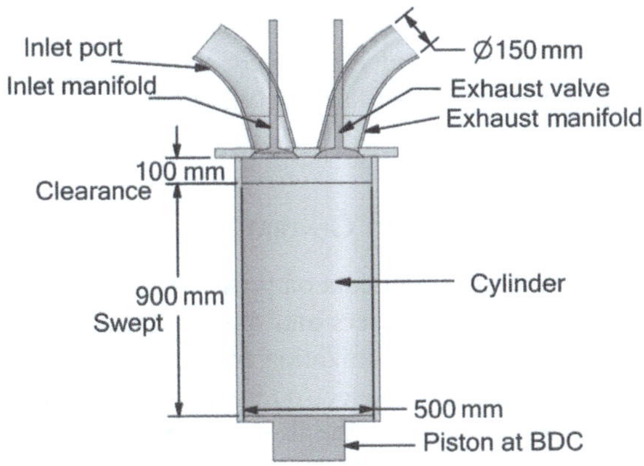

FIGURE 6.11 Cross section of a single cylinder internal combustion engine.

Table 6.4) and, in certain situations, unburned H_2 or O_2 slippage. These emissions impact both system efficiency and environmental sustainability.

CFD offers the dominant method for determining emission mechanisms and designing control plans for H_2 burners, IC engines, and electrolyzer vent gases. According to reviews, real-time emission estimation, optimization, and maintenance are made possible by integrating CFD with detailed chemical kinetics and machine-learning surrogates [43–45].

Chemical reaction models must be coupled with flow and heat transfer solvers to accurately predict emissions. Current practice is dominated by three primary frameworks: (i) steady-based CFD with post-processing NO_x models, (ii) transient LES or DNS models, and (iii) hybrid CFD-machine learning surrogates.

A machine learning-assisted CFD workflow was demonstrated by Zhang et al. [46], which reduced simulation time by more than 80% by using convolutional neural networks to predict instantaneous NO_x fields inside an H_2 burner. Such hybrid techniques are increasingly being incorporated into digital-twin environments for real-time monitoring of emissions.

CFD supports emission control and mitigation. Typical hydrogen-engine strategies consist of: (i) lean-burn operation and Exhaust Gas Recirculation (EGR), (ii) water or steam injection, micromix and staged combustion, and (iii) electrolyzer vent management [47–50]. The lean-burn operation and EGR lower flame temperature and $NO_{x,}$ but run the danger of misfiring, which CFD aids in defining lean-limit boundaries.

6.6 OPTIMIZATION IN HYDROGEN GENERATION AND COMBUSTION SYSTEMS

CFD modeling naturally leads to optimization, which bridges simulation and system design. Identifying the operating parameters and geometric arrangements of combustion and H_2-generating systems that optimize conversion efficiency, H_2 yield, or combustion stability while reducing energy losses and emissions is paramount.

As a result of the computational cost of conventional trial-and-error parametric studies, CFD-based optimization is widely adopted, which is supplemented by metamodels like artificial neural networks (ANNs), response-surface methods, and genetic algorithms [51, 52].

6.6.1 Optimization in Hydrogen Generation Systems

CFD provides comprehensive spatial data on pressure, current density, temperature, and bubble distribution for electrolyzers and thermochemical reactors. These outcomes from CFD are utilized in optimization frameworks to adjust design parameters, including flow-channel shape, electrode gaps, and operating pressure. Such optimization frameworks adopted are the Response Surface Methodology (RSM), Genetic Algorithms (GA), and Neural-Network-Assisted CFD.

The RSM builds a quadratic regression that connects design parameters (such as flow rate, temperature, and electrolyte concentration) to design objectives like energy efficiency or hydrogen production rate [53]. Jabbari et al. [54] employed CFD-RSM optimization on membrane gas separation. The RSM was employed in the study to identify an optimal combination of variables, resulting in fewer CFD runs compared to a full parametric sweep.

In the GA algorithm, stochastic search approaches imitate natural evolution to identify optimal combinations. In neural-network-assisted CFD, such as the application of ANNs to CFD data, nearly instantaneous predictions of meaningful outputs are made for new operating conditions. Such optimizations greatly accelerate the maturity of technology by directly influencing process control strategies and material selection.

6.6.2 Optimization in Hydrogen Combustion Systems

In combustion systems, performance-emission trade-offs are the primary focus. CFD provides the thermodynamic field, and algorithms guide parameter adjustments to meet targets such as stable lean-limit operation, minimum NO_x emissions, or maximum indicated efficiency. The following are typical design variables: (i) equivalence ratio and dilution fraction, (ii) Swirl number, staging ratio, and burner geometry, especially for turbines, and (iii) injection pressure, orientation, and timing, especially for engines.

Future hydrogen-system optimization will use cloud-deployed digital twins, multi-objective evolutionary algorithms, and physics-informed neural networks [55, 56]. Some of the key areas for future research will be on integrating CFD-generated datasets into a continual-learning framework [57]. Another area will be on quantifying uncertainties to ensure robust optimization under fluctuating ambient conditions [58].

Additionally, another area of focus is the use of reinforcement learning to control combustion and generation in dynamic power grids autonomously. The convergence of CFD outputs, optimization, and AI is anticipated to enable fully autonomous and shortened design cycles, self-optimizing engines, and hydrogen plants, which are features of Industry 5.0 energy systems.

REFERENCES

1. J. G. Segovia-Hernández, S. Hernández, E. Cossío-Vargas, M. Juarez-García, and E. Sánchez-Ramírez, 'Green hydrogen production for sustainable development: A critical examination of barriers and strategic opportunities', *RSC Sustainability*, vol. 3, no. 1, pp. 134–157, 2025. doi: 10.1039/D4SU00630E.
2. M. Jaradat, S. Almashaileh, C. Bendea, A. Juaidi, G. Bendea, and T. Bungau, 'Green hydrogen in focus: A review of production technologies, policy impact, and market developments', *Energies*, vol. 17, no. 16, p. 3992, 2024. doi: 10.3390/en17163992.
3. A. Sirat, S. Ahmad, I. Ahmad, N. Ahmed, and M. Ahsan, 'Integrative CFD and AI/ML-based modeling for enhanced alkaline water electrolysis cell performance for hydrogen production', *International Journal of Hydrogen Energy*, vol. 83, pp. 1120–1131, 2024. doi: 10.1016/j.ijhydene.2024.08.184.
4. C. Huang, J. L. R. Torres, Y. Zong, S. You, and X. Jin, 'A review of alkaline electrolyzer technology modeling and applications for decision-making optimization in energy systems', *Renewable and Sustainable Energy Reviews*, vol. 224, p. 116005, 2025. doi: 10.1016/j.rser.2025.116005.
5. Z. Zhang, W. Liu, and M. L. Free, 'Phase-field modeling and simulation of gas bubble coalescence and detachment in a gas-liquid two-phase electrochemical System', *Journal of the Electrochemical Society*, vol. 167, no. 1, p. 013532, 2020. doi: 10.1149/2.0322001JES.
6. J. Park, and G. Son, 'Numerical study on mass transfer and hydrogen bubble generation in water electrolysis', *Journal of Computational Fluids Engineering*, vol. 29, no. 3, pp. 173–185, 2024. doi: 10.6112/kscfe.2024.29.3.173.
7. M. A. Riaz *et al*, 'Water electrolysis technologies: The importance of new cell designs and fundamental modelling to guide industrial-scale development', *Energy & Environmental Science*, vol. 18, no. 11, pp. 5190–5214, 2025. doi: 10.1039/D4EE05559D.
8. M. Saleh Shahreza, S. Sogbaike, A. Archibong-Eso, I. M. Albayati, and A. M. Aliyu, 'Multiphase computational fluid dynamics simulation-based performance investigation of hydrogen production with patterned electrodes in alkaline-water electrolysis', *Journal of Energy Resources Technology, Part A*, vol. 1, no. 6, p. 061703, 2025. doi: 10.1115/1.4069226.
9. A. Bayat, P. K. Das, and S. C. Saha, 'Modeling porosity distribution strategies in PEM water electrolyzers: A comparative analytical and numerical study', *Mathematics*, vol. 13, no. 13, p. 2077, 2025. doi: 10.3390/math13132077.
10. K. M. Busam, S. P. Vudata, W. Li, and F. V. Lima, 'Macro-level electrochemical modeling strategies of low-temperature and high-temperature electrolyzers for hydrogen production: A review', *Industrial & Engineering Chemical Research*, vol. 64, no. 23, pp. 11129–11152, 2025. doi: 10.1021/acs.iecr.4c05008.
11. L. Deng *et al.*, 'Bubble evolution dynamics in alkaline water electrolysis', *eScience*, vol. 5, no. 4, p. 100353, 2025. doi: 10.1016/j.esci.2024.100353.
12. J. Rodríguez, and E. Amores, 'CFD modeling and experimental validation of an alkaline water electrolysis cell for hydrogen production', *Processes*, vol. 8, no. 12, p. 1634, 2020. doi: 10.3390/pr8121634.
13. M. Ali *et al*, 'Recent progress in underground hydrogen storage', *Energy & Environmental Science*, vol. 18, no. 12, pp. 5740–5810, 2025. doi: 10.1039/D4EE04564E.
14. Y. Khalili, S. Yasemi, M. Bagheri, and A. Sanati, 'Advancements in hydrogen storage technologies: Integrating with renewable energy and innovative solutions for a sustainable future', *Energy Geoscience*, vol. 6, no. 2, p. 100408, 2025. doi: 10.1016/j.engeos.2025.100408.

15. M. E. Okolo, D. S. Adebayo, and C. F. Oduoza, 'Numerical modelling of non-premixed hydrogen blended combustion in a 3D-combustor with jet optimisation', *Reaction Chemistry and Engineering*, vol. 10, no. 5, pp. 1114–1130, 2025. doi: 10.1039/D4RE00537F.
16. R. Al-ajmi, A. H. Qazak, A. M. Sadeq, M. Al-Shaghdari, S. F. Ahmed, and A. K. Sleiti, 'Numerical investigation of the potential of using hydrogen as an alternative fuel in an industrial burner', *Fuel*, vol. 385, p. 134194, 2025. doi: 10.1016/j.fuel.2024.134194.
17. B. Zhang, H. Wang, and S. Wang, 'Computational investigation of combustion, performance, and emissions of a diesel-hydrogen dual-fuel engine', *Sustainability*, vol. 15, no. 4, p. 3610, 2023. doi: 10.3390/su15043610.
18. A. Ballatore, D. Q. Reyes, H. Bao, and J. van Oijen, 'LES-FGM modelling of non-premixed auto-igniting turbulent hydrogen flames including preferential diffusion', 2024. doi: 10.48550/ARXIV.2411.08505.
19. C. J. Ramsay, and K. K. J. R. Dinesh, 'Numerical modelling of a heavy-duty diesel-hydrogen dual-fuel engine with late high pressure hydrogen direct injection and diesel pilot', *International Journal of Hydrogen Energy*, vol. 49, pp. 674–696, 2024. doi: 10.1016/j.ijhydene.2023.09.019.
20. M. C. Cameretti, R. De Robbio, E. Mancaruso, and M. Palomba, 'CFD study of dual fuel combustion in a research diesel engine fueled by hydrogen', *Energies*, vol. 15, no. 15, p. 5521, 2022. doi: 10.3390/en15155521.
21. F. S. Almutairi, K. K. J. R. Dinesh, and J. A. Van Oijen, 'Modelling of hydrogen-blended dual-fuel combustion using flamelet-generated manifold and preferential diffusion effects', *International Journal of Hydrogen Energy*, vol. 48, no. 4, pp. 1602–1624, 2023. doi: 10.1016/j.ijhydene.2022.10.078.
22. S. Ruiz, S. Valencia, F. Illacanchi, A. Mendiburu, L. Bravo, P. Khare, and C. Celis, 'Numerical Assessment of a Partially Stirred Reactor-Based Combustion Model Using a High-Fidelity CFD Tool', in *Volume 8: Fluids Engineering*, American Society of Mechanical Engineers, Nov. 2024, p. V008T10A003. doi: 10.1115/IMECE2024-145408.
23. 'Accelerate Hydrogen Adoption Using Simulation', ANSYS part of synopsis. [Online]. Available: https://www.ansys.com/blog/accelerate-hydrogen-adoption-using-simulation
24. B. T. Sangtam, and H. Park, 'Review on bubble dynamics in proton exchange membrane water electrolysis: Towards optimal green hydrogen yield', *Micromachines*, vol. 14, no. 12, p. 2234, 2023. doi: 10.3390/mi14122234.
25. Hydrox Holdings, 'DEFTTM (Divergent Electrode Flow Through) Electrolysis', 2023. [Online]. Available: https://hydroxholdings.co.za/wp-content/uploads/2020/04/DEFT-Technology-Showcase-Presentation_Website_RevB-1.pdf
26. H. Tüysüz, 'Alkaline water electrolysis for green hydrogen production', *Accounts of Chemical Research*. doi: 10.1021/acs.accounts.3c00709.
27. S. Hu *et al.*, 'A comprehensive review of alkaline water electrolysis mathematical modeling', *Applied Energy*, vol. 327, p. 120099, 2022. doi: 10.1016/j.apenergy.2022.120099.
28. Hydrox Holdings, 'Hydrox Technology', 2025. [Online]. Available: https://hydroxholdings.co.za/technology
29. F. Khalighi, N. G. Deen, Y. Tang, and A. W. Vreman, 'Hydrogen bubble growth in alkaline water electrolysis: An immersed boundary simulation study', *Chemical Engineering Science*, vol. 267, p. 118280, 2023. doi: 10.1016/j.ces.2022.118280.
30. ANSYS, 'ANSYS Fluent Theory Guide: 4.12.4 Enhanced Wall Treatment', 2009. [Online]. Available: https://www.afs.enea.it/project/neptunius/docs/fluent/html/th/node101.htm
31. ANSYS, 'ANSYS Fluent 12.0 User's Guide: 4.12.4 Enhanced Wall Treatment', 2009. [Online]. Available: https://www.afs.enea.it/project/neptunius/docs/fluent/html/ug/node410.htm

32. Zulfhazli, A. R. Keeley, S. Takeda, and S. Managi, 'A systematic review of the techno-economic assessment of various hydrogen production methods of power generation', *Frontiers in Sustainability*, vol. 3, p. 943145, 2022. doi: 10.3389/frsus.2022.943145.
33. N. Ghasem, 'A review of the CFD modeling of hydrogen production in catalytic steam reforming reactors', *International Journal of Molecular Sciences*, vol. 23, no. 24, p. 16064, 2022. doi: 10.3390/ijms232416064.
34. Y. Han, Z. Zhang, Z. Wang, and G. Zhang, 'Numerical investigation of hydrogen production via methane steam reforming in tubular packed bed reactors integrated with annular metal foam gas channels', *Energies*, vol. 18, no. 17, p. 4758, 2025. doi: 10.3390/en18174758.
35. L. D'Souza, 'Thermochemical hydrogen production from water using reducible oxide materials: A critical review', *Materials for Renewable and Sustainable Energy*, vol. 2, no. 1, p. 7, 2013. doi: 10.1007/s40243-013-0007-0.
36. J. Wang, Z. Liu, C. Ji, and L. Liu, 'Heat transfer and reaction characteristics of steam methane reforming in a novel composite packed bed microreactor for distributed hydrogen production', *Energies*, vol. 16, no. 11, p. 4347, 2023. doi: 10.3390/en16114347.
37. T. Zhao, C. Chen, and H. Ye, 'CFD simulation of hydrogen generation and methane combustion inside a water splitting membrane reactor', *Energies*, vol. 14, no. 21, p. 7175, 2021. doi: 10.3390/en14217175.
38. S. Verhelst, and T. Wallner, 'Hydrogen-fueled internal combustion engines', *Progress in Energy and Combustion Science*, vol. 35, no. 6, pp. 490–527, 2009. doi: 10.1016/j.pecs.2009.08.001.
39. M. A. Habib, A. Q. Abdulrahman, A. B. S. Alquaity, and A. Qasem, 'Hydrogen combustion, production, and applications: A review', *Alexandria Engineering Journal*, vol. 100, pp. 182–207, 2024. doi: 10.1016/j.aej.2024.05.030.
40. E. Giacomazzi *et al.*, 'Hydrogen combustion: Features and barriers to its exploitation in the energy transition', *Energies*, vol. 16, no. 20, p. 7174, 2023. doi: 10.3390/en16207174.
41. G. Li, M. Niu, J. Jian, and Y. Lu, 'Recent progress and prospects of hydrogen combustion chemistry in the gas phase', *Renewable and Sustainable Energy Reviews*, vol. 213, p. 115411, 2025. doi: 10.1016/j.rser.2025.115411.
42. Z. Wang, and X. Yang, 'NO_x formation mechanism and emission prediction in turbulent combustion: A review', *Applied Sciences*, vol. 14, no. 14, p. 6104, 2024. doi: 10.3390/app14146104.
43. N. Swaminathan, and A. Parente, Eds, *Machine Learning and Its Application to Reacting Flows: ML and Combustion*, in Lecture Notes in Energy, vol. 44, Springer International Publishing, 2023. doi: 10.1007/978-3-031-16248-0.
44. J. L. Wilk-Jakubowski, L. Pawlik, D. Frej, and G. Wilk-Jakubowski, 'Data-driven computational methods in fuel combustion: A review of applications', *Applied Sciences*, vol. 15, no. 13, p. 7204, 2025. doi: 10.3390/app15137204.
45. T. Kumar, A. Kumar, and P. Pal, 'A Posteriori Evaluation of a Physics-Constrained Neural Ordinary Differential Equations Approach Coupled with CFD Solver for Modeling Stiff Chemical Kinetics', 2023. doi: 10.48550/ARXIV.2312.00038.
46. M. Zhang, Q. Cui, M. Xiao, and G. Ni, 'Deep learning accelerated numerical simulation for three-dimensional compressible fluids', *Physics in Fluids*, vol. 36, no. 7, p. 076124, 2024. doi: 10.1063/5.0214669.
47. P. Abinash, 'Formation, kinetics and control strategies of NO_x emission in hydrogen fueled IC engine', *International Journal of Engineering Research & Technology*, vol. 9, no. 01, 2020. doi: 10.17577/IJERTV9IS010081.
48. W. Gao, Z. Fu, Y. Li, Y. Li, and J. Zou, 'Progress of performance, emission, and technical measures of hydrogen fuel internal-combustion engines', *Energies*, vol. 15, no. 19, p. 7401, 2022. doi: 10.3390/en15197401.

49. N. Zhao, Z. Liu, L. Wang, J. Pan, and Z. Huang, 'Experimental study on performance and emission of an electric turbocharged hydrogen direct injection engine', *ACS Omega*, vol. 9, no. 25, pp. 27407–27414, 2024. doi: 10.1021/acsomega.4c02222.
50. L. Wang, X. Li, C. Hong, P. Guo, S. Guo, and Z. Yang, 'Research and development of hydrogen-fueled internal combustion engines in China', *ACS Omega*, vol. 8, no. 51, pp. 48590–48612, 2023, doi: 10.1021/acsomega.3c05397.
51. H. Wang *et al.*, 'Multi-objective optimization of a hydrogen-fueled Wankel rotary engine based on machine learning and genetic algorithm', *Energy*, vol. 263, p. 125961, 2023. doi: 10.1016/j.energy.2022.125961.
52. B. Suatean, G. Cican, S. Guilain, and G. De-Paz-Alcolado, 'Optimization of hydrogen combustion in diesel engines: A CFD-based approach for efficient hydrogen mixing and emission reduction', *Fuels*, vol. 6, no. 2, p. 27, 2025. doi: 10.3390/fuels6020027.
53. O. O. Olabinjo, 'Response surface techniques as an inevitable tool in optimization process', in *Response Surface Methods – Theory, Applications and Optimization Techniques*, V. Silva and J. Sousa Cardoso, Eds, IntechOpen, 2024. doi: 10.5772/intechopen.1004575.
54. B. Jabbari, E. Jalilnejad, K. Ghasemzadeh, and A. Iulianelli, 'Modeling and optimization of a membrane gas separation based bioreactor plant for biohydrogen production by CFD–RSM combined method', *Journal of Water Process Engineering*, vol. 43, p. 102288, 2021. doi: 10.1016/j.jwpe.2021.102288.
55. Z. Ren, S. Zhou, D. Liu, and Q. Liu, 'Physics-informed neural networks: A review of methodological evolution, theoretical foundations, and interdisciplinary frontiers toward next-generation scientific computing', *Applied Sciences*, vol. 15, no. 14, p. 8092, 2025. doi: 10.3390/app15148092.
56. N. Wang, Y. Chen, and D. Zhang, 'A comprehensive review of physics-informed deep learning and its applications in geoenergy development', *Innovation Energy*, vol. 2, p. 100087, 2025. doi: 10.59717/j.xinn-energy.2025.100087.
57. S. Kumar, P. Kumar, and K. S. Grewal, 'Machine learning-driven multi-objective optimization of microchannel reactors for CO_2 conversion', *Advanced Sustainable Systems*, vol. 9, no. 6, p. 2500064, 2025. doi: 10.1002/adsu.202500064.
58. L. Li, L. Zhang, B. Zhang, H. Liu, and Z. Zheng, 'A review of uncertainty quantification and its applications in numerical simulation of scramjet combustor', *International Journal of Thermofluids*, vol. 18, p. 100351, 2023. doi: 10.1016/j.ijft.2023.100351.

APPENDIX

```
#include "udf.h"
                                         /* ---------- USER INPUT SECTION ---------- */
#define CURRENT_DENSITY     1000.0                        /* A/m2 */
#define ELECTRODE_AREA      0.025                         /* m2 */
#define FARADAY_CONST       96485.0                       /* C/mol */
#define MOLAR_MASS_H2       0.002016                      /* kg/mol */
#define MOLAR_MASS_O2       0.03200                       /* kg/mol */
#define ELEC_PER_MOL_H2     2.0
#define ELEC_PER_MOL_O2     4.0

                                     /* Porous-zone volumes (adjust to your geometry) */
#define VOL_H2_ZONE         5.0e-6                        /* m3 */
#define VOL_O2_ZONE         5.0e-6                        /* m3 */

                    /* ---------- CALCULATED MASS SOURCE RATES ---------- */
#define H2_RATE  (CURRENT_DENSITY * ELECTRODE_AREA * MOLAR_MASS_H2 / (ELEC_PER_
MOL_H2 * FARADAY_CONST * VOL_H2_ZONE))
```

```
#define O2_RATE  (CURRENT_DENSITY * ELECTRODE_AREA * MOLAR_MASS_O2 / (ELEC_PER_
MOL_O2 * FARADAY_CONST * VOL_O2_ZONE))
#define LIQ_RATE (-(H2_RATE + O2_RATE))

                        /* ---------- UDFs FOR EACH PHASE ---------- */
           /* H2 phase: positive source (mass added) */
DEFINE_SOURCE(h2_mass_source, c, t, dS, eqn)
{
dS[eqn] = 0.0;
return (real)H2_RATE;                                        /* kg/m3-s */
}

       /* O2 phase: positive source (mass added) */
DEFINE_SOURCE(o2_mass_source, c, t, dS, eqn)
{
dS[eqn] = 0.0;
return (real)O2_RATE;                                        /* kg/m3-s */
}

      /* Liquid phase: negative source (mass removed) */
DEFINE_SOURCE(liquid_mass_sink, c, t, dS, eqn)
{
dS[eqn] = 0.0;
return (real)LIQ_RATE;                                  /* negative value */
}
```

7 The Studies of Hydrogen Storage and Safety Using CFD

7.1 MASS AND HEAT TRANSFER IN HYDROGEN STORAGE SYSTEMS

Some of the various technologies employed in hydrogen (H_2) storage systems include liquefied H_2 vessels, high-pressure gaseous tanks, cryogenic compressed storage, absorbents, metal hydrides, and chemical carriers. The effective mass transfer (such as H_2 absorption and release) and heat (such as compression/expansion, exothermic/endothermic sorption, convection/conduction, and boil-off) inside the storage medium and its supporting systems is a significant challenge shared by all these technologies. In this section, the basic physics of heat and mass transfer, specific problems with H_2 storage systems, modeling considerations for CFD, and recommendations for researchers and energy practitioners are provided in detail.

The twin engines of H_2 storage performance are heat and mass transfer. Heat transfer is the corresponding conduction, convection, and occasionally radiation that accompany the uptake or release of H_2, or low-temperature/high-pressure storage operations. Mass transfer is the movement of H_2 atoms (or molecules) into, within, and out of the storage medium [1]. The following are the fundamental equations that govern the mass and heat transfer (detailed derivations of the equations are covered in Chapter 5).

The mass conservation, continuity equation is given by:

$$\frac{\partial \rho}{\partial t} + \nabla \cdot (\rho \boldsymbol{u}) = S_m \tag{7.1}$$

where $\boldsymbol{u}$ is the velocity vector of the three coordinates, ρ is the density, and S_m is a mass source for possible H_2 sorption and desorption.

The Navier-Stokes (momentum conservation) equation is expressed as:

$$\rho\left(\frac{\partial \boldsymbol{u}}{\partial t} + \boldsymbol{u}.\nabla \boldsymbol{u}\right) = -\nabla p + \mu \nabla^2 u + \rho g + \mathbf{F} \tag{7.2}$$

where p is pressure, μ is the fluid viscosity, g is the gravitational acceleration, and $\mathbf{F}$ is the sum of other external body forces, including centrifugal/Coriolis forces in rotating systems, electromagnetic forces (if an electric or magnetic field acts on the

DOI: 10.1201/9781003727194-7

fluid), and drag in porous media. If H_2 flow is modeled through a porous medium, it may include the Darcy-Forchheimer drag [2], i.e.,

$$\mathbf{F} = -\frac{\mu}{K}\boldsymbol{u} - C_F \rho |\boldsymbol{u}| \boldsymbol{u} \tag{7.3}$$

where K is permeability and C_F is an inertial resistance coefficient.

Heat transfer is given from the energy equation, i.e.,

$$\rho c_p \left(\frac{\partial (T)}{\partial t} + \boldsymbol{u}\nabla \mathrm{T} \right) = k\ \nabla^2 T + \Phi \tag{7.4}$$

where c_p is the specific heat, T is the temperature, Φ represent the viscous dissipation, which is the volumetric heat generation or absorption term.

Species transport for the mass fraction of H_2 (Y_{H_2}), is expressed as:

$$\frac{\partial \left(\rho Y_{H_2} \right)}{\partial t} + \nabla \left(\rho Y_{H_2}),\ \boldsymbol{u} \right) = \nabla \left(\rho \Gamma_{H_2} \nabla Y_{H_2} \right) + S_{Y_{H_2}} \tag{7.5}$$

where Γ_{H_2} is the H_2 mass-diffusion coefficient, and $S_{Y_{H_2}}$ is the volumetric source term.

The coupling of heat and mass transfer is crucial in many H_2 storage applications. The sorption/desorption process produces or absorbs heat through a reaction, which requires it to be efficiently transported away; meanwhile, the mass uptake rate frequently depends on local temperature through equilibrium or kinetics. Ineffective heat removal reduces system performance by slowing the uptake or release of H_2 [2].

7.2 HYDROGEN STORAGE CATEGORIES AND THEIR ASSOCIATED MASS AND HEAT TRANSFER PROBLEMS

For mass and heat transfer, H_2 storage systems may be grouped into: (i) cryo-compressed and liquefied H_2 systems, (ii) cryogenic or high-pressure gaseous tanks, and (iii) solid-state material-based storage, such as adsorbents or hydrides (Figure 7.1 depicts this category). Each of the groups has unique mass transfer and heat transfer problems.

7.2.1 Storage of High-Pressure Gases

For stationary or transportation applications, H_2 is stored at approximately 70 MPa in compressed gaseous H_2 tanks [3]. The tanks' discharging and filling constitute a comparatively straightforward mass transfer. Heat transfer, however, is a significant problem since compression raises the temperature of the gas (through adiabatic compression and Joule heating), which must be managed. Also, the temperature rise during compression increases the internal pressure and decreases the mass of H_2 that can be achieved at a given pressure [4, 5].

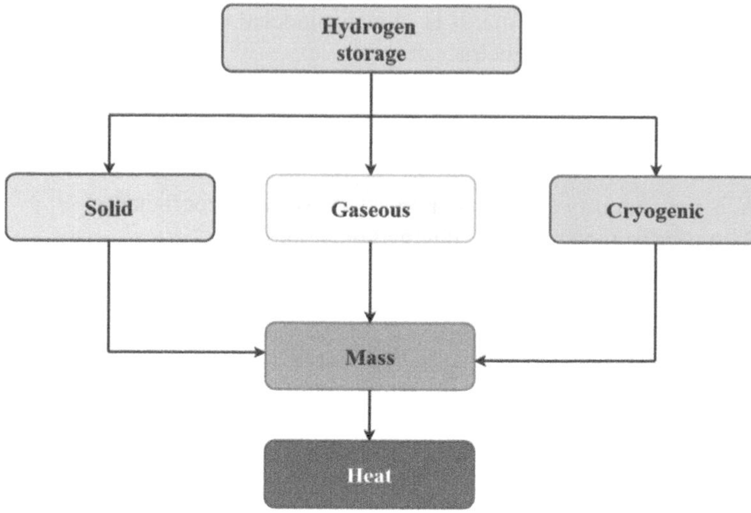

FIGURE 7.1 Mass and heat transfer coupling in hydrogen storage systems (solid/gas/cryogenic).

The gas cools and expands upon discharging, which may lead to condensation and embrittlement issues. Conductive and convective heat transfer, compressible flow, and occasionally phase change (if liquefaction or supercritical conditions are approached) must all be considered while modeling the discharging and filling processes using coupled transient CFD [6, 7].

According to Luo [8] study on heat transfer analysis during discharging-charging cycles of H_2 storage tanks, the convective heat transfer coefficient and vessel wall conduction determine the temperature excursion and, therefore, safe operation. Also, a similar study conducted by Melideo et al. [9] stated that when filling an H_2 tank (2000 l), pre-cooling guarantees thermal safety.

7.2.2 Liquefied/cryogenic and Cryo-Compressed Hydrogen Storage

Cryo-compressed H_2 and liquefied H_2 systems increase volumetric density but introduce significant heat transfer problems, such as stratification of cold H_2, boil-off from heat ingress, and two-phase flow [10]. While heat transfer must be addressed, convection within the tank, conduction through insulation, radiation loads, and complex fluid dynamics all play a role. In this case, mass transfer is linked to venting and phase change.

Multiphase and Conjugate Heat Transfer (CHT) CFD are often utilized in computational modeling to represent the dynamics as pointed out. The significant mass and heat transfer problems in cryogenic systems, especially maintaining low temperatures and managing boil-off with minimal losses, have been highlighted in recent H_2 storage technology studies [11].

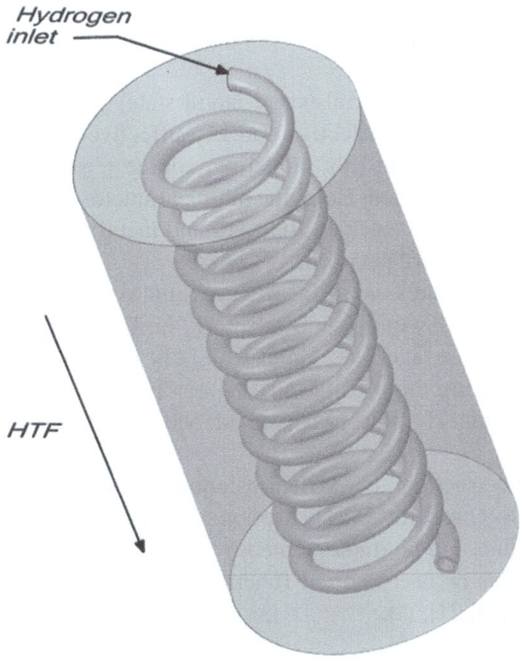

FIGURE 7.2 A schematic of a metal hydride storage bed indicating internal cooling tubes, heat transfer fluid flow (HTF), and hydrogen inlet/outlet.

7.2.3 Hydrogen Storage Based on Materials

Material-based storage, such as in metal and complex hydrides, provides H_2 densities [12]. However, this storage poses unique problems from the standpoint of mass and heat transfer. H_2 transport, which involves diffusion through the solid lattice, also involves sorption and desorption reactions that absorb or generate large amounts of heat. The effective thermal conductivity of the hydride bed is often low, resulting in thermal gradients that slow down kinetics and reduce storage capacity [13]. Figure 7.2 shows a metal hydride storage bed with cooling tubes.

According to Larpruenrudee et al. [14], fast H_2 release and uptake in metal hydride systems remain hindered by inadequate heat removal and the hydride bed's thermal conductivity. Also, a review which covers a wide range of advanced thermal-management techniques in metal hydride H_2 systems has been reported [15]. These strategies include metal forms, phase change materials, internal cooling tubes, high-conductivity additives, and others.

7.3 CFD GOVERNING MODELING CONSIDERATIONS IN HYDROGEN STORAGE

From the standpoint of CFD, several factors must be considered when simulating H_2 storage systems with significant heat and mass coupling. Such factors include porous media and effective properties, multiphase and compressible flow, reaction kinetics and heat generation coupling, thermal boundary conditions and coupling, and mesh and solver strategies.

The H_2 storage bed in material-based (solid-state) systems is frequently modeled as a porous medium with effective porosity, effective thermal conductivity, and effective permeability [16]. In this modeling, source terms (such as mass source terms, for instance) are included. Here, the main parameters are the effective thermal conductivity of packed beds (such as additives and metal hydrides) and the effective diffusivity of H_2 via the pore network.

Compressibility becomes essential in cryogenic or gaseous H_2 storage. When liquid changes phase to vapor, such as in H_2 boiling off, a multiphase model (Mixture, VOF, or more accurate Euler-Euler) is needed. Boundary layers and transients in discharging and filling must be resolved by meshing. Studies applying CFD to cryogenic H_2 flow and packed bed heat transfer have been documented [17–20].

Reaction kinetics (such as the Arrhenius equation) and the related heat reaction are often utilized to characterize H_2 sorption and desorption in metal hydride systems. These couplings can be implemented by using CFD codes such as the built-in multiphysics modules or user-defined functions (UDFs). Such coupling is demonstrated in Eq. (7.6) [14, 21], i.e.,

$$\dot{m}_{H_2} = k_a exp\left(\frac{E_a}{RT}\right)\frac{P_{H_2} - P_{eq}}{P_{eq}}(1 - X) \tag{7.6}$$

where $\dot{m}_{H_2}$ is the mass flow rate of H_2 that is absorbed or desorbed, k_a is the absorption rate constant, E_a is the activation energy for the absorption of H_2, P_{H_2} is the H_2 pressure supplied, P_{eq} is the equilibrium pressure, and X is the fraction of hydride reacted. In the case of desorption, the term $(1 - X)$ in Eq. (7.6) is replaced by (X).

The couplings between structural components (such as insulation and the tank wall), the H_2 bed, heat transfer fluid loops (for heating and cooling), and the ambient environment must be considered for a precise simulation. Time-dependent (i.e., transient) CFD is essential and dominates. Such transient operations are sorption/desorption and filling/discharging cycles.

High-fidelity simulations require fine meshing for H_2 jets or two-phase flows, as well as near walls and tubes in storage beds, to resolve conduction paths. Tracking bubbles and vapor in cryogenic tanks may be assisted by dynamic mesh or adaptive mesh refinement.

7.4 CRYOGENIC TANKS AND PRESSURE VESSEL SIMULATIONS IN CFD

A key component of modern H_2 energy systems is the use of cryogenic and high-pressure storage vessels. Predicting internal thermodynamic behavior, evaluating safety margins, and assisting with advanced tank design all strongly depend on the accuracy of CFD modeling of these vessels.

The applications of these vessels include fuel cell pressure vessels, liquid and gaseous H_2 tanks, and cryogenic storage thermal management. CFD is beneficial in offering a non-intrusive method for evaluating flow fields, temperature, and pressure inside these enclosures under a range of operational and transient conditions.

In this section, a computational simulation case study of static H_2 gas in a closed vertical cylinder was conducted, demonstrating fundamental aspects of pressure vessel simulation and validating the static equilibrium behavior anticipated under uniform thermal and boundary conditions.

The geometry used for the simulation is depicted in Figure 7.3. The figure represents a vertical pressure vessel with smooth, solid walls. It has an internal diameter

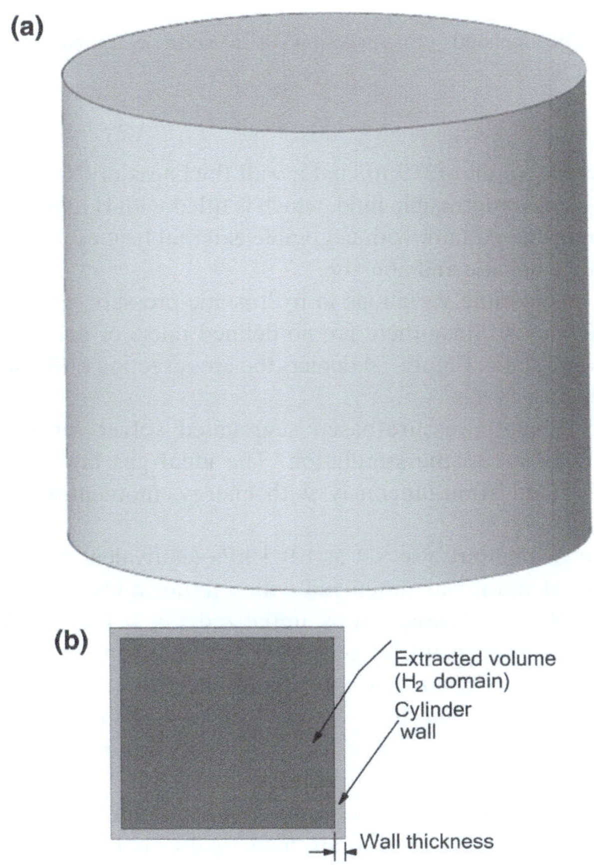

FIGURE 7.3 Hydrogen gas in a vertical cylindrical pressure tank. (a) A vertical cylindrical tank filled with hydrogen gas. (b) A cross section of the tank indicating the extracted volume of the gas.

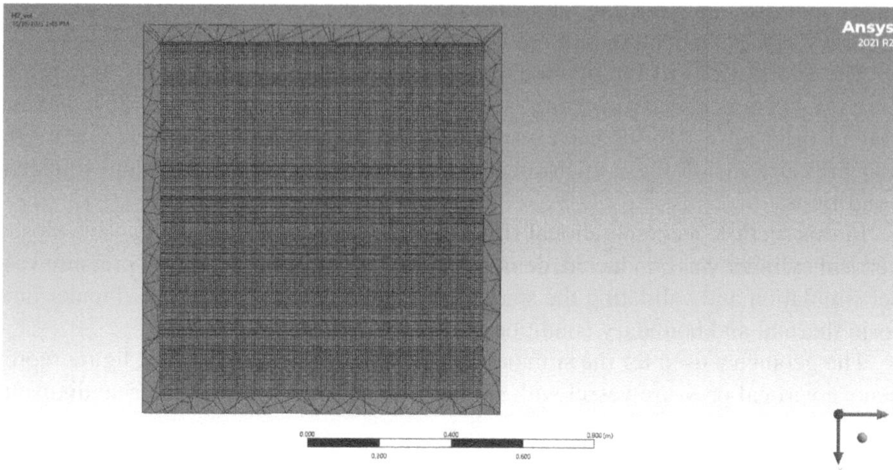

FIGURE 7.4 The meshed cross section of a solid cylindrical tank filled with hydrogen gas.

of 1 m, an internal height of 0.9 m, and a wall thickness of 0.005 m. The vessel is modeled as an ideal compressible fluid, which is filled with H_2 gas. The vessel is also simulated as an insulated tank with negligible external heat exchange; all the vessel walls were made adiabatic and non-slip.

To account for possible variations in hydrostatic pressure, gravity was included along the vertical axis. Since there are no defined inlets or outlets, the vessel was entirely sealed and static. Figure 7.4 depicts the cross section of the mesh of the vessel filled with hydrogen gas.

An ANSYS Fluent pressure-based segregated solver for steady state settings was utilized to run the simulation. The ideal gas law governing density variation was solved simultaneously with energy, momentum, and continuity equations.

The operational pressure was set at 101.3 kPa (equivalent to atmospheric pressure), and the H_2 domain was initialized with a uniform temperature of 288.16 K. The initial equilibrium condition was initialized via hybrid (hydrostatic-aware) initialization. Neither internal heat generation nor volumetric energy sources were specified as the tank was assumed to be fully insulated.

A completely static equilibrium was converged by the simulation. The static pressure contour, as depicted in Figure 7.5, was essentially uniform, with values ranging from -5.44×10^{-15} Pa at the top of the tank to 7.66×10^{-17} Pa at the bottom of the tank. The gas is in hydrostatic equilibrium with negligible gravitational influence, as confirmed by these values, which are numerical round-off variations about the reference operating pressure.

The temperature field stayed perfectly uniform at 288 K, which is consistent with the imposed isothermal boundary conditions (Figure 7.6). Likewise, the velocity field was almost zero throughout the domain, confirming that no

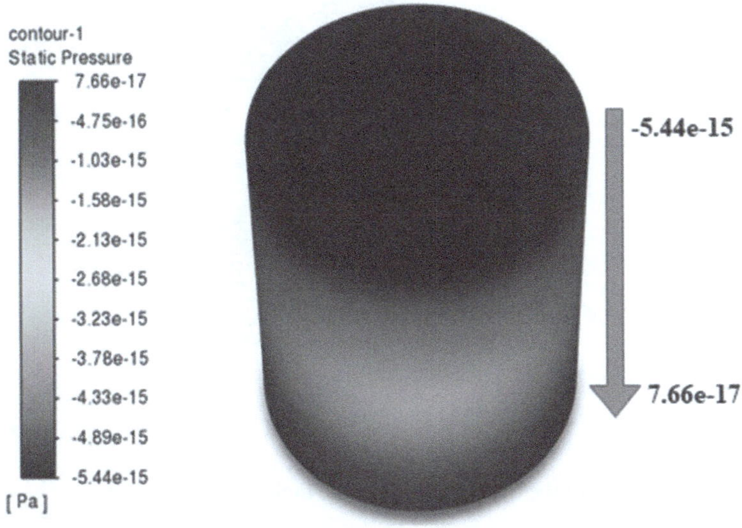

FIGURE 7.5 Static pressure of the hydrogen tank.

buoyant circulation or convective motion developed under the given conditions (Figure 7.7).

Wall shear stress and wall heat flux were effectively zero since the gas remained quiescent. These demonstrative outcomes show a correct physical behavior for a thermally uniform, sealed H_2 vessel and offer a reliable reference case for more intricate simulations involving phase change, pressurization, and thermal stratification. These validation cases are crucial benchmarks before applying CFD studies to non-isothermal or dynamic cryogenic tank simulations.

FIGURE 7.6 Static temperature of the hydrogen tank.

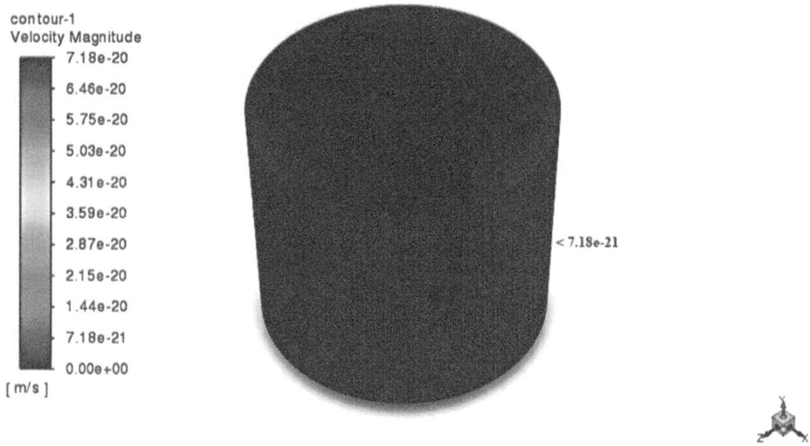

FIGURE 7.7 Velocity of the hydrogen fluid in the tank.

This baseline case study demonstrates that when no external gradients exist, equilibrium thermodynamic states are accurately reproduced by CFD. It is also shown that the application of ideal gas and incompressibility assumptions for small-scale storage tanks is validated by the negligible hydrostatic pressure variation (i.e., less than 1 Pa over a 0.9 m vessel). Achieving physically consistent steady states requires accurate initialization (hydrostatic or hybrid). The approach in this case may be extended to model boil-off, venting, pressurization, and heat leakage in cryogenic tanks.

7.5 DETECTION OF LEAKS AND DISPERSION MODELING

Leak detection and dispersion modeling are essential elements of H_2 safety engineering. In the event of H_2 release (due to valve leakage, storage tank system failure, tank rupture, or piping failure), subsequent mixing, transport, accumulation, or dilution of H_2 in confined spaces or the ambient environment defines the hazard footprint [22]. Figure 7.8 depicts three hydrogen scenarios from an orifice, a leak with ambient wind, and an indoor confined space leak.

In this section, H_2 leak physics, CFD modeling techniques for dispersion, detection, and sensor strategies, as well as best practices for incorporating CFD into safety assessment and mitigation design, are discussed.

The initial behavior of the release of H_2 is determined by several physical phenomena when a H_2 containment system breaches. Such physical phenomena include orifice or jet flow regimes, buoyancy and lighter-than-air behavior, dilution and entrainment, accumulation in semi-confined or confined spaces, and geometry and leak source duration. These are all discussed in the following:

i. Jet or orifice flow regime:
   The leak generally appears as a high-velocity jet (sometimes choked or sonic) until the pressure decreases (or the vena contracta forms) if the

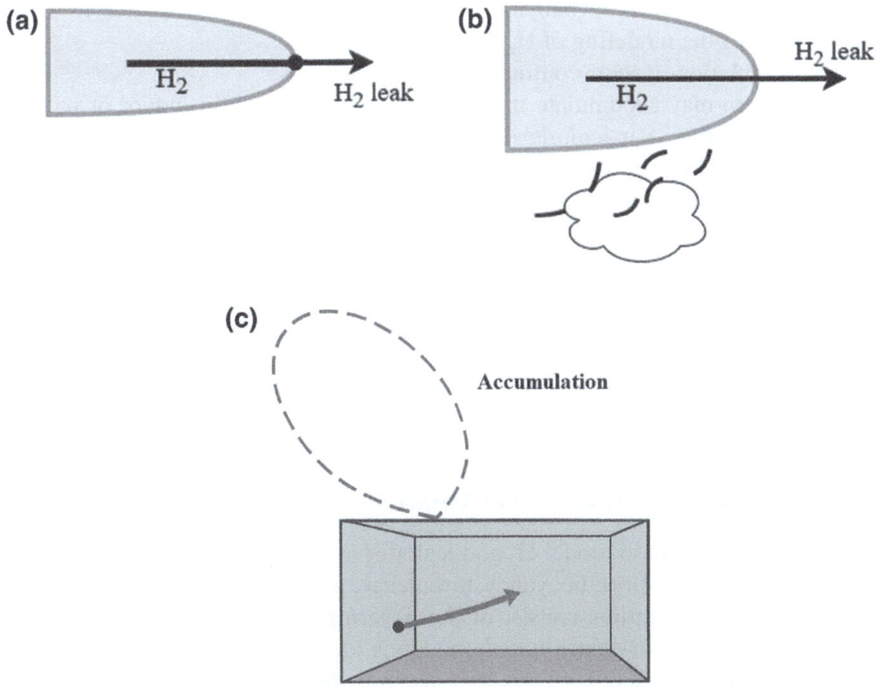

FIGURE 7.8 Schematic of hydrogen leak scenarios: (a) high-pressure orifice jet, (b) horizontal leak with ambient wind, and (c) indoor confined space leak showing rise, entrainment, and accumulation close to the ceiling.

ambient pressure is significantly lower than that of the storage system's pressure. Cryogenic/orifice flow relation can be used to estimate the mass flow rate ($\dot{m}$) [23, 24], i.e.,

$$\dot{m} = C_d A \sqrt{\gamma \rho_u P_u \left(\frac{2}{\gamma + 1} \right)^{\frac{\gamma+1}{\gamma-1}}} \tag{7.7}$$

where C_d is the discharge coefficient, A is the discharge hole area, γ is the specific heat ratio, ρ_u is the real gas upstream density, and P_u is the absolute upstream pressure.

ii. Buoyancy and lighter-than-air behavior:
Hydrogen has a substantially lower density than air, which is approximately 0.089 kg/m³ at 273 K and 1 atm. The buoyant force has a significant impact on the plume rise, accumulation, and dispersion close to the vent/ceiling regions [25].

iii. Entrainment and dilution:
Following the initial momentum-driven jet phase, the H_2 combines with ambient air, causing the plume to slow down and dilute, and may potentially form

a neutrally buoyant or buoyant cloud. This diluted cloud is treated in a large portion of the modeling of H_2 dispersion.

iv. Accumulation in semi-confined or confined spaces:
Hydrogen may accumulate instead of dispersing freely in indoor or partially confined areas, particularly if there is limited ventilation, corners that allow pooling, or obstructions blocking transport. A CFD research pointed out that in a confined industrial building with insufficient ventilation, H_2 concentration approached the lower flammable limit in less than 8 s [26].

v. Geometry or leak source duration:
The dispersion behavior is influenced by the location (i.e., ceiling vs. flow), orientation, duration, leak hole size, and ambient wind/ventilation conditions. Higher pressure differentials, larger orifice diameters, or downward-facing leaks result in more rapid accumulation and increased hazard distance, according to numerous recent modeling studies [27, 28].

7.5.1 Hydrogen Dispersion CFD Modeling Approaches

CFD is widely utilized to model H_2 and leak dispersion phenomena, with its ability to capture species transport, buoyancy, turbulence, geometry, and ventilation effects. Essential modeling features consist of the governing equations, buoyancy modeling, turbulence-chemistry/entrainment modeling, defining boundary conditions and domain, meshing, ventilation effects, and accumulation.

As discussed earlier in Section 7.1, the flow is governed by the Navier-Stokes (momentum), species transport, continuity, and energy (optional for temperature effects) equations. In the turbulent modeling, Reynolds-Averaged Navier-Stokes (RANS), i.e., k-ω SST, k-ε, or Large Eddy Simulation (LES), are often employed in dispersion models to resolve turbulence caused by the jet and ambient mixing. A recent study, for instance, utilized the k-ε model to study hydrogen jet dispersion in a pipeline leak scenario [28].

Buoyant acceleration results from density gradients because H_2 is less dense than air [29]. In the buoyancy modeling, for a slight temperature difference, the Boussinesq approximation may be applied, i.e.,

$$\rho \approx \rho_0\left[1-\beta\left(T-T_0\right)\right] \tag{7.8}$$

Using the gravity term, Eq. (7.8) may be given by:

$$\rho g \approx \rho_0 g\left[1-\beta\left(T-T_0\right)\right] \tag{7.9}$$

where β is the coefficient of thermal expansion. However, H_2 in ambient air with large concentration gradients, full variable density is often utilized.

In meshing, grid dependence is crucial; studies reveal that coarse grids tend to underestimate the dispersion distance. A comparative study conducted by Guan et al. [27] selected 0.1 m for accuracy when comparing 0.1 m, 0.2 m, and 0.4 m grid sizes in an H_2 pipeline leak simulation.

For the effects of ventilation and accumulation, forced or natural ventilation must be incorporated into indoor modeling. According to a finding, placing exhaust outlets near the ceiling decreased the volume of flammable cloud up to 25% [26].

Beyond modeling, safety depends on effective detection. CFD helps determine the optimal placement and H_2 concentration that triggers alarm thresholds. Essential aspects of the strategies include the Lower Flammability Limit (LFL) of H_2 in air, sensor placement, ventilation and obstacle effects, as well as response time and detection instrumentation.

The LFL of H_2 in air is approximately 4% by volume. In safety studies, alarm thresholds are often set at 20% and 40% of LFL (i.e., at 0.8% and 1.6% below the H_2 LFL, respectively). CFD studies can indicate a minor diameter delay approaching alarm concentration and increase the duration to unsafe concentrations [27].

Sensor placement for detecting H_2 is crucial. In small, confined rooms, H_2 buildup tends to occur at corners or near the ceiling. According to the findings of Yassin et al. [26], the highest concentrations of H_2 were found in corner areas with poor ventilation. Sensors placed close to ventilation outlets and accumulation regions achieve faster detection. Magneto-electronic H_2 sensors are gaining popularity. They can detect 0.06% H_2 by volume, and they have a response time of about 1 s [30].

7.5.2 Safety Design Implications for Hydrogen Systems

When leak detection design and dispersion modeling are integrated, designers can achieve the following: (i) sensor placement optimization, (ii) ventilation layout design, (iii) hazard footprint estimation, (iv) mitigation strategies evaluation, and (v) risk assessment and regulatory compliance. Early detection and optimized sensor placement are crucial to avoid accumulation beyond the LFL.

CFD simulations demonstrate how exhaust placement near the ceiling significantly reduces the volume of hazardous clouds. Dispersion modeling defines hazard footprints, such as safe distances, exclusion zones, and emergency response times. CFD aids in the development of mitigation strategies by comparing the effects of reducing leak size, placing barriers, or improving ventilation on dispersion and accumulation. According to pipeline leak studies, reducing the leak diameter causes a delay in approaching critical concentration thresholds [27, 28]. CFD data contributes to Quantitative Risk Assessments (QRA) and helps meet safety requirements for H_2 infrastructure. For modeling H_2 dispersion and leaks in CFD for systems, the workflow as shown in the flowchart in Figure 7.9 may be followed.

7.6 RISK ASSESSMENTS OF EXPLOSION AND FIRE

This section covers a thorough treatment of QRA approaches for hydrogen-related explosions and fire, along with their practical implications, the physics underlying the relationship between consequence estimation and leak/dispersion modeling, and suggested workflows and mitigations. The goal of this section is to provide the theoretical and practical modeling processes required to conduct a convincing risk assessment of H_2 explosion and fire for plant facilities, storage, or refueling.

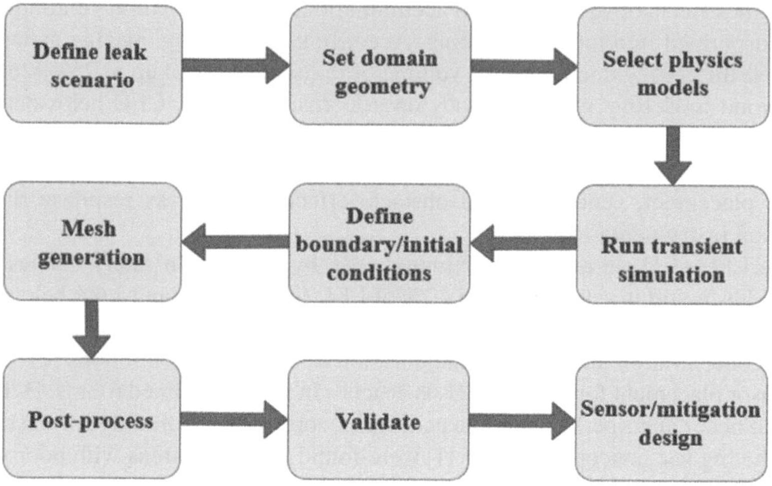

FIGURE 7.9 Workflow for modeling hydrogen leaks and dispersion in CFD.

Typically, risk is expressed as the product of the frequency (i.e., the probability per unit of time) of an initiating event and the impact (or repercussion) if the event were to occur. The usual outputs for the assessments of H_2 fire and explosion include Individual Risk (IR), societal or group risk, and consequence metrics [31].

The IR represents the annual probability of fatality for a representative individual at a specified location, such as a member of the public or even an adjacent worker. The group or societal risk is often plotted as a frequency (*F*) vs. number of fatalities (*N*) curve, which usually shows the annual *F* surpassing *N*. The metrics of consequence include building damage probabilities, flammable cloud volume/time, overpressure, exposure time, and thermal radiation. Figure 7.10 is a flowchart used to illustrate a conventional QRA for the H_2 risk procedure [32, 33].

A QRA can basically be expressed as [31]:

$$\text{Risk} = \sum_i P_{ig,i} \times C_i \tag{7.8}$$

where i is a scenario, $P_{ig,i}$ is the probability of ignition occurrence, which may lead to explosion or fire, and C_i is the consequence metric or expected number of fatalities. Detailed QRA uses fault trees and event trees to break down leak events or equipment failures into these terms; Hydrogen Risk Assessment Models (HyRAM) and several QRA frameworks formalize these steps for H_2 [34].

Pipes or vessel fractures, connector failures, leaks through valves, and significant releases from storage (such as tank ruptures) are key initiating events for H_2 explosive risk. Human-error probabilities, failure-rate data, and likelihoods of external events (such as vehicle collisions) are required for estimating *F*. Familiar sources used in practice include the industry failure database and code guidance, such as the HyRAM and SAND reports [34, 35]; event data and incident databases [36]; and reliability models.

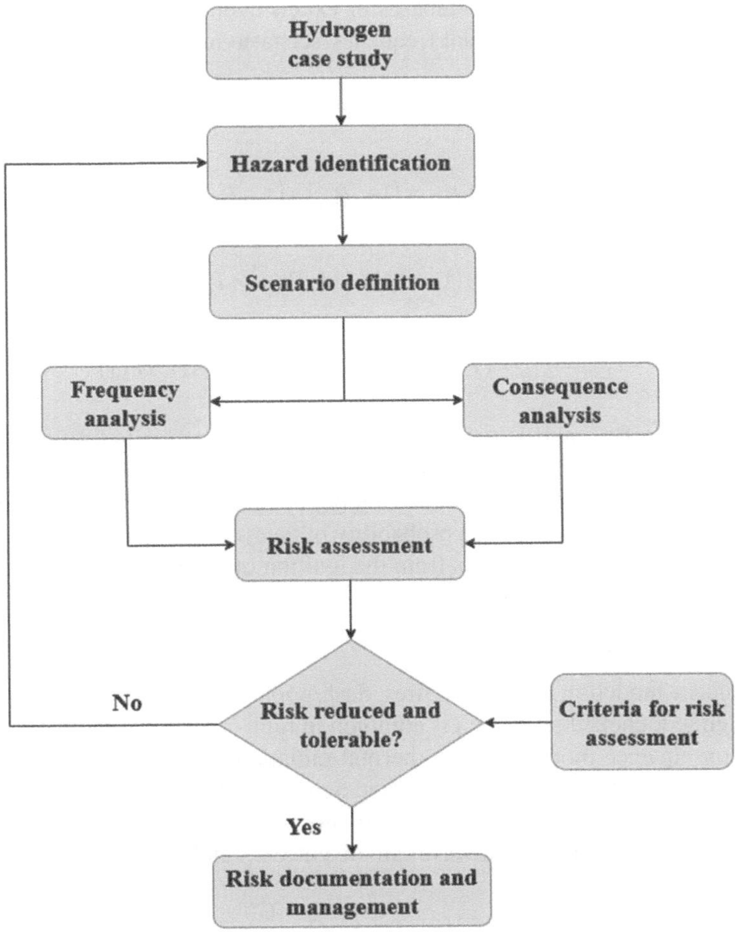

FIGURE 7.10 Framework for qualitative risk management in hydrogen production systems, outlining hazard identification and assessment.

7.6.1 Dispersion and Ignition Modeling

A release scenario (such as temperature, mass flow, or momentum) is the first step in this modeling. The CFD and dispersion outcomes feed this model. For compressible high-pressure leaks, choked relations are utilized to evaluate mass, while for cryogenic or liquid spills, the two-phase vaporization models are used. Afterward, the leak mass flow is then utilized as a source in reduced models or dispersion CFD [34, 37]. CFD turbulence models, such as the RANS and LES, aid in predicting concentration fields, thresholds, and exposure duration at the locations of the receptor [38].

The probability of ignition depends significantly on gas concentration (which must be within the range of the lower and upper flammability limits), ignition sources (such as hot sources or electrical sparks), turbulence, and the temporal coincidence of these factors. Conditional ignition probability is usually used in practical

QRA, which is derived from databases or experiments. The ignition probability is expressed in the following annual frequency scenario models used in HyRAM [35]:

$$F_{\text{uig}} = F_{\text{H}_2} \times P_{\text{iso}} + F_{\text{H}_2} \times (1 - P_{\text{iso}}) \times \left(1 - P_{\text{im}_{\text{ig}}}\right) \times \left(1 - P_{\text{de}_{\text{ig}}}\right) \tag{7.9}$$

$$F_{\text{Jfire}} = F_{\text{H}_2} \times (1 - P_{\text{iso}}) \times \left(1 - P_{\text{im}_{\text{ig}}}\right) \tag{7.10}$$

$$F_{\text{Ffire}} = F_{\text{H}_2} \times (1 - P_{\text{iso}}) \times \left(1 - P_{\text{im}_{\text{ig}}}\right) \times P_{\text{de}_{\text{ig}}} \times P_{\text{TP}} \tag{7.11}$$

$$F_{\text{Exp}} = F_{\text{H}_2} \times (1 - P_{\text{iso}}) \times \left(1 - P_{\text{im}_{\text{ig}}}\right) \times P_{\text{de}_{\text{ig}}} \times (1 - P_{\text{TP}}) \tag{7.12}$$

where F_{uig}, F_{Jfire}, F_{Ffire}, and F_{Exp} are the frequencies of the unignited, jet fire, flash fire, and explosion, respectively. F_{H_2} is the annual frequency of a gaseous release, P_{iso} is the probability of leak detection and isolation for the ignition, $P_{\text{im}_{\text{ig}}}$ is the probability of immediate ignition, $P_{\text{de}_{\text{ig}}}$ is the probability of delayed ignition, and P_{TP} is the probability that the dominant impacts from the ignition of accumulated gas are thermal.

7.6.2 Consequence Modeling

Consequence modeling includes jet fires, flash/pool fires, and vapor cloud explosions. For an ignited, choked jet, which is common in high-pressure gaseous H_2, the focus of the consequence modeling is on thermal radiation and flame length. A common engineering model developed by Houf and Schefer [39] is adopted for predicting the radiant heat flux from high-pressure turbulent jet flames. The radiant heat flux, q_{rad}, is expressed as a function of downstream distance, x, and radial distance, r, i.e.,

$$q_{\text{rad}}(x, r) = \chi_{\text{rad}} Q_f \frac{C^*}{4\pi r^2} \tag{7.13}$$

where χ_{rad} is the radiant fraction, Q_f is the heat release rate, and C^* is the normalized radiative heat flux. χ_{rad} is given by:

$$\chi_{\text{rad}} = 0.082737 \log\left(\tau_f\right) - 0.080435 \tag{7.14}$$

τ_f is the flame residence time, i.e.,

$$\tau_f = \frac{\rho_f w_f^2 L_{vis} f_s}{3 \rho_j d_j^2 u_j} \tag{7.15}$$

ρ_f, f_s, w_f, and L_{vis} are the flame density, mass fraction of H_2 (from stoichiometric), width, and visible length, respectively. ρ_j, u_j, and d_j are the density, velocity, and diameter of the jet at the orifice. Q_f in Eq. (7.13) is given by:

$$Q_f = \dot{m}_{fu} \times \Delta H_c \tag{7.16}$$

$\dot{m}_{fu}$ is the fuel mass flow rate and ΔH_c is the heat of combustion. C^* (Eq. (7.13)) is given by:

$$C^* = 0.85985\exp\left(-2.75791\left|\frac{x}{L_{vis}} - 0.6352\right|\right) \tag{7.17}$$

Rapidly vaporizing liquified releases can create rich vapor clouds that are vulnerable to flash fires. A dispersed cloud may deflagrate or transition to detonate when it ignites. Overpressure is the main consequence metric for vapor cloud explosions. Trinitrotoluene-equivalent (TNT-equivalent) approaches and empirical blast scaling are examples of engineering vapor cloud explosion models. However, these models are approximations that need careful justification.

The state-of-the-art QRA provides worst-case overpressure vs. distance by linking CFD and dispersion outcomes to semi-empirical vapor cloud explosion models. A recent study conducted by Ma et al. [38] cautions that, in contrast to hydrocarbon clouds, blast scaling is altered by hydrogen's high diffusivity and rapid flame speed.

7.6.3 Quantification of Consequences

Converting physical metrics to human and population consequences involves employing vulnerability functions. Such functions include thermal radiation vulnerability and overpressure vulnerability. In thermal radiation vulnerability, the probability of fatality vs. radiative flux and exposure time utilizes logistic or probit functions, such as the probit models from HyRAM. Overpressure vulnerability involves the likelihood of fatality vs. overpressure.

For scenario i, the expected number of fatalities, J_i, is expressed by:

$$J_i = \sum_{k \in \text{receptors}} P_{\text{fatal}}(m_k) \times N_k \tag{7.18}$$

where N_k is the number of people at receptor k, and $P_{\text{fatal}}(m_k)$ is the vulnerability at metric m_k (such as overpressure, flux) at that receptor. Standard vulnerability curves for H_2 hazards are supplied by HyRAM, which are frequently utilized in code development [35].

7.6.4 Validation, Mitigation, and Design Implications

Ignition probabilities, dispersion model fidelity, leaf frequency, and vulnerability curves all affect QRA outcomes. Best practices include model validation, sensitivity analysis, and standardized tools. Consequence and dispersion models should be validated against experiments [37, 38]. Sensitivity studies are also crucial in identifying the most influential parameters, such as ignition probability and wind speed, among others. Standardized tools (such as the HyRAM and code guidance) aid in harmonizing assumptions and offer traceable model inputs [35].

QRA informs mitigation in terms of reducing leak frequency, decreasing ignition probability, and minimizing consequences. In terms of reducing leak frequency, components and quality control are improved, and maintenance is reduced. In reducing ignition probability, the spark is being controlled, and safe electrical equipment is produced. Consequence reduction involves building blast-resistant structures, improving ventilation, and maintaining separation distances, among other measures. CFD can demonstrate how ventilation locations, such as ceiling exhausts close to accumulation zones, lower flammable cloud volumes and, therefore, their consequences [36,37].

7.7 IMPROVEMENT IN SAFETY THROUGH CFD STUDIES

Safety concerns, including hydrogen's high diffusivity, broad flammability limits, and low ignition energy, limit its potential as a clean, renewable energy carrier. Because of these properties, H_2 is more likely to diffuse, leak, and ignite in confined spaces such as refueling stations, pipelines, and storage tanks. Conventional empirical safety design approaches are frequently conservative and expensive. CFD has become a transformative tool for enhancing H_2 safety via detailed analysis of dispersion, release, explosion, and ignition phenomena under realistic geometries and boundary conditions.

CFD offers physics-based, high-resolution insights that allow researchers and engineers to [40]:

i. Quantify ignition sources and patterns of flame propagation.
ii. Evaluate and visualize gas temperature and concentration fields.
iii. Improve the sensor locations, venting, and mitigation of explosion systems.
iv. Develop safety standards based on performance to reduce experimental expenses.

The study of safety in CFD depends on solving the LES or RANS forms of the conservation equations. The conservation equations, along with their source terms, are discussed in the earlier section. The turbulent combustion process can be modeled utilizing a partially stirred reactor or eddy dissipation concept methods, while discrete phase models (DPM) may be adopted for cryogenic liquid H_2 evaporation [41]. Studies have applied these models to H_2 safety analysis [42].

7.7.1 CFD Applications for Safety Enhancement

By optimizing H_2 leakages and ventilation, CFD aids in determining the optimal vent positions and ventilation rates to regulate H_2 concentrations below 25% of the LFL. In flame acceleration and explosion mitigation, installing baffles or adjusting the size of vents helps reduce the overpressure of an explosion. Using detailed combustion kinetics, CFD can simulate deflagration-to-detonation transition [43–45].

In detection and sensor placement enhancement, CFD can help to identify hotspots through H_2 dispersion simulations [46–48]. This technique helps in designing smart sensor networks, which are integrated with the Internet of Things (IoT) and AI-based H_2 leak prediction systems [49].

One of the most vital uses of CFD-based safety studies is for H_2 refueling stations. CFD helps in evaluating the effects of pressure relief valves, blast radius, and jet flame impingement. In structural reinforcement, guided by CFD, thermal and pressure load transfer from CFD to structural solvers (such as ANSYS Mechanical) is possible by linking CFD and Finite Element Analysis (FEA). This aids in identifying fatigue and stress hotspot regions in pipelines and vessels during accidental ignition [50, 51]. Identifying the stress hotspot regions, the Hoop stress equation is given by:

$$\sigma_{\mathrm{h}} = \frac{P \times D}{2t} \tag{7.19}$$

where t is the wall thickness, P is the internal pressure, and D is the diameter. CFD-predicted transient pressure profiles may be coupled into the hoop stress for dynamic safety margin assessment.

CFD enhances the accuracy of QRA by providing realistic frequency and consequence data. In the HyRAM framework, CFD data feeds into probabilistic consequence modules.

Some of the emerging trends in modern CFD safety research involve integrating real-time CFD monitoring in digital twins of H_2 facilities [52, 53]; AI-assisted surrogate models for rapid hazard prediction [54, 55]; multiphase coupling (i.e., CFD-thermal-electrical) for H_2 storage batteries and cryogenic systems [56]; and adaptive mesh refinement to capture flame acceleration and shock-flame interactions [57, 58]. Future directions should emphasize CFD validation with large-scale experiments such as HyTunnel and HySEA, benchmarking across codes (such as OpenFOAM, Fluent, FLACS, etc.) [59, 60], standardization of CFD safety protocols in NFPA 2:2023 and ISO/TR 15916:2023 [61, 62].

REFERENCES

1. S. Srinivasa Murthy, 'Heat and mass transfer in solid state hydrogen storage: A review', *Journal of Heat Transfer*, vol. 134, no. 3, p. 031020, 2012. doi: 10.1115/1.4005156.
2. H. Mao, Z. Chen, L. Cheng, and K. Wang, 'The activity and reversibility of intramolecular B/N Frustrated Lewis Pairs in the hydrogenation', *International Journal of Hydrogen Energy*, vol. 46, no. 1, pp. 110–118, 2021. doi: 10.1016/j.ijhydene.2020.09.248.
3. A. Magliano, C. Perez Carrera, C. M. Pappalardo, D. Guida, and V. P. Berardi, 'A comprehensive literature review on hydrogen tanks: Storage, safety, and structural integrity', *Applied Sciences*, vol. 14, no. 20, p. 9348, 2024. doi: 10.3390/app14209348.
4. J.-Q. Li, J.-C. Li, K. Park, S.-J. Jang, and J.-T. Kwon, 'An analysis on the compressed hydrogen storage system for the fast-filling process of hydrogen gas at the pressure of 82 MPa', *Energies*, vol. 14, no. 9, p. 2635, May 2021. doi: 10.3390/en14092635.
5. J.-C. Li, H. Xu, K. Zhou, and J.-Q. Li, 'A review on the research progress and application of compressed hydrogen in the marine hydrogen fuel cell power system', *Heliyon*, vol. 10, no. 3, p. e25304, 2024. doi: 10.1016/j.heliyon.2024.e25304.
6. J. M. Monteiro, L. Ribeiro, J. Monteiro, A. Baptista, and G. F. Pinto, 'Computational fluid dynamics simulation of filling a hydrogen type 3 tank at a constant mass flow rate', *Energies*, vol. 17, no. 6, p. 1375, 2024. doi: 10.3390/en17061375.

7. S. Li, J. Guo, X. Lv, T. Deng, B. Cao, and J. Wang, 'Research on high-pressure hydrogen pre-cooling based on CFD technology in fast filling process', *Processes*, vol. 9, no. 12, p. 2208, 2021. doi: 10.3390/pr9122208.
8. H. Luo, 'Modelling, Optimization and Protocol Study for Efficient and Safe Hydrogen Refuelling System', PhD, Université du Québec à Trois-Rivières, Canada, 2024. [Online]. Available: https://depot-e.uqtr.ca/id/eprint/11524/1/eprint11524.pdf
9. D. Melideo, U. Desideri, J. Andreas, and T. Kalisch, 'Zero-dimensional modeling and optimization of hydrogen refueling for type IV tanks: From validation to large-scale applications', *Energy*, vol. 332, p. 137254, 2025. doi: 10.1016/j.energy.2025.137254.
10. M. Aziz, 'Liquid hydrogen: A review on liquefaction, storage, transportation, and safety', *Energies*, vol. 14, no. 18, p. 5917, 2021. doi: 10.3390/en14185917.
11. A. S. Mekonnin, K. Wacławiak, M. Humayun, S. Zhang, and H. Ullah, 'Hydrogen storage technology, and its challenges: A review', *Catalysts*, vol. 15, no. 3, p. 260, 2025. doi: 10.3390/catal15030260.
12. G. Chen, D. Liang, Z. Kang, J. Fan, S. Fan, and X. Zhou, 'Review of hydrogen storage in solid-state materials', *Energies*, vol. 18, no. 11, p. 2930, 2025. doi: 10.3390/en18112930.
13. I. P. Jain, C. Lal, and A. Jain, 'Hydrogen storage in Mg: A most promising material', *International Journal of Hydrogen Energy*, vol. 35, no. 10, pp. 5133–5144, 2010. doi: 10.1016/j.ijhydene.2009.08.088.
14. P. Larpruenrudee *et al.*, 'A review on the overall performance of metal hydride-based hydrogen storage systems', *Energies*, vol. 18, no. 5, p. 1291, 2025. doi: 10.3390/en18051291.
15. V. K. Kukkapalli, S. Kim, and S. A. Thomas, 'Thermal management techniques in metal hydrides for hydrogen storage applications: A review', *Energies*, vol. 16, no. 8, p. 3444, 2023. doi: 10.3390/en16083444.
16. K. Lahmer, R. Bessaïh, A. Scipioni, and M. El Ganaoui, 'Thermal modeling of hydrogen storage by absorption in a magnesium hydrides tank', *International Journal of Simulation and Multidisciplinary Design Optimization*, vol. 5, p. A21, 2014. doi: 10.1051/smdo/2014002.
17. X. Quan, Y. Xia, and Q. Zhang, 'CFD-DEM modeling of cryogenic hydrogen flow and heat transfer in packed bed', *Sustainability*, vol. 16, no. 11, p. 4522, 2024. doi: 10.3390/su16114522.
18. Q. Ba *et al.*, 'Modeling of cryogenic compressed hydrogen jet flames', *International Journal of Hydrogen Energy*, vol. 51, pp. 917–927, 2024. doi: 10.1016/j.ijhydene.2023.06.265.
19. K. B. Minko, M. V. Lototskyy, I. E. Bessarabskaya, and B. P. Tarasov, 'CFD simulation of heat and mass transfer processes in a metal hydride hydrogen storage system, taking into account changes in the bed structure', *International Journal of Hydrogen Energy*, vol. 137, pp. 861–869, 2025. doi: 10.1016/j.ijhydene.2024.05.083.
20. Q. Zhang, Y. Xia, Z. Cheng, and X. Quan, 'Numerical investigation of heat transfer characteristics of hydrogen flow in a randomly packed bed', *Case Studies in Thermal Engineering*, vol. 61, p. 105143, 2024. doi: 10.1016/j.csite.2024.105143.
21. C. A. Chung, and C.-S. Lin, 'Prediction of hydrogen desorption performance of Mg2Ni hydride reactors', *International Journal of Hydrogen Energy*, vol. 34, no. 23, pp. 9409–9423, 2009. doi: 10.1016/j.ijhydene.2009.09.061.
22. T. Yang *et al.*, 'Characteristics of hydrogen leakage and dissipation from storage tanks in an integrated hydrogen production and refueling station', *Fire*, vol. 7, no. 9, p. 306, 2024. doi: 10.3390/fire7090306.
23. D. W. Green, and M. Z. Southard, *Perry's Chemical Engineers' Handbook*, 9th edition, McGraw-Hill Education, 2019.

24. S. T. Le, T. N. Nguyen, S. Linforth, and T. D. Ngo, 'Safety investigation of hydrogen energy storage systems using quantitative risk assessment', *International Journal of Hydrogen Energy*, vol. 48, no. 7, pp. 2861–2875, 2023. doi: 10.1016/j.ijhydene.2022.10.082.
25. Z. Shu *et al.*, 'Dispersion characteristics of hydrogen leakage: Comparing the prediction model with the experiment', *Energy*, vol. 236, p. 121420, 2021. doi: 10.1016/j.energy.2021.121420.
26. K. Yassin, S. Kelm, and E.-A. Reinecke, 'Numerical simulation of dispersion and ventilation of hydrogen clouds in case of leakage inside a large-scale industrial building', *Hydrogen*, vol. 6, no. 2, p. 40, 2025. doi: 10.3390/hydrogen6020040.
27. W. Guan, J. Chen, L. Chen, J. Cao, and H. Fan, 'Safe design of a hydrogen-powered ship: CFD simulation on hydrogen leakage in the fuel cell room', *Journal of Marine Science and Engineering*, vol. 11, no. 3, p. 651, 2023. doi: 10.3390/jmse11030651.
28. Y. Lin *et al.*, 'Modeling of hydrogen dispersion, jet fires and explosions caused by hydrogen pipeline leakage', *Fire*, vol. 7, no. 1, p. 8, 2023. doi: 10.3390/fire7010008.
29. D. Brzezińska, 'Hydrogen dispersion and ventilation effects in enclosures under different release conditions', *Energies*, vol. 14, no. 13, p. 4029, 2021. doi: 10.3390/en14134029.
30. I. S. Maksymov, and M. Kostylev, 'Magneto-electronic hydrogen gas sensors: A critical review', 2021. doi: 10.48550/ARXIV.2112.07296.
31. S. Kashkarov, M. Dadashzadeh, S. Sivaraman, and V. Molkov, 'Quantitative risk assessment methodology for hydrogen tank rupture in a tunnel fire', *Hydrogen*, vol. 3, no. 4, pp. 512–530, 2022. doi: 10.3390/hydrogen3040033.
32. H.-R. Gye, S.-K. Seo, Q.-V. Bach, D. Ha, and C.-J. Lee, 'Quantitative risk assessment of an urban hydrogen refueling station', *International Journal of Hydrogen Energy*, vol. 44, no. 2, pp. 1288–1298, 2019. doi: 10.1016/j.ijhydene.2018.11.035.
33. J. Zhao, M. Wu, H. Lu, G. Li, Y. Xu, and M. Tian, 'Quantitative risk assessment of an oil-gas-hydrogen-electricity integrated energy station in China', *ACS Omega*, vol. 9, no. 37, pp. 38887–38896, 2024. doi: 10.1021/acsomega.4c05050.
34. K. M. Groth, E. S. Hecht, and J. T. Reynolds, 'Methodology for assessing the safety of hydrogen systems: HyRAM 1.0 technical reference manual', Sandia National Laboratories, California, SAND2015-10216, 2015. [Online]. Available: https://h2tools.org/sites/default/files/SAND2015-10216-HyRAM-1.0-Technical-Reference-Manual.pdf
35. K. Groth, E. Hecht, and J. T. Reynolds, 'Methodology for assessing the safety of Hydrogen Systems: HyRAM 1.0 technical reference manual', 2015. doi: 10.2172/1814840.
36. M. Calabrese *et al.*, 'Hydrogen safety challenges: A comprehensive review on production, storage, transport, utilization, and CFD-based consequence and risk assessment', *Energies*, vol. 17, no. 6, p. 1350, 2024. doi: 10.3390/en17061350.
37. X. Fu *et al.*, 'hydrogen jet flame simulation and thermal radiation damage estimation for leakage accidents in a hydrogen refueling station', *Fire*, vol. 7, no. 7, p. 210, 2024. doi: 10.3390/fire7070210.
38. Q. Ma, Y. He, J. You, J. Chen, and Z. Zhang, 'Probabilistic risk assessment of fire and explosion of onboard high-pressure hydrogen system', *International Journal of Hydrogen Energy*, vol. 50, pp. 1261–1273, 2024. doi: 10.1016/j.ijhydene.2023.10.157.
39. W. Houf, and R. Schefer, 'Predicting radiative heat fluxes and flammability envelopes from unintended releases of hydrogen', *International Journal of Hydrogen Energy*, vol. 32, no. 1, pp. 136–151, 2007. doi: 10.1016/j.ijhydene.2006.04.009.
40. A. G. Venetsanos, D. Baraldi, P. Adams, P. S. Heggem, and H. Wilkening, 'CFD modelling of hydrogen release, dispersion and combustion for automotive scenarios', *Journal of Loss Prevention in the Process Industries*, vol. 21, no. 2, pp. 162–184, 2008. doi: 10.1016/j.jlp.2007.06.016.

41. D. Veynante, and L. Vervisch, 'Turbulent combustion modeling', *Progress in Energy and Combustion Science*, vol. 28, no. 3, pp. 193–266, 2002. doi: 10.1016/S0360-1285(01)00017-X.
42. H. Li, X. Cao, H. Du, L. Teng, Y. Shao, and J. Bian, 'Numerical simulation of leakage and diffusion distribution of natural gas and hydrogen mixtures in a closed container', *International Journal of Hydrogen Energy*, vol. 47, no. 84, pp. 35928–35939, 2022. doi: 10.1016/j.ijhydene.2022.08.142.
43. D. Baraldi, and E. Kommission, Eds, *Gap Analysis of CFD Modelling of Accidental Hydrogen Release and Combustion*, in EUR, no. 24399, Publications Office of the European Union, 2010. doi: 10.2790/2090.
44. H. Sun *et al.*, 'Effects of vent size and pressure on hydrogen explosion dynamic characteristics', *ACS Omega*, 2024. doi: 10.1021/acsomega.4c04972.
45. Y. Cao, F. Qian, P. Xi, J. Wang, and Y. An, 'Influence of obstacles on the vented hydrogen deflagration in an explosion tube', *ACS Omega*, vol. 10, no. 27, pp. 29827–29835, 2025. doi: 10.1021/acsomega.5c04428.
46. X. Deng, J. Sun, F. Yang, and M. Ouyang, 'Simulation and quantitative assessment of sensor placement in a hydrogen bus for risk mitigation', *Hydrogen*, vol. 5, no. 4, pp. 976–986, 2024. doi: 10.3390/hydrogen5040052.
47. Q. He, F. Kong, R. Sun, R. Li, J. Yang, and Q. Min, 'Experimental study on dynamic response performance of hydrogen sensor in confined space under ceiling', *Frontiers in Energy Research*, vol. 12, p. 1456316, 2024. doi: 10.3389/fenrg.2024.1456316.
48. X. Wang *et al.*, 'Influence of longitudinal wind on hydrogen leakage and hydrogen concentration sensor layout of fuel cell vehicles', *Sustainability*, vol. 15, no. 13, p. 10712, 2023. doi: 10.3390/su151310712.
49. H.-J. Hwang, and S.-H. Hwang, 'Developing an internet of things system for hydrogen leak detection at hydrogen refueling stations integrating LoRa and global positioning system', *Edelweiss Applied Science and Technology*, vol. 8, no. 5, pp. 1582–1593, 2024. doi: 10.55214/25768484.v8i5.1874.
50. G. Kim *et al.*, 'Finite element analysis framework for structural safety evaluation of type IV hydrogen storage vessel', *Hydrogen*, vol. 6, no. 3, p. 44, 2025. doi: 10.3390/hydrogen6030044.
51. Z. Chen, H. Hui, S. Huang, Z. Du, G. Xue, and F. Meng, 'Fitness-for-service assessment of hoop-wrapped vessel with metal liner in high-pressure hydrogen environment', *Energies*, vol. 18, no. 17, p. 4463, 2025. doi: 10.3390/en18174463.
52. D. A. Paternina-Verona, O. E. Coronado-Hernández, M. Pérez-Sánchez, and H. M. Ramos, 'Real-time analysis and digital twin modeling for CFD-based air valve control during filling procedures', *Water*, vol. 16, no. 21, p. 3015, 2024. doi: 10.3390/w16213015.
53. Z. Feng, I. Eiubovi, Y. Shao, Z. Fan, and R. Tan, 'Review of digital twin technology applications in hydrogen energy', *CHAIN*, vol. 1, no. 1, pp. 54–74, 2024. doi: 10.23919/CHAIN.2024.000001.
54. M. M. Peksen, 'CFD and artificial intelligence-based machine learning synergy for the assessment of syngas-utilizing pre-reformer in r-SOC technology advancement', *Applied Sciences*, vol. 14, no. 22, p. 10181, 2024. doi: 10.3390/app142210181.
55. G. Yang, D. Kong, X. He, X. Yu, and K. Jiang, 'Prediction of hydrogen leakage location and intensity in hydrogen refueling stations based on deep learning', *International Journal of Hydrogen Energy*, vol. 68, pp. 209–220, 2024. doi: 10.1016/j.ijhydene.2024.04.234.
56. L. Zhang, E. Wang, and J. Shi, 'Dynamic thermal performance of cryogenic liquid hydrogen heating from gas-liquid stratification to supercritical under a constant heat flux condition: A two-step simulation', *International Journal of Hydrogen Energy*, vol. 136, pp. 1242–1250, 2025. doi: 10.1016/j.ijhydene.2024.02.025.

57. B. Vanbersel, R. Meziat, A. Francis, P. Mohanamuraly, G. Staffelbach, T. Jaravel, Q. Douasbin, O. Dounia, and O. Vermorel, 'A systematic adaptive mesh refinement method for large eddy simulation of turbulent flame propagation', *Flow, Turbulence and Combustion*, vol. 112, no. 4, pp. 1127–1160, 2024, doi. 10.1007/s10494-024-00534-6.
58. A. Pappa, A. Verhaeghe, P. Bénard, W. De Paepe, and L. Bricteux, 'Adaptive mesh refinement towards optimized mesh generation for large eddy simulation of turbulent combustion in a typical micro gas turbine combustor', *Energy*, vol. 301, p. 131550, 2024. doi: 10.1016/j.energy.2024.131550.
59. V. Molkov, and W. Dery, 'The blast wave decay correlation for hydrogen tank rupture in a tunnel fire', *International Journal of Hydrogen Energy*, vol. 45, no. 55, pp. 31289–31302, 2020. doi: 10.1016/j.ijhydene.2020.08.062.
60. A. W. Lach, A. V. Gaathaug, and K. Vaagsaether, 'Pressure peaking phenomena: Unignited hydrogen releases in confined spaces – Large-scale experiments', *International Journal of Hydrogen Energy*, vol. 45, no. 56, pp. 32702–32712, 2020. doi: 10.1016/j.ijhydene.2020.08.221.
61. NFPA 2, 'NFPA 2, Hydrogen Technologies Code (2023)', 2024. [Online]. Available: https://www.nfpa.org/product/nfpa-2-code/p0002code
62. ISO/TR 15916:2015, 'Basic considerations for the safety of hydrogen systems', 2023. [Online]. Available: https://www.iso.org/standard/56546.html

8 Green Hydrogen Integration in Renewable and Combined Energy Systems

8.1 PREAMBLE

Green hydrogen has become a key element in the global community's efforts to move toward sustainable energy systems, with deep decarbonization, improved energy resilience, and sectoral integration being the main goals [1, 2]. Green hydrogen is a clean, adaptable, and scalable energy carrier that has the potential to revolutionize the current energy environment. It is created by electrolyzing water with electricity generated from renewable sources like solar and wind. Its incorporation into hybrid and renewable energy systems offers a game-changing chance to solve the intermittent nature of renewable power, guarantee steady grid operations, and facilitate cross-sectoral decarbonization, especially in difficult-to-abate industries like transportation, heating, and industry [3, 4].

Green hydrogen's incorporation into energy systems solves a number of significant issues related to renewable energy. Despite being plentiful and sustainable, solar and wind power are erratic and unpredictable by nature because they rely on the weather [5, 6]. This unpredictability frequently results in imbalances between the supply and demand for energy, which limits the use of renewable energy sources and causes grid management inefficiencies. By absorbing excess renewable power during times of surplus generation and turning it back into electricity or usable fuels when demand outpaces supply, green hydrogen acts as a long-duration energy storage medium. This capability lessens dependency on backup systems that rely on fossil fuels, promotes energy arbitrage, and improves grid stability [7, 8].

Additionally, because of its adaptability, green hydrogen may be used in a variety of industries, which promotes sectoral integration. For heavy-duty and long-distance vehicles, hydrogen fuel cells offer a zero-emission substitute in the transportation industry [9, 10]. Green hydrogen has the potential to replace carbon-intensive fuels and feedstocks in industrial operations like ammonia manufacturing and steelmaking. Hydrogen may be integrated into current natural gas networks and used for heating purposes in both residential and commercial settings. By extending the advantages of renewable energy outside the electrical grid, these multi-sector applications enable a comprehensive strategy to decarbonization [11, 12].

This section's goal is to examine the many approaches and systems for integrating green hydrogen into diverse energy systems. Hydrogen is sectionalized into

DOI: 10.1201/9781003727194-8

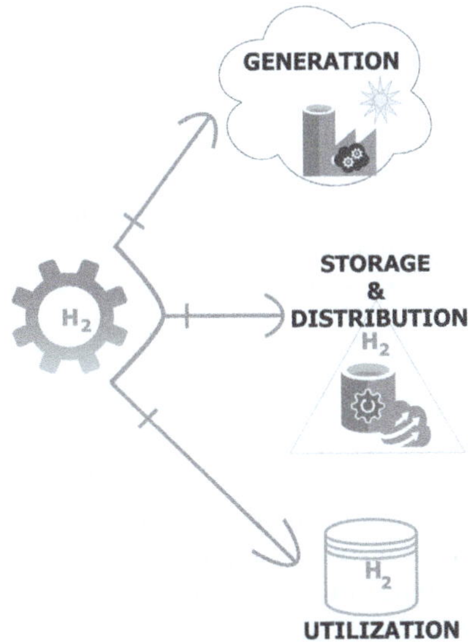

FIGURE 8.1 Green hydrogen.

generation, storage and distribution, and utilization as shown in Figure 8.1. This involves looking at how hydrogen can be used in Power-to-Gas (P2G) systems, microgrids, and hybrid topologies, in conjunction with solar and wind energy systems, and to improve demand response and grid services. Additionally, the part will include system architectures and illustrative examples that demonstrate the theoretical and practical frameworks for successful integration. In order to contribute to a more adaptable, robust, and decarbonized energy future, the section seeks to offer a thorough grasp of how green hydrogen might close the gap between renewable generation and end use demands.

In addition to being a technological advancement, integrating green hydrogen into renewable and integrated energy systems is strategically necessary to meet long-term energy and climate goals. By promoting the use of renewable energy, guaranteeing energy security, and allowing the coupling of energy sectors, it is a crucial enabler in the shift to a low-carbon economy [13, 14]. Therefore, the success of the global energy transition will depend on comprehending and putting into practice efficient integration strategies.

8.2 INTEGRATION IN SOLAR AND WIND ENERGY SYSTEMS

An important step in the global endeavor to reduce the carbon footprint of energy production while improving energy storage and grid flexibility is the incorporation of green hydrogen into solar and wind energy systems. Thanks to regulatory backing and technological improvements, solar and wind power have become the most popular renewable energy sources [15, 16]. Their inherent unreliability, however, presents problems for grid stability and a steady supply of electricity. Combining these resources with water electrolysis to produce hydrogen provides a workable approach that makes it possible to turn surplus electrical energy into chemical energy that can be stored.

Green hydrogen is created using electrolyzers, which use electricity to split water into hydrogen and oxygen. They can also be fueled by solar or wind energy. When supply outpaces demand during times of high renewable generation, these systems work very well. Instead of reducing the excess energy, which results in waste and financial losses, it can be used to create hydrogen in electrolyzers as shown in Eq. (8.1).

$$2H_2O(l) \rightarrow 2H_2(g) + O_2(g) \tag{8.1}$$

By generating an extra source of income, this strategy not only makes renewable projects more economically viable but also aids in load control and grid balancing. A key component of hydrogen integration's value proposition is addressing intermittency. Grid operators frequently have to make decisions about curtailment during periods of peak output due to the daily and seasonal fluctuations in renewable energy generation. Electrolyzer systems function as dynamic loads that can increase or decrease in response to surplus renewable energy. This load flexibility enables a greater penetration of renewables into the energy mix and reduces the requirement for peaking plants based on fossil fuels.

Electrolyzer sizing in relation to the renewable energy source's capacity must be carefully considered when designing efficient hydrogen production systems for solar and wind installations. Higher capital expenditures per unit of hydrogen could result from oversized electrolyzers being underutilized during times of low renewable output [17, 18]. However, at times of high generation, undersized systems might not be able to fully capture excess energy. Therefore, cost, utilization rate, and hydrogen demand projections must all be balanced in order to achieve optimal sizing.

The decision between grid-tied and off-grid configurations is another crucial element. Electrolyzers work in tandem with grid signals in grid-tied systems, allowing them to take part in demand response initiatives and supplementary services. In markets where electricity prices fluctuate, this arrangement is beneficial since it enables operators to reduce energy expenses. On the other hand, off-grid technologies improve energy resilience and autonomy, particularly in isolated or distant areas with inconsistent or restricted grid connectivity. When there is a shortage of renewable energy, these systems are frequently used in conjunction with battery storage to guarantee uninterrupted operation.

Pilot projects and commercial endeavors to combine hydrogen with solar and wind power have been developed in a number of nations with strong renewable energy penetration rates. In order to generate green hydrogen for industrial usage and gas grid injection, the Schleswig-Holstein region of Germany's "WESTKÜSTE100" project combines a 30 MW electrolyzer with onshore wind farms. In an energy system with a high wind capacity, this project is an excellent example of sector coupling and the use of hydrogen as a storage and conversion medium.

The "Hydrogen Park South Australia" project in Australia uses surplus solar energy to create green hydrogen, which is then combined with natural gas to heat homes. The area is a leader in the integration of renewable hydrogen due to its plentiful solar resources and dedication to decarbonization [19, 20]. The first solar-powered green hydrogen plant in the Middle East is located in the "Mohammed bin Rashid Al Maktoum Solar Park" in the United Arab Emirates. In order to assist regional energy transition goals, Siemens Energy, DEWA, and Expo 2020 Dubai collaborated to develop this pilot, which shows how concentrated solar power can be used to manufacture hydrogen in desert settings.

The intermittency of renewable energy presents a realistic and scalable solution through the incorporation of green hydrogen into solar and wind energy systems. Energy that would normally be limited can be stored and used for a variety of purposes by combining electrolyzers with changeable renewable energy sources. Green hydrogen has the potential to significantly contribute to energy system stabilization, renewable energy usage, and the acceleration of the transition to a low-carbon energy future with the right system design and supportive policies [21, 22].

8.3 MICROGRIDS AND HYBRID ENERGY SYSTEMS

Particularly in isolated or off-grid areas, microgrids and hybrid energy systems are becoming more widely acknowledged as essential tools for incorporating renewable energy sources and improving energy dependability. According to Darani *et al.* [23] and Cagnano *et al.* [24], a microgrid is a localized energy network that can function independently or in tandem with the main power grid. Because of their localized nature, microgrids can lower transmission losses, promote the integration of various energy technologies, such as solar photovoltaics (PV), wind turbines, battery storage, and green hydrogen production units, and offer energy security. These elements work together to optimize the flexibility, efficiency, and resilience of hybrid energy systems.

Microgrids are made up of distributed energy resources (DERs) and interconnected loads that work together to provide electricity to a specific location, like a community, industrial park, campus, or isolated village. The ability to work in two modes—grid-connected, where it communicates with the main grid, and islanded, or off-grid, where it operates independently during grid failures or in areas without grid access—is what distinguishes a microgrid.

Energy storage systems (batteries and increasingly hydrogen-based storage), loads (residential, commercial, or industrial energy users), control and communication infrastructure, and renewable energy producing units (wind turbines, solar panels) are the main parts of microgrids. Microgrids have taken on a new dimension with the incorporation of electrolysis-based hydrogen production. Hydrogen serves both as a long-duration storage medium and a flexible energy carrier, complementing short-term storage solutions like batteries.

In off-grid or islanded microgrids, green hydrogen plays a crucial role in maintaining energy supply reliability. Due to the intermittent nature of solar and wind resources, energy availability can fluctuate significantly. While batteries provide effective short-term storage for hours or days, hydrogen's chemical energy storage capabilities enable seasonal and extended backup solutions [25, 26]. Electrolyzers convert surplus renewable electricity into hydrogen, which can be stored in tanks for long periods without significant losses.

Fuel cells turn stored hydrogen back into electricity to fulfill demand when renewable energy declines because of weather fluctuations or at night, guaranteeing a steady supply of electricity. During extended outages, periods of high demand, or protracted renewable downtime, this cycle enables microgrids to remain resilient and stable. Because of its high energy density and adaptability in storage, hydrogen is especially well-suited for remote settlements, islanded systems, and areas with poor access to fuel logistics.

As illustrated in Figure 8.2, the integration of microgrids and green hydrogen production provides a number of noteworthy advantages, especially in terms of improving energy access, resilience, and autonomy. Microgrids overcome the difficulties and expenses involved in expanding centralized grid infrastructure by offering off-grid and remote locations a decentralized energy solution that can be customized to local resources and requirements. By facilitating longer-term energy storage and fuel flexibility, hydrogen integration enhances this autonomy even further.

The system's ability to function autonomously in the event of natural disasters or main grid failures enhances energy resilience by lowering outage vulnerability. By acting as an energy reserve that may be used during prolonged times of low renewable generation, hydrogen storage provides an additional degree of security [27, 28]. For vital infrastructure like hospitals, communication networks, and emergency services in remote areas, this resilience is essential.

Another significant benefit is decentralization. Microgrids enable communities and businesses to efficiently manage their energy resources by fostering local control over energy production and use. By generating jobs in hydrogen and renewable

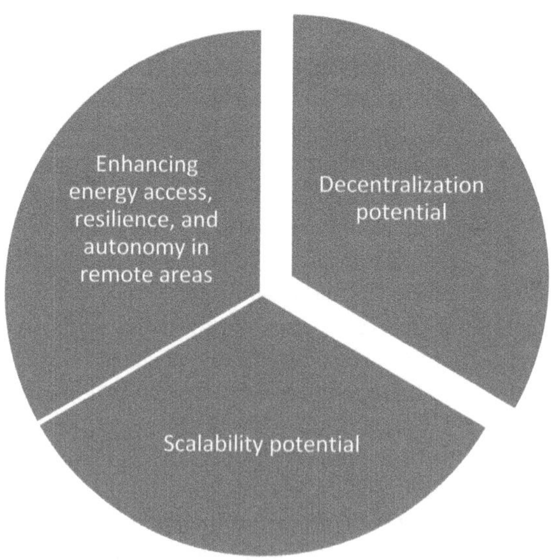

FIGURE 8.2 Benefits of microgrids and hybrid energy systems.

energy technologies, this decentralization can boost local economic growth and strengthen energy democracy. Microgrids with integrated hydrogen systems also provide flexibility and scalability. Systems can begin small, servicing a few homes or a small community, and develop as demand increases or the cost of technology decreases. Because of its modularity, green hydrogen adoption is more possible in a variety of socioeconomic circumstances by enabling incremental investment and technology enhancements.

Green hydrogen production and storage in microgrids and hybrid energy systems is a forward-thinking solution to the problems of energy access and integration of renewable energy. These systems offer improved resilience, dependability, and sustainability by fusing localized energy autonomy with the adaptability of hydrogen as a long-duration storage medium, particularly in isolated or off-grid areas. Microgrids' scalability and decentralization help the world's clean energy transition and emphasize how important green hydrogen will be to the energy landscape of the future.

8.4 POWER-TO-GAS AND OTHER ENERGY SYSTEMS INTEGRATION

As P2G and associated technologies advance, green hydrogen's incorporation into energy systems is changing quickly. Energy storage, sector coupling, and decarbonization across several industries are made possible by these strategies, which use renewable electricity to generate hydrogen and synthetic gases. In order to integrate green hydrogen at scale, P2G, power-to-liquid, and power-to-chemicals pathways are disruptive strategies that leverage current infrastructure and industrial processes [29, 30].

The Power-to-Gas is the process of turning excess or off-peak renewable electricity into gaseous fuels, mainly hydrogen, by electrolysis and then methanation to produce synthetic methane. Green hydrogen is created by electrolyzers that use renewable electricity to break water molecules into hydrogen and oxygen. This hydrogen can either be stored for later use or injected straight into natural gas pipes. In methanation reactors, hydrogen can also be mixed with collected carbon dioxide to create synthetic methane, also known as renewable natural gas, which is chemically comparable to regular natural gas.

P2G's interoperability with the current natural gas infrastructure is one of its main advantages. With very slight adjustments, hydrogen or synthetic methane can be transported and stored via gas pipelines, storage facilities, and distribution networks, providing an affordable option for long-distance and seasonal energy storage. By transforming electrical energy into transportable, storable gases, this method gets around some of the drawbacks of electrical grids, like transmission bottlenecks and battery storage issues.

In addition to producing gas, green hydrogen is a fundamental component of power-to-liquid and power-to-chemicals technologies. The process of power-to-liquid entails combining green hydrogen with biomass or captured carbon feedstocks to create liquid fuels like methanol, synthetic diesel, or jet fuel [31, 32].

These synthetic fuels are carbon-neutral, renewable substitutes for fuels derived from fossil fuels that work with current fuel delivery infrastructure and combustion engines.

Green hydrogen is used by power-to-chemicals to produce vital industrial chemicals like methanol and ammonia. Traditionally, ammonia—which is mostly used in fertilizers—is made with hydrogen that comes from fossil fuels, which releases a lot of carbon dioxide into the atmosphere. These emissions are significantly reduced by switching to green hydrogen-based ammonia synthesis, which supports environmentally friendly chemical production and agriculture. Likewise, green hydrogen-derived methanol provides decarbonization pathways for the petrochemical sectors by serving as a fuel and chemical feedstock.

The idea of sector linking and industrial symbiosis is embodied in the incorporation of green hydrogen into industrial operations. The optimization of resource utilization across several businesses, when surplus energy or byproducts from one operation are used as inputs for another, is known as industrial symbiosis. Because of its adaptability, hydrogen may connect the transportation, chemical, power, and heating industries, promoting cross-sector integration and decarbonization. Green ammonia produced using electrolyzed hydrogen helps fertilizer production by removing the need for fossil fuels. Green hydrogen allows for a cleaner production chain in the petrochemical industry by replacing grey hydrogen, which is utilized in chemical synthesis and refining.

Additionally, sector coupling improves the resilience and flexibility of the energy system. Depending on changes in demand, hydrogen generated during times of strong renewable electricity output can be stored and diverted across sectors [33, 34]. This integrated strategy minimizes curtailment, optimizes the use of renewable energy, and speeds up the shift to a low-carbon energy economy.

Scaling green hydrogen in the global energy transition requires the integration of P2G and related energy systems. P2G technologies facilitate effective energy storage and transportation by transforming renewable electricity into synthetic gases and hydrogen that are compatible with current infrastructure. While sector coupling and industrial symbiosis encourage cross-sector decarbonization and resource efficiency, power-to-liquid and power-to-chemicals expand the role of hydrogen into sustainable fuel and chemical manufacturing. By combining these strategies, green hydrogen's full potential as an industrial feedstock and flexible energy vector is unlocked, promoting a resilient and sustainable energy future.

8.5 INTEGRATION IN GRID AND DEMAND RESPONSE

For contemporary power networks, the growing use of variable renewable energy sources like wind and solar presents serious issues. These resources' erratic and sporadic nature can lead to supply-demand imbalances, frequency variations, and grid instability. As illustrated in Figure 8.3, green hydrogen, which is created by electrolysis driven by renewable energy sources, presents promising options for demand response and grid balancing, improving grid dependability and enabling a sustainable energy transition.

Controlling variations in the supply and demand for electricity is one of the main obstacles to incorporating renewables. Peak shaving, frequency management,

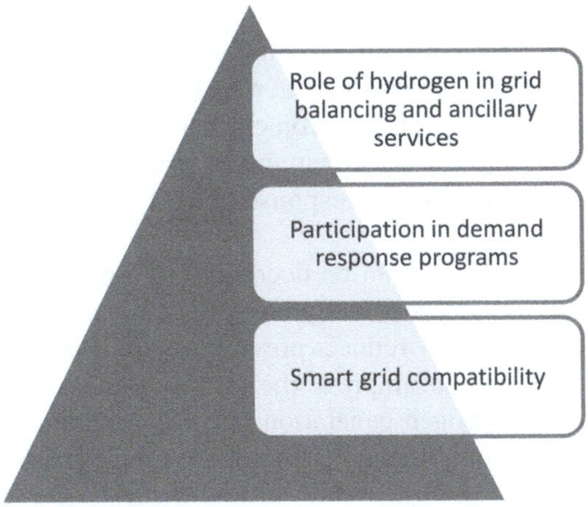

FIGURE 8.3 Integration in grid and demand response.

and load leveling are made possible by green hydrogen's versatility as an energy storage medium [35, 36]. Electrolyzers keep renewable energy from being curtailed and ease the burden on the grid by turning excess renewable electricity into hydrogen during times of oversupply, as shown in Eqs. (8.3) and (8.4). When demand spikes or renewable energy declines, this stored hydrogen can be transformed back into power using fuel cells or turbines, thereby mitigating supply fluctuations.

$$2H_2O(l) \rightarrow 2H_2(g) + O_2(g) \tag{8.2}$$

This hydrogen can then be stored and later reconverted to electricity in fuel cells when solar and wind outputs are low:

$$2H_2(g) + O_2(g) \rightarrow 2H_2O(l) + \text{Electricity} \tag{8.3}$$

In order to preserve stability in frequency regulation, the grid must react quickly to slight imbalances between generation and consumption. Hydrogen systems can contribute to frequency management by dynamically ramping up or down the electrolyzer load in a matter of seconds. Peak shaving is a similar strategy that lowers grid demand during times of high consumption. By allowing electricity supply to complement the system, hydrogen storage lessens grid congestion and dependency on fossil fuel peaking plants.

While traditional batteries are only capable of short-term storage, hydrogen storage offers a long-term buffer against varying renewable supplies. Given that renewable generation fluctuates greatly during the months, this feature is essential for seasonal balancing. Large amounts of hydrogen may be stored for a long time, guaranteeing system continuity and stability even amid protracted shortages in renewable energy.

Demand response (DR) programs encourage customers or energy users to modify their patterns of power use in response to price signals or grid conditions. Because of their operating flexibility, green hydrogen generation electrolyzers are ideally suited for involvement in DR projects. Electrolyzers can move energy consumption away from times of peak demand or grid stress events by dynamically modifying hydrogen production in response to grid constraints or electricity market prices [37, 38].

On the other hand, demand can be decreased by limiting hydrogen production at times of high prices or grid stress. By leveling demand peaks and filling valleys, this dynamic operation not only reduces power costs for hydrogen producers but also offers useful grid services that improve the overall efficiency and dependability of the grid. By including hydrogen generation into DR programs, financial incentives are matched with grid requirements, encouraging greater use of renewable energy sources and lowering the need for costly infrastructure improvements. This adaptability helps to close the gap between stable energy demand and the unpredictability of renewable output by allowing hydrogen production facilities to function as both energy consumers and grid stabilizers.

Smart grid technologies harness the full potential of integrating green hydrogen into demand response and grid balancing. Smart grids optimize energy flows and coordinate DERs in real-time by utilizing sophisticated digital control, communication, and automation systems. Intelligent control and optimization of hydrogen production and consumption depending on grid circumstances, power costs, and renewable availability are made possible by the integration of electrolyzers and hydrogen storage inside smart networks.

The production of hydrogen can be synchronized with other energy storage technologies, including thermal storage, pumped hydro, and batteries, using digital control systems. By directing energy flows to the most economical and technically sound storage or generation alternative at any given time, this multi-resource optimization improves grid flexibility. Additionally, predictive analytics and machine learning algorithms are made possible by smart grid platforms to forecast market conditions for hydrogen, energy demand, and renewable generation [39, 40]. By using these projections to guide electrolyzer scheduling, hydrogen production is able to maximize financial returns while still meeting grid demands. In order to preserve grid stability in the face of growing renewable integration, real-time monitoring and automated control enhance system responsiveness and reliability.

A low-carbon energy future is made possible in large part by green hydrogen's function in demand response and grid integration. Hydrogen storage reduces renewable intermittency and improves grid stability by offering load leveling, frequency management, and peak shaving services. By taking part in demand response programs, hydrogen production can be dynamically adjusted to grid circumstances and energy prices, maximizing resource use and cutting expenses. Advanced digital control, cooperation with other storage and production assets, and predictive optimization are made possible by hydrogen technologies' connection with smart grid systems. When taken as a whole, these qualities establish green hydrogen as a strategic and adaptable part of contemporary, sustainable, and robust power systems.

8.6 ILLUSTRATIONS OF INTEGRATED ENERGY SYSTEMS

Energy storage, flexible energy carriers like green hydrogen, and the integration of various renewable energy technologies are critical to the shift to sustainable energy systems. As illustrated in Figure 8.4, visualizing and modeling these integrated energy systems using energy flow routes, system architectural diagrams, and simulation results offers important insights into their effectiveness, efficiency, and advantages [41, 42]. With the help of case studies and comparative analysis, this explores illustrated instances that show how green hydrogen may be smoothly integrated with solar, wind, batteries, and the electrical grid.

To see how green hydrogen interacts with other renewable energy sources in an integrated energy system, system architecture diagrams are crucial tools. These diagrams usually show wind turbines and solar PV arrays as the main sources of electricity that feed into a hybrid system that also contains electrolyzers and batteries. Hydrogen is produced by electrolyzers and can be stored in underground caves or tanks. Later, the hydrogen can be repurposed into energy using fuel cells or combustion turbines, fed into the gas grid, or used directly for industrial activities.

Multiple energy vectors are present in a single architecture: electrolyzers for the production of hydrogen, batteries for short-term storage, and electricity flowing from renewable sources to direct consumption loads. In turn, the hydrogen can be used for chemical synthesis, power generation, heating, transportation, and long-term storage. These graphs demonstrate the flexibility and robustness that hydrogen integration adds while assisting stakeholders in understanding the relationships and energy flows between components.

The conversion of energy from generation to final consumption in integrated systems is explained using illustrative energy flow paths. A solar-wind-hydrogen microgrid is a prime example; it uses solar panels and wind turbines to provide electricity to meet urgent demands, such those of homes or businesses [43, 44]. Green

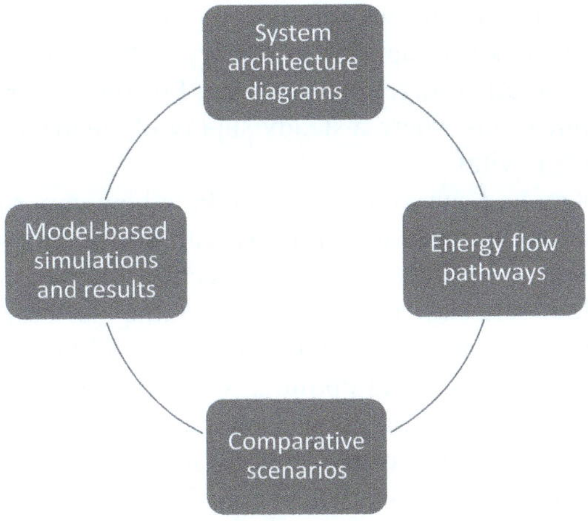

FIGURE 8.4 Illustrations of integrated energy systems.

hydrogen is produced by electrolyzers powered by electricity when demand for renewable energy surpasses supply. This hydrogen is kept for later usage and serves as a buffer.

Fuel cells or turbines can be used to turn stored hydrogen back into electricity during times of high demand or poor renewable output, assisting isolated loads or boosting system power. Hydrogen can also be used in industry for things like steel manufacture and ammonia synthesis, or it can be used as fuel in transportation. This method emphasizes the function of hydrogen in addressing temporal discrepancies between fluctuating or constant demand across sectors and changeable renewable supply.

Detailed evaluation and optimization of integrated energy systems that incorporate green hydrogen are made possible by computational models and simulations. System sizing, operational tactics, and economic trade-offs are examined in case studies that use optimization techniques. Models balance capital costs and operating advantages by revealing the ideal electrolyzer capacity in relation to renewable energy characteristics. Additionally, simulations measure how hydrogen storage lowers greenhouse gas emissions by reducing dependency on fossil fuel backup. Diversified energy vectors have been shown to improve system resilience and reduce the risk of supply shortages during inclement weather or periods of high demand.

Dynamic simulations demonstrate how hydrogen can improve grid flexibility by supplying power during demand peaks and absorbing excess renewable power during off-peak hours [45, 46]. These insights direct investment and design choices, enabling the implementation of integrated systems that are sustainable and reasonably priced. Analyses of comparative scenarios demonstrate the observable advantages of integrating hydrogen as opposed to systems that just employ batteries and direct renewable electricity. Research contrasting "with hydrogen" and "without hydrogen" situations shows that adding hydrogen significantly lowers system costs, emissions, and operational restrictions.

Hydrogen storage also improves resilience indicators like the capacity to sustain supplies amid protracted renewable droughts. The energy density and scalability of hydrogen make it superior to battery-only systems, which might not be able to meet seasonal storage requirements. For remote or islanded networks and industrial applications where a steady supply of electricity is necessary, this enhanced resilience is vital.

Examples of integrated energy systems highlight how important green hydrogen is as a supplementary energy source in grids with a high proportion of renewable energy. Energy flow pathways and system architecture diagrams show how hydrogen connects solar, wind, batteries, and the grid, providing sectoral and temporal flexibility [47, 48]. These advantages are confirmed by model-based simulations, which also show efficiency increases and optimize system performance. The benefits of hydrogen in lowering expenses, emissions, and improving resilience are further supported by comparative scenario assessments. In support of the worldwide shift to a sustainable, dependable, and low-carbon energy future, these examples collectively offer strong arguments for integrating green hydrogen into future energy system designs [49, 50].

8.7 CONCLUSION

A revolutionary chance to improve the adaptability, durability, and sustainability of upcoming energy infrastructures is presented by the incorporation of green hydrogen into renewable and hybrid energy systems. Green hydrogen serves as a flexible energy carrier that makes it possible to efficiently handle the unpredictability of solar and wind energy, allowing for long-term energy storage and sector coupling between industry, transportation, and power. This integration promotes a greener and more secure energy future by lowering dependency on fossil fuels and significantly reducing greenhouse gas emissions. When hydrogen is combined with batteries, grid systems, and various renewable energy sources, a synergistic framework is created that dynamically balances supply and demand, enhancing grid stability and increasing access to energy, especially in isolated or off-grid areas.

However, substantial investment and strong legislative backing are needed to fully realize the potential of integrated hydrogen systems. In order to overcome the present financial obstacles and technological difficulties related to electrolyzers, storage, and fuel cells, strategic frameworks that encourage research, development, and implementation are essential. Commercial viability and scale-up will be accelerated by policies that support market mechanisms, standardization, and infrastructure development. Pilot projects and demonstration plants can be sparked by public-private partnerships and targeted finance, which can also boost stakeholder confidence and provide useful data.

Going forward, green hydrogen is expected to be a key component of next-generation energy networks that are decentralized, digitized, and decarbonized. Optimized hydrogen production and consumption will be made possible by developments in smart grid technologies and digital control systems, which will be smoothly connected with demand response plans and variable renewable energy sources. Furthermore, new developments in sector coupling, P2G, and power-to-liquid will increase the use of hydrogen and promote a more thorough decarbonization of difficult-to-abate industries like heavy manufacturing and long-distance transportation. Green hydrogen integration is a crucial tenet in creating resilient, sustainable, and low-carbon energy systems that will satisfy tomorrow's energy demands as international efforts to tackle climate change grow.

REFERENCES

1. M. J. B. Kabeyi, and O. A. Olanrewaju, 'Sustainable energy transition for renewable and low carbon grid electricity generation and supply', *Frontiers in Energy Research*, vol. 9, p. 743114, 2022.
2. B. Zakeri, K. Paulavets, L. Barreto-Gomez, L. Gomez Echeverri, S. Pachauri, J. Rogelj, F. Creutzig, D. Urge-Vorsatz, D. Victor, B. Boza-Kiss, and C. Zimm, 2021. Transformations within reach: Pathways to a sustainable and resilient world - Rethinking energy solutions. https://pure.iiasa.ac.at/id/eprint/16820/
3. I. E. Atawi, A. Q. Al-Shetwi, A. M. Magableh, and O. H. Albalawi, 'Recent advances in hybrid energy storage system integrated renewable power generation: Configuration, control, applications, and future directions', *Batteries*, vol. 9, no. 1, p. 29, 2022.

4. M. P. Bakht, Z. Salam, M. Gul, W. Anjum, M. A. Kamaruddin, N. Khan, and A. L. Bukar, 'The potential role of hybrid renewable energy system for grid intermittency problem: A techno-economic optimisation and comparative analysis', *Sustainability*, vol. 14, no. 21, p. 14045, 2022.
5. S. Asiaban, N. Kayedpour, A. E. Samani, D. Bozalakov, J. D. De Kooning, G. Crevecoeur, and L. Vandevelde, 'Wind and solar intermittency and the associated integration challenges: A comprehensive review including the status in the Belgian power system', *Energies*, vol. 14, no. 9, p. 2630, 2021.
6. J. Li, S. Chen, Y. Wu, Q. Wang, X. Liu, L. Qi, X. Lu, and L. Gao, 'How to make better use of intermittent and variable energy? A review of wind and photovoltaic power consumption in China', *Renewable and Sustainable Energy Reviews*, vol. 137, p. 110626, 2021.
7. J. W. McNamara, V. DeAngelis, R. H. Byrne, A. Benson, B. R. Chalamala, and R. Masiello, 'Long-duration energy storage in a decarbonized future: Policy gaps, needs, and opportunities', *MRS Energy & Sustainability*, vol. 9, no. 2, pp. 142–170, 2022.
8. M. Wei, D. Black, J. Elliott, H. Goudey, S. E. Greenberg, and A. Weber, 2020. Improving Energy System Resilience at Berkeley Lab and Beyond. https://eta-publications.lbl.gov/sites/default/files/wei_-_improving_energy_resilience_at_berkeley_final_v2.pdf
9. C. Cunanan, M. K. Tran, Y. Lee, S. Kwok, V. Leung, and M. Fowler, 'A review of heavy-duty vehicle powertrain technologies: Diesel engine vehicles, battery electric vehicles, and hydrogen fuel cell electric vehicles', *Clean Technologies*, vol. 3, no. 2, pp. 474–489, 2021.
10. I. Staffell, D. Scamman, A. V. Abad, P. Balcombe, P. E. Dodds, P. Ekins, N. Shah, and K. R. Ward, 'The role of hydrogen and fuel cells in the global energy system', *Energy & Environmental Science*, vol. 12, no. 2, pp. 463–491, 2019.
11. X. Hou, S. Zhong, and J. A. Zhao, 'A critical review on decarbonizing heating in China: Pathway exploration for technology with multi-sector applications', *Energies*, vol. 15, no. 3, p. 1183, 2022.
12. R. Kakodkar, G. He, C. D. Demirhan, M. Arbabzadeh, S. G. Baratsas, S. Avraamidou, D. Mallapragada, I. Miller, R. C. Allen, E. Gençer, and E. N. Pistikopoulos, 'A review of analytical and optimization methodologies for transitions in multi-scale energy systems', *Renewable and Sustainable Energy Reviews*, vol. 160, p. 112277, 2022.
13. A. Long, M. B. Mokhtar, M. F. Ahmed, and C. K. Lim, 'Enhancing sustainable development via low carbon energy transition approaches', *Journal of Cleaner Production*, vol. 379, p. 134678, 2022.
14. M. Murshed, Z. Ahmed, M. S. Alam, H. Mahmood, A. Rehman, and V. Dagar, 'Reinvigorating the role of clean energy transition for achieving a low-carbon economy: Evidence from Bangladesh', *Environmental Science and Pollution Research*, vol. 28, pp. 67689–67710, 2021.
15. A. S. Darwish, and R. Al-Dabbagh, 'Wind energy state of the art: Present and future technology advancements', *Renewable Energy and Environmental Sustainability*, vol. 5, p. 7, 2020.
16. T. Lehtola, and A. Zahedi, 'Solar energy and wind power supply supported by storage technology: A review', *Sustainable Energy Technologies and Assessments*, vol. 35, pp. 25–31, 2019.
17. G. Bristowe, and A. Smallbone, 'The key techno-economic and manufacturing drivers for reducing the cost of power-to-gas and a hydrogen-enabled energy system', *Hydrogen*, vol. 2, no. 3, pp. 273–300, 2021.
18. M. J. Ginsberg, M. Venkatraman, D. V. Esposito, and V. M. Fthenakis, 'Minimizing the cost of hydrogen production through dynamic polymer electrolyte membrane electrolyzer operation', *Cell Reports Physical Science*, vol. 3, no. 6, pp 1–19, 2022. DOI: 10.1016/j.xcrp.2022.100935

19. D. Hayashi, 'Harnessing innovation policy for industrial decarbonization: Capabilities and manufacturing in the wind and solar power sectors of China and India', *Energy Research & Social Science*, vol. 70, p. 101644, 2020.
20. L. Hermwille, 'Hardwired towards transformation? Assessing global climate governance for power sector decarbonization', *Earth System Governance*, vol. 8, p. 100054, 2021.
21. M. Frankowska, A. Rzeczycki, M. Sowa, and W. Drożdż, 'Functional model of power grid stabilization in the green hydrogen supply chain system—Conceptual assumptions', *Energies*, vol. 16, no. 1, p. 154, 2022.
22. A. H. Schrotenboer, A. A. Veenstra, M. A. uit het Broek, and E. Ursavas, 'A Green Hydrogen Energy System: Optimal control strategies for integrated hydrogen storage and power generation with wind energy', *Renewable and Sustainable Energy Reviews*, vol. 168, p. 112744, 2022.
23. S. A. Darani, C. D. Majhor, W. W. Weaver, R. D. Robinett, and O. Abdelkhalik, 'Optimal positioning of energy assets in autonomous robotic microgrids for power restoration', *IEEE Transactions on Industrial Informatics*, vol. 15, no. 7, pp. 4370–4380, 2019.
24. A. Cagnano, E. De Tuglie, and P. Mancarella, 'Microgrids: Overview and guidelines for practical implementations and operation', *Applied Energy*, vol. 258, p. 114039, 2020.
25. N. Khan, S. Dilshad, R. Khalid, A. R. Kalair, and N. Abas, 'Review of energy storage and transportation of energy', *Energy Storage*, vol. 1, no. 3, p. e49, 2019.
26. Z. Zhu, T. Jiang, M. Ali, Y. Meng, Y. Jin, Y. Cui, and W. Chen, 'Rechargeable batteries for grid scale energy storage', *Chemical Reviews*, vol. 122, no. 22, pp. 16610–16751, 2022.
27. A. M. Elberry, J. Thakur, A. Santasalo-Aarnio, and M. Larmi, 'Large-scale compressed hydrogen storage as part of renewable electricity storage systems', *International Journal of Hydrogen Energy*, vol. 46, no. 29, pp. 15671–15690, 2021.
28. S. R. Thiyagarajan, H. Emadi, A. Hussain, P. Patange, and M. Watson, 'A comprehensive review of the mechanisms and efficiency of underground hydrogen storage', *Journal of Energy Storage*, vol. 51, p. 104490, 2022.
29. J. A. Schaidle, R. G. Grim, L. Tao, M. Ruth, K. Harrison, N. Dowe, C. McMillan, S. Pless, and D. J. Arent., 2022. Power Conversion Technologies: The Advent of Power-to-Gas, Power-to-Liquid, and Power-to-Heat. *Technologies for Integrated Energy Systems and Networks*, pp. 41–70.
30. M. Sterner, and M. Specht, 'Power-to-gas and power-to-X—The history and results of developing a new storage concept', *Energies*, vol. 14, no. 20, p. 6594, 2021.
31. T. N. Do, Y. G. Hur, H. E. Jeong, J. W. Chung, W. Won, and J. Kim, 'Rethinking of conventional gas-to-liquid via dimethyl ether intermediate incorporating renewable energy against power-to-liquid', *Energy Conversion and Management*, vol. 261, p. 115643, 2022.
32. R. J. White, 2022. Power-to-Liquids-Conversion of CO_2 and Renewable H_2 to Methanol. *Advances in Energy Storage: Latest Developments from R&D to the Market*, pp. 489–520. https://doi.org/10.1002/9781119239390.ch22
33. A. I. Osman, N. Mehta, A. M. Elgarahy, M. Hefny, A. Al-Hinai, A. A. H. Al-Muhtaseb, and D. W. Rooney, 'Hydrogen production, storage, utilisation and environmental impacts: A review', *Environmental Chemistry Letters*, vol. 20, no. 1, pp. 153–188, 2022.
34. M. G. Rasul, M. A. Hazrat, M. A. Sattar, M. I. Jahirul, and M. J. Shearer, 'The future of hydrogen: Challenges on production, storage and applications', *Energy Conversion and Management*, vol. 272, p. 116326, 2022.
35. A. Alzahrani, S. K. Ramu, G. Devarajan, I. Vairavasundaram, and S. Vairavasundaram, 'A review on hydrogen-based hybrid microgrid system: Topologies for hydrogen energy storage, integration, and energy management with solar and wind energy', *Energies*, vol. 15, no. 21, p. 7979, 2022.

36. W. Sun, and G. P. Harrison, 'Active load management of hydrogen refuelling stations for increasing the grid integration of renewable generation', *IEEE Access*, vol. 9, pp. 101681–101694, 2021.
37. A. Mayyas, M. Wei, and G. Levis, 'Hydrogen as a long-term, large-scale energy storage solution when coupled with renewable energy sources or grids with dynamic electricity pricing schemes', *International Journal of Hydrogen Energy*, vol. 45, no. 33, pp. 16311–16325, 2020.
38. C. Schnuelle, T. Wassermann, D. Fuhrlaender, and E. Zondervan, 'Dynamic hydrogen production from PV & wind direct electricity supply – Modeling and techno-economic assessment', *International Journal of Hydrogen Energy*, vol. 45, no. 55, pp. 29938–29952, 2020.
39. S. Koshy, S. Rahul, R. Sunitha, and E. P. Cheriyan, 'Smart grid–based big data analytics using machine learning and artificial intelligence: A survey', *Artificial Intelligence and Internet of Things for Renewable Energy Systems*, vol. 12, p. 241, 2021.
40. N. Mostafa, H. S. M. Ramadan, and O. Elfarouk, 'Renewable energy management in smart grids by using big data analytics and machine learning', *Machine Learning with Applications*, vol. 9, p. 100363, 2022.
41. D. Huppmann, M. J. Gidden, Z. Nicholls, J. Hörsch, R. Lamboll, P. N. Kishimoto, T. Burandt, O. Fricko, E. Byers, J. Kikstra, and M. Brinkerink, 'Pyam: Analysis and visualisation of integrated assessment and macro-energy scenarios', *Open Research Europe*, vol. 1, p. 74, 2021.
42. D. Song, W. Meng, M. Dong, J. Yang, J. Wang, X. Chen, and L. Huang, 'A critical survey of integrated energy system: Summaries, methodologies and analysis', *Energy Conversion and Management*, vol. 266, p. 115863, 2022.
43. O. M. Babatunde, J. L. Munda, and Y. Hamam, 'Off-grid hybrid photovoltaic–micro wind turbine renewable energy system with hydrogen and battery storage: Effects of sun tracking technologies', *Energy Conversion and Management*, vol. 255, p. 115335, 2022.
44. T. Hinokuma, H. Farzaneh, and A. Shaqour, 'Techno-economic analysis of a fuzzy logic control based hybrid renewable energy system to power a university campus in Japan', *Energies*, vol. 14, no. 7, p. 1960, 2021.
45. G. Buffo, D. Ferrero, M. Santarelli, and A. Lanzini, 'Energy and environmental analysis of a flexible power-to-x plant based on reversible solid oxide cells (rSOCs) for an urban district', *Journal of Energy Storage*, vol. 29, p. 101314, 2020.
46. P. Colombo, A. Saeedmanesh, M. Santarelli, and J. Brouwer, 'Dynamic dispatch of solid oxide electrolysis system for high renewable energy penetration in a microgrid', *Energy Conversion and Management*, vol. 204, p. 112322, 2020.
47. J. D. Fonseca, M. Camargo, J. M. Commenge, L. Falk, and I. D. Gil, 'Trends in design of distributed energy systems using hydrogen as energy vector: A systematic literature review', *International Journal of Hydrogen Energy*, vol. 44, no. 19, pp. 9486–9504, 2019.
48. Y. Zhou, 'Low-carbon transition in smart city with sustainable airport energy ecosystems and hydrogen-based renewable-grid-storage-flexibility', *Energy Reviews*, vol. 1, no. 1, p. 100001, 2022.
49. A. Z. Arsad, M. A. Hannan, A. Q. Al-Shetwi, M. Mansur, K. M. Muttaqi, Z. Y. Dong, and F. Blaabjerg, 'Hydrogen energy storage integrated hybrid renewable energy systems: A review analysis for future research directions', *International Journal of Hydrogen Energy*, vol. 47, no. 39, pp. 17285–17312, 2022.
50. A. Safari, N. Das, O. Langhelle, J. Roy, and M. Assadi, 'Natural gas: A transition fuel for sustainable energy system transformation?', *Energy Science & Engineering* vol. 7, no. 4, pp. 1075–1094, 2019.

9 Analysis of Exergy, Exergoeconomic, and Exergoenvironmental Studies in Hydrogen Energy Systems

9.1 EXERGY ANALYSIS BASICS

The second law of thermodynamics describes the maximum potential work obtainable from a system nearing equilibrium with a reference environment. Exergy measures the quality of energy and highlights losses due to irreversibilities, unlike energy, which is conserved by the first law. In hydrogen (H_2) energy systems—involving electrolysis, compression, storage, and fuel cell conversion—exergy analysis serves as a valuable tool for diagnosis and optimization [1].

Hydrogen is an emerging, clean energy vector that can, in theory, deliver very high energy quality, primarily due to its high chemical exergy-to-energy ratio. However, there are inherent inefficiencies in every conversion step. Exergy analysis is robust at identifying the locations of significant losses, whereas conventional energy analysis is unable to do so.

The exergy analysis is essential in extending its outcomes to the fields of exergoeconomic and exergoenvironmental analyses, because it offers a rigorous framework for improving system designs and connecting thermodynamic performance with economic and environmental factors [1–4]. Figure 9.1 illustrates the typical exergy flow for a H_2 system, which highlights the partitioning of input exergy into useful work, losses, and exergy destruction.

Exergy is the maximum useful work obtainable from a system as it interacts with a reference environment until it approaches equilibrium. It is composed of kinetic, potential, physical, and chemical exergies. The kinetic and potential exergies are fully convertible to work (ordered forms of energy) [5]. For a steady-flow system, the specific physical exergy (due to deviations in temperature and pressure) is given by [5]:

$$\varepsilon_{ph}(T, P) = (h - h_0) - T_0(s - s_0) \tag{9.1}$$

where ε_{ph} is the specific physical exergy and h and s are the specific enthalpy and entropy of a stream, respectively. h_0 and s_0 are the corresponding values of the

DOI: 10.1201/9781003727194-9

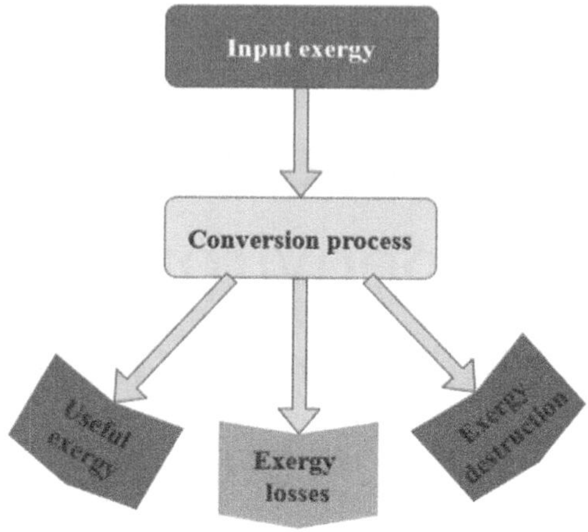

FIGURE 9.1 General schematic of exergy flows in a hydrogen energy system.

specific enthalpy and entropy at the reference environment (T_0, P_0). The reference environment (also known as dead state) is characterized by fixed temperature T_0, pressure P_0, and composition (chemical), usually representing natural atmospheric conditions.

For an ideal gas flow, the specific physical exergy stream may be given by [5]:

$$\varepsilon_{ph}(T, P) = \underbrace{\int_{T_0}^{T} c_{p,i}(T)dT}_{h_i - h_{i0}} - T_0 \underbrace{\int_{T_0}^{T} \frac{C_{p,i}(T)}{T} dT}_{s_i - s_{i0}} + RT_0 \ln\left(\frac{P}{P_0}\right) \tag{9.2}$$

where $c_{p,i}(T)$ is the temperature-dependent specific heat capacity at ith gas and R is the specific gas constant.

The specific chemical exergy per mole of a substance i is computed as

$$\varepsilon_{ch} = \sum_i x_i \bar{\varepsilon}_i^{ch} + RT_0 \sum_i x_i \ln(x_i) \tag{9.3}$$

Here, $\bar{\varepsilon}_i^{ch}$ is the standard chemical exergy and x_i is the mole fraction of the combustion species.

Thus, the overall exergy rate for a given stream is expressed as:

$$\dot{\Xi} = \dot{m}\left(\varepsilon_{ph} + \varepsilon_{ch}\right) \tag{9.4}$$

where $\dot{m}$ is the stream's mass flow rate.

For a control volume, the exergy balance under the steady state conditions is given by:

$$\dot{\Xi}_{in} + \dot{\Xi}_q = \dot{\Xi}_{out} + \dot{W} + \dot{\Xi}_D \tag{9.5}$$

where $\dot{\Xi}_{in}$ and $\dot{\Xi}_{out}$ are exergy into and out of the control volume, respectively. $\dot{\Xi}_q$ and $\dot{W}$ are the rates of heat and work in the control volume, respectively. $\dot{\Xi}_D$ is the exergy destruction rate in the control volume.

The exergy balance may be represented as fuel and product terms, i.e.,

$$\dot{\Xi}_F = \dot{\Xi}_P + \dot{\Xi}_D \tag{9.6}$$

$\dot{\Xi}_F$ and $\dot{\Xi}_P$ are the fuel and product exergy rates of the control volume, respectively. In the entire system, the exergy balance may be given by:

$$\dot{\Xi}_{F,tot} = \dot{\Xi}_{P,tot} + \sum \dot{\Xi}_D + \dot{\Xi}_{L,tot} \tag{9.7}$$

where $\dot{\Xi}_{F,tot}$ and $\dot{\Xi}_{P,tot}$ are the total fuel and product exergy rates in the system. $\dot{\Xi}_{L,tot}$ is the total exergy loss in the system.

Following the exergy balance formulation, the exergy efficiency (second-law efficiency) of the control volume, ϵ, is expressed as:

$$\epsilon = \frac{\dot{\Xi}_{output}}{\dot{\Xi}_{input}} = \frac{\dot{\Xi}_P}{\dot{\Xi}_F} = \frac{\dot{\Xi}_P}{\dot{\Xi}_P + \dot{\Xi}_D} \tag{9.8}$$

Exergy efficiency provides a realistic measure of how effectively a process converts available energy into useful work.

9.1.1 Sources of Exergy Destruction

Exergy destruction comes from irreversibilities within conversion processes. The primary sources of energy loss in a thermodynamic system, such as hydrogen energy systems, include thermal irreversibility, electrochemical losses, mechanical losses, mixing and separation, and chemical reaction irreversibility. Thermal irreversibility results from thermal transfer across finite temperature differences in fuel cell stacks, reformers, and heat exchangers.

In electrolyzers and fuel cells, electrochemical losses result from ohmic, concentration, and activation overpotentials. Mechanical losses occur due to friction in pumps, compressors, and turbines. In mixing and separation techniques, there are diffusion and recombination losses, such as in H_2 blending and gas purification systems. Incomplete reactions or kinetic limitations, such as reforming and combustion processes, result in chemical reaction irreversibilities.

Reducing irreversibilities leads to both increased thermodynamic efficiency and lower environmental impacts because the rate of exergy destruction is directly proportional to entropy generation [6]. Table 9.1 summarizes the irreversibilities in some components of H_2 systems and their mitigation strategies.

9.1.2 Exergy in Hydrogen Energy Systems

Both physical and chemical exergies are crucial for H_2 energy systems. Hydrogen has a chemical exergy of roughly 120 MJ/kg, which is approximately 1.04 times its lower

TABLE 9.1
Typical Exergy Destruction Sources in Hydrogen Energy Systems

Component	Irreversibility	Major Cause	Mitigation Approach	Source
Storage tank	Thermal	Heat leakage	Insulation, cryogenic optimization	[7–10]
Compressor	Mechanical	Heat loss, friction	Multistage compression, intercooling	[11]
Electrolyzer	Electrochemical	overpotentials	Catalyst optimization, lower resistance	[12]
Fuel cell	Electrochemical	Activation and ohmic losses	Improved membrane design	[13]
Heat exchangers	Thermal	Temperature gradient	Pinch point optimization	[1]

heating value (LHV) [1]. Exergy analysis makes it possible for detailed evaluation of H_2 systems, such as H_2 storage systems (comprising of pressure and thermal exergy changes), electrolyzers (comprising of electricity to chemical exergy changes), and fuel cells (consisting of chemical to electrical and thermal exergy changes).

9.2 CASE STUDY: ELECTROLYZER TO FUEL CELL SYSTEM

In this section, an illustration of the exergy analysis for an H_2 system is undertaken. Figure 9.2 represents a H_2 energy conversion chain which consists of four main subsystems, such as:

i. Electrolyzer
ii. Compressor
iii. H_2 storage tank
iv. Fuel cell

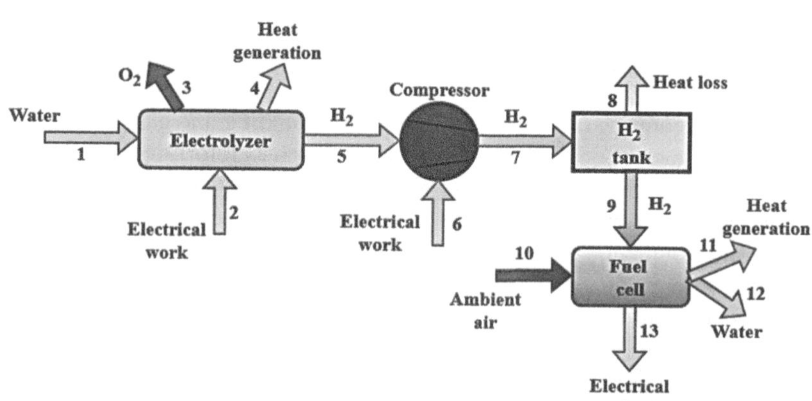

FIGURE 9.2 Hydrogen generation and conversion chain.

The system operates as an open cycle. Water is input into the electrolyzer (as stream 1), and it undergoes an electrochemical reaction to produce H_2 (as the primary product, in stream 5) and O_2 (stream 3) with the electrical supply (as stream 2).

A compressor, which requires electrical power, compresses the H_2 generated by the electrolyzer and stored in the H_2 storage tank. The H_2 stored in the tank is then used as the fuel input to the fuel cell, which combines it with O_2 from ambient air. The fuel cell generates electrical power as its primary product, along with water and heat as byproducts.

9.2.1 Exergy Analysis of the Hydrogen System

The H_2 conversion chain in Figure 9.2 is a theoretical system explicitly tailored for exergy analysis. The following discussion examines the exergy flows, exergy balances, and exergy efficiencies of each system's component and the entire system.

The electrolyzer input or fuel exergy is the water exergy and the electrical work input. Hydrogen exergy (and O_2 exergy if not vented to the atmosphere) is the primary output or product exergy (and possible heat generation), while there is exergy destruction due to irreversibilities, such as activation losses, ohmic losses, and heat losses [14]. The electrolyzer basically converts electricity into chemical exergy stored in H_2.

The exergy balance for the electrolyzer is formulated from the fuel and product exergy concept, such as:

$$\dot{\Xi}_1 + \dot{\Xi}_2 = \dot{\Xi}_3 + \dot{\Xi}_4 + \dot{\Xi}_5 + \dot{\Xi}_{D,Elect} \tag{9.9a}$$

$$\dot{\Xi}_1 + \dot{W}_2 = \dot{\Xi}_3 + \dot{\Xi}_{q,4} + \dot{\Xi}_5 + \dot{\Xi}_{D,Elect} \tag{9.9b}$$

$$\dot{\Xi}_1 + \dot{W}_2 = \dot{\Xi}_3 + \dot{Q}_{Elect}\left(1 - \frac{T_0}{T_4}\right) + \dot{\Xi}_5 + \dot{\Xi}_{D,Elect} \tag{9.9c}$$

where $\dot{\Xi}_1$ and $\dot{\Xi}_2$ are the exergy rates of water and electrical work ($\dot{W}_2$). $\dot{\Xi}_3$ is the O_2 exergy rate, $\dot{\Xi}_4$ is the heat exergy rate generated in the electrolyzer, $\dot{Q}_{Elect}$ is the heat loss rate (heat generated, which is lost to the cooling system) in the electrolyzer, $\dot{\Xi}_5$ is the H_2 exergy rate, and $\dot{\Xi}_{D,Elect}$ is the exergy destruction in the electrolyzer. For this electrolyzer, the exergy rates of water and electrical work are the fuel exergy into the electrolyzer, while the H_2 exergy rate is the product exergy rate.

From the exergy balance, the exergy destruction in the electrolyzer is given by:

$$\dot{\Xi}_{D,Elect} = \dot{\Xi}_1 + \dot{W}_2 - \left[\dot{\Xi}_3 + \dot{Q}_{Elect}\left(1 - \frac{T_0}{T_4}\right) + \dot{\Xi}_5\right] \tag{9.10}$$

Following Eqs. (9.9) and (9.10), the exergy efficiency of the electrolyzer is given by:

$$\epsilon_{Elect} = \frac{\dot{\Xi}_3 + \dot{Q}_{Elect}\left(1 - \frac{T_0}{T_4}\right) + \dot{\Xi}_5}{\dot{\Xi}_1 + \dot{W}_2} = \frac{\dot{\Xi}_3 + \dot{Q}_{Elect}\left(1 - \frac{T_0}{T_4}\right) + \dot{\Xi}_5}{\dot{\Xi}_3 + \dot{Q}_{Elect}\left(1 - \frac{T_0}{T_4}\right) + \dot{\Xi}_5 + \dot{\Xi}_{D,Elect}} \tag{9.11}$$

Eq. (9.11) shows the quantity of the input exergy into the electrolyzer that is useful for producing H_2.

Hydrogen from the electrolyzer is the input to the compressor. Electrical power drives the compressor to mechanically compress the H_2 exiting the electrolyzer. Essentially, the compressor increases the pressure of the H_2 for storage. With this, the input exergies to the compressor are the H_2 exergy rate from the electrolyzer, electrical work to the compressor, and a possible small exergy associated with ambient air (which may be negligible). The output exergy is the compressed H_2 exergy exiting the compressor. There is exergy destruction due to compression work, heat dissipation, and mechanical inefficiencies [6].

The exergy balance of the compressor is formulated similarly as:

$$\dot{\Xi}_6 = (\dot{\Xi}_7 - \dot{\Xi}_5) + \dot{\Xi}_{D,Comp} \tag{9.12a}$$

$$\dot{W}_6 = (\dot{\Xi}_7 - \dot{\Xi}_5) + \dot{\Xi}_{D,Comp} \tag{9.12b}$$

where $\dot{\Xi}_6$ or $\dot{W}_6$ is the exergy rate of the compressor work, $\dot{\Xi}_7$ is the compressed H_2 exergy rate, while $\dot{\Xi}_{D,Comp}$ is the exergy destruction in the compressor, which may be given as:

$$\dot{\Xi}_{D,Comp} = \dot{W}_6 - \left(\dot{\Xi}_7 - \dot{\Xi}_5\right) \tag{9.13}$$

The exergy efficiency of the compressor is thus given by:

$$\epsilon_{Comp} = \frac{\dot{\Xi}_7 - \dot{\Xi}_5}{\dot{W}_6} \tag{9.14}$$

The compressed H_2 is stored in the H_2 storage tank, which is the only input to the tank. Therefore, the compressed H_2 exergy is the input (fuel) exergy to the storage tank. The output from the storage tank is the H_2 exergy delivered to the fuel cell, along with any possible heat-rejection (or addition) exergy. There is exergy destruction due to thermal losses, pressure decay, and tank material interactions [15]. The tank acts as an exergy buffer [16].

The exergy balance of the storage tank is given by the following:

$$\dot{\Xi}_7 = \dot{\Xi}_8 + \dot{\Xi}_9 + \dot{\Xi}_{D,Tank} \tag{9.15a}$$

$$\dot{\Xi}_7 = \dot{\Xi}_{q,8} + \dot{\Xi}_9 + \dot{\Xi}_{D,Tank} \tag{9.15b}$$

$$\dot{\Xi}_7 = \dot{Q}_{Tank}\left(1 - \frac{T_0}{T_8}\right) + \dot{\Xi}_9 + \dot{\Xi}_{D,Tank} \tag{9.15c}$$

where $\dot{\Xi}_8 = \dot{\Xi}_{q,8} = \dot{Q}_{Tank}\left(1 - \frac{T_0}{T_8}\right)$ is the exergy rate of the possible heat rejection to the environment or addition to the tank, and $\dot{Q}_{Tank}$ is the heat loss rate in the tank. $\dot{\Xi}_9$ is the exergy rate of the H_2 output to the fuel cell.

When throttling losses are added to the tank, the exergy balance of Eq. (9.15c) becomes:

$$\dot{\Xi}_7 = \dot{Q}_{Tank}\left(1 - \frac{T_0}{T_8}\right) + \dot{\Xi}_{th} + \dot{\Xi}_{D,Tank} \tag{9.15d}$$

where $\dot{\Xi}_{th}$ is the exergy rate exiting the throttle (i.e., the compressed H_2 in the tank that has undergone the throttling process). There may also be leak losses from the tank, which will be accounted for in the event of a leakage. $\dot{\Xi}_{D,Tank}$ is the exergy destruction in the tank, given by:

$$\dot{\Xi}_{D,Tank} = \dot{\Xi}_7 - \left[\dot{Q}_{Tank}\left(1 - \frac{T_0}{T_8}\right) + \dot{\Xi}_{th}\right] \tag{9.16}$$

The exergy efficiency of the tank is computed as:

$$\epsilon_{Tank} = \frac{\dot{Q}_{Tank}\left(1 - \frac{T_0}{T_8}\right) + \dot{\Xi}_{th}}{\dot{\Xi}_7} \tag{9.17}$$

The fuel cell converts the chemical exergy of H_2 into electrical power. The input exergies are the ambient air exergy and H_2 exergy from the storage tank (throttled). Additional compressors may be required for both the H_2 and air inputs, depending on the fuel cell inlet pressures. The compressors are required when the storage tank pressure is lower than the fuel cell H_2 inlet pressure, and the air inlet pressure required is higher than the ambient pressure.

The exergy balance of the fuel cell generating electrical power while producing water as a byproduct and generating heat is given by:

$$\dot{\Xi}_9 + \dot{\Xi}_{10} = \dot{\Xi}_{11} + \dot{\Xi}_{12} + \dot{\Xi}_{13} + \dot{\Xi}_{D,FC} \tag{9.18a}$$

$$\dot{\Xi}_9 + \dot{\Xi}_{10} = \dot{\Xi}_{q,11} + \dot{\Xi}_{12} + \dot{W}_{13} + \dot{\Xi}_{D,FC} \tag{9.18b}$$

$$\dot{\Xi}_9 + \dot{\Xi}_{10} = \dot{Q}_{FC}\left(1 - \frac{T_0}{T_{11}}\right) + \dot{\Xi}_{12} + \dot{W}_{13} + \dot{\Xi}_{D,FC} \tag{9.18c}$$

where $\dot{\Xi}_{10}$ is the ambient air exergy rate. $\dot{\Xi}_{11}$ ($\dot{\Xi}_{q,11}$) is the heat exergy rate generated in the fuel cell, and $\dot{Q}_{FC}$ is the heat generation in the fuel cell. $\dot{\Xi}_{13}$ ($\dot{W}_{13}$) is the exergy rate of the electrical power generation, while $\dot{\Xi}_{D,FC}$ is the exergy destruction in the fuel cell, which is computed as:

$$\dot{\Xi}_{D,FC} = \dot{\Xi}_9 + \dot{\Xi}_{10} - \left[\dot{Q}_{FC}\left(1 - \frac{T_0}{T_{11}}\right) + \dot{\Xi}_{12} + \dot{W}_{13}\right] \tag{9.19}$$

while the exergy efficiency of the fuel cell is computed as:

$$\epsilon_{FC} = \frac{\dot{Q}_{FC}\left(1 - \frac{T_0}{T_{11}}\right) + \dot{\Xi}_{12} + \dot{W}_{13}}{\dot{\Xi}_9 + \dot{\Xi}_{10}} \tag{9.20}$$

9.2.2 The System Exergy Efficiency

The H_2 system exergy efficiency for solely generating electricity in the fuel cell without considering O_2 generated (i.e., when O_2 is vented) in the electrolyzer is computed as:

$$\epsilon_{sys} = \frac{\dot{W}_{13} + \dot{\Xi}_{11}}{\dot{\Xi}_1 + \dot{W}_2 + \dot{W}_6 + \dot{\Xi}_{10}} \tag{9.21}$$

However, when the O_2 generated in the electrolyzer is recovered, the system's exergy efficiency is thus given by:

$$\epsilon_{sys} = \frac{\dot{W}_{13} + \dot{\Xi}_3 + \dot{\Xi}_{11}}{\dot{\Xi}_1 + \dot{W}_2 + \dot{W}_6 + \dot{\Xi}_{10}} \tag{9.22}$$

9.2.3 The Hydrogen System Exergy Analysis Illustration

Following Eqs. (9.9) to (9.22), the exergy analysis is further illustrated by computing the exergy destruction and exergy efficiencies of the H_2 system. Table 9.2 presents the parameters used to illustrate the system's components for calculating fuel exergy, product exergy, exergy destruction, component exergy efficiency, and system exergy efficiency. The parameters in Table 9.2 are for illustration purposes only and do not represent a real H_2 system.

For this illustration, Eqs. (9.9) to (9.22) coupled with the parameters given in Table 9.2 are solved in Scilab computational software. The Scilab code for the system computation is given in Appendix.

The outcomes of the exergy analysis are presented in Figures 9.3–9.5. Figure 9.3 shows the fuel and product exergies in the components of the H_2 system.

The fuel exergy in the components is the resources expended to generate the product. The product exergy represents the purpose of purchasing the components. Fuel exergy, as depicted in the figure, is all higher in all the components of the system,

TABLE 9.2
Parameters for Hydrogen Conversion System Work Example

Parameter	Values	Unit
Reference Temperature (T_0)	298.15	K
Reference pressure (P_0)	101.325	kPa
Electrolyzer electrical power ($\dot{W}_2$)	100	kW
Hydrogen mass flow rate ($\dot{m}_{H2}$)	0.000582	kg/s
Electrolyzer stack temperature (T_4)	300	K
Storage tank temperature (T_8)	300	K
Electrolyzer operating pressure (P_4)	3	MPa
Compressor outlet pressure (P_7)	20	MPa
Hydrogen specific gas constant (R_{H2}) at T_4	4.124	kJ/(kg K)
Oxygen specific gas constant (R_{O2}) at T_4	0.2598	kJ/(kg K)
Storage tank heat loss ($\dot{Q}_{tank}$)	0.5	kW
Fuel cell hydrogen inlet pressure (P_9)	0.2	MPa
Hydrogen molar mass (M_{H2})	2.016	g/mol
Oxygen molar mass (M_{O2})	32	g/mol
Hydrogen specific chemical exergy (ε_{H2})	120	MJ/kg
Oxygen specific chemical exergy (ε_{O2})	0.124	MJ/kg
Electrolyzer enthalpy change (ΔH)	285.83	kJ/mol
Hydrogen heat capacity ratio (k)	1.41	
Compressor isentropic efficiency (η_{com})	75	%
Molar mass of water (M_{H2O})	18.015	g/mol
Fuel cell temperature (T_{11})	353	K
Heat capacity of water at T_{11}	4.181	kJ (kg K)

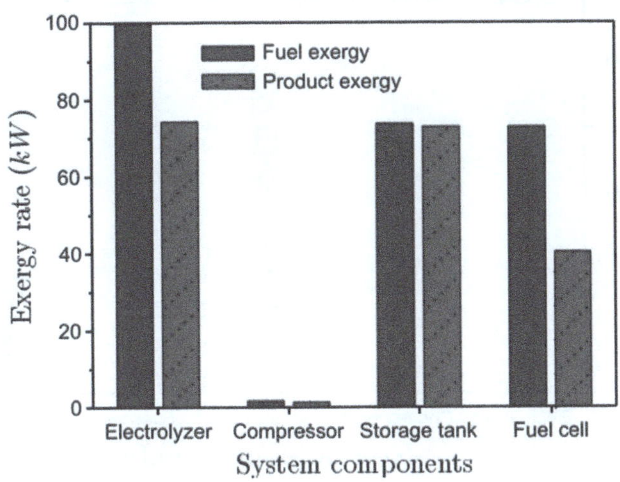

FIGURE 9.3 Fuel and product exergy rates in the components.

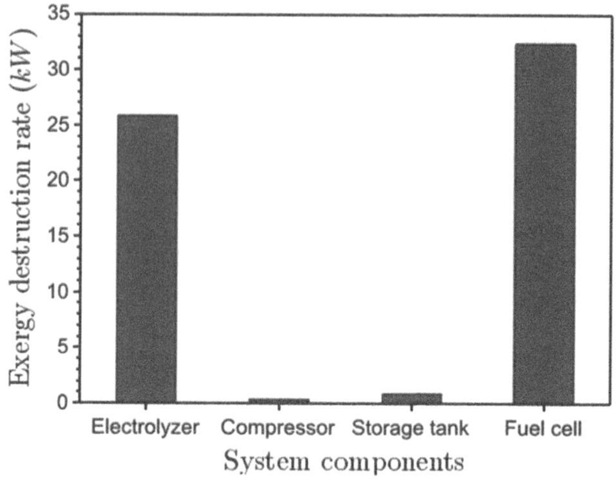

FIGURE 9.4 Exergy destruction in the components.

indicating the electrolyzer has the highest for both the fuel and product exergies when compared to other components.

This shows that higher fuel exergy (electrical energy) is required to generate H_2 and O_2 in the electrolyzer. The product exergy rate obtained in the electrolyzer indicates that from the 100 kW required by the electrolyzer, less than 80 kW is used to generate the products, while that of the fuel cell is less than 45 kW. The compressor has the lowest rates for both fuel and product exergy.

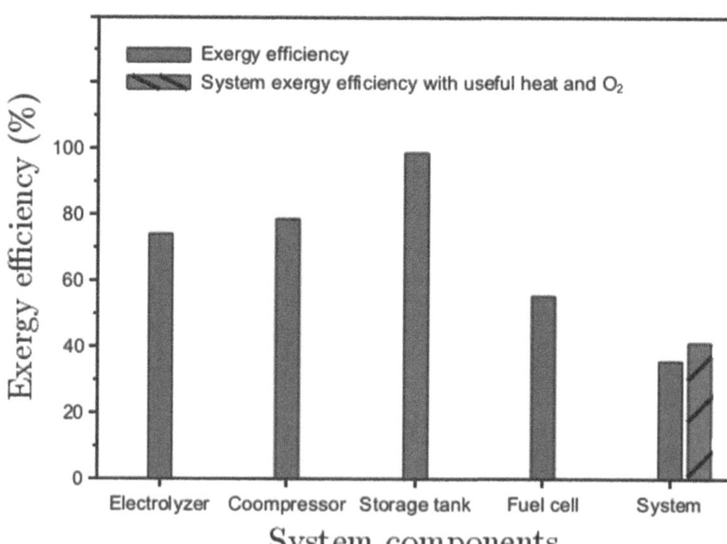

FIGURE 9.5 Exergy efficiency of the components and the overall hydrogen system.

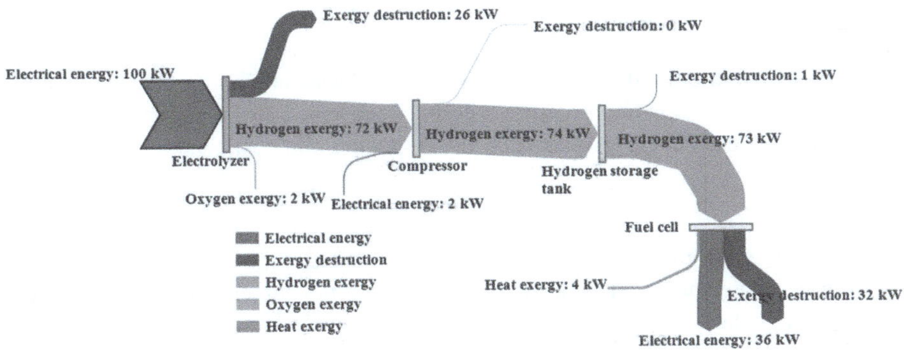

FIGURE 9.6 Exergy flow rates of the hydrogen system components.

The components' exergy destruction shown in Figure 9.4 indicates that, despite the electrolyzer having the highest product exergy rate, exergy destruction is highest in the fuel. This is primarily due to the fuel cell's higher operating temperature. Low exergy destruction rates are obtained for the compressor and storage tank.

Following the fuel exergy, product exergy, and exergy destruction in the system components, the exergy efficiency is presented in Figure 9.5. As expected, the exergy efficiencies obtained for both the compressor and storage tank are higher than for the fuel cell and electrolyzer. The high exergy efficiencies in the compressor and storage tank result from low exergy destruction in the components; reducing exergy destruction increases exergy efficiency.

The overall system exergy efficiencies for the two cases (i.e., from Eqs. (9.21) and (9.22)) are lower compared to those of the components also shown in Figure 9.5. For the second case (Eq. (9.22)), where O_2 is recovered, the exergy efficiency increased by approximately 6%. Figure 9.6 shows the exergy flow rates of the components of the hydrogen system. The figure clearly indicates that exergy destruction rates in the compressor and storage tank are negligible compared to those of the electrolyzer and fuel cell.

9.3 EXERGOECONOMIC ANALYTICAL APPROACHES

Exergoeconomics is the union of thermodynamics with economic rationality, which creates a framework in which each exergy carries a cost and each exergy destruction of useful work potential is reflected as a financial penalty. Exergoeconomics serves as a cost compass for H_2 energy systems, revealing how inefficiencies accrue financial costs. Two objectives in exergoeconomics are to quantify exergy flows within components and assign unit costs to these flows to determine optimal design, operation, and investment strategies.

Hydrogen systems, like the H_2 conversion chain described earlier—including the electrolyzer, compressor, storage tank, and fuel cell—show complex interactions between chemical exergy, physical exergy, and economic costs. Consequently, exergoeconomic analysis helps identify thermodynamic inefficiencies that are costly, highlighting design improvements that can lead to tangible financial benefits [6, 17, 18].

Exergoeconomics analysis combines two main ideas: attributing thermodynamic costs to exergy streams and creating a component-wise cost balance. In thermodynamic costing, each exergy stream is given a specific unit cost, i.e.,

$$\dot{C} = c\dot{\Xi} \tag{9.23}$$

where $\dot{C}$ is the cost rate, c is the unit exergy cost, and $\dot{\Xi}$ is the exergy flow rate.

In the component-based cost balance, every component satisfies a general exergoeconomic balance. The general cost balance of a component is formulated as [4]:

$$\dot{C}_{in} + \dot{C}_q + \dot{Z} = \dot{C}_{out} + \dot{C}_w \tag{9.24}$$

where $\dot{C}_{in}$ and $\dot{C}_{out}$ are the cost rates of input and output exergy streams, respectively. $\dot{C}_q$ and $\dot{C}_w$ are the cost rates associated with possible heat and work in the component, respectively. $\dot{Z}$ is the cost rate for the investment term associated with the component's operation and maintenance. The cost balance in Eq. (9.24) is expressed further as:

$$c_{in}\dot{\Xi}_{in} + c_q\dot{\Xi}_q + \dot{Z} = c_{out}\dot{\Xi}_{out} + c_w\dot{\Xi}_w \tag{9.25}$$

where c_{in} and c_{out} are the unit costs of exergy input and output of the control volume, respectively. c_q and c_w are the unit costs of heat and work, respectively.

The cost balance in Eqs. (9.24) and (9.25) may similarly be written in terms of the fuel and product exergies, i.e.,

$$\dot{C}_F + \dot{Z} = \dot{C}_P \tag{9.26}$$

$$c_F\dot{\Xi}_F + \dot{Z} = c_P\dot{\Xi}_P \tag{9.27}$$

where $\dot{C}_F$ and $\dot{C}_P$ are the fuel and product exergy cost rates, respectively, while c_F and c_P are the unit costs of the respective exergies.

Following the cost balances, the exergy destruction cost rate ($\dot{C}_D$) of the component is given by:

$$\dot{C}_D = c_F\dot{\Xi}_D \tag{9.28}$$

Exergoeconomic performance metrics include the total cost rate, exergoeconomic factor, and relative cost difference. The total cost rate, $\dot{C}_{tot}$, is given by:

$$\dot{C}_{tot} = \dot{Z} + \dot{C}_D \tag{9.29}$$

The exergoeoconomic factor, f, is given by:

$$f = \frac{\dot{Z}}{\dot{Z} + \dot{C}_D} \tag{9.30}$$

The exergoeconomic factor reveals the contribution of a component's investment cost to its total cost rate. The relative cost difference, r, is computed as:

$$r = \frac{c_P - c_F}{c_F} \tag{9.31}$$

The relative cost difference determines the actual cost sources in the component.

9.3.1 Exergoeconomic Methods

Exergoeconomic analysis may be performed utilizing a variety of exergoeconomic methods. The earliest approaches are the Exergy Economic Approach (EEA) [19], Thermodynamic Functional Analysis (TFA) [20], First Exergoeconomic Approach (FEA) [21], Exergy Costing Theory (ECT) [22], Moran method [23], Last-In-First-Out (LIFO) [24], Structural Analysis Approach (SAA) [25], and the Average-Cost Exergy Costing (AVCO) [26]. The most recent approach, known as Specific Exergy Costing (SPECO), was developed in 2006 [27].

The SPECO approach offers a methodical framework for the tracking of component-level exergy costs utilizing the concepts known as the fuel-product rules, which are provided as auxiliary equations. This approach has been validated, demonstrating that it is the most effective compared to other validated methods [28]. The approach is especially appealing for H_2 systems because it explicitly separates the economic penalties associated with electrical input from those related to thermodynamic losses. The advantages of the H_2 chain are the intuitive implementation, suitability for chemical exergy flows, and its provision for assigning a direct unit cost of H_2 [28].

9.3.2 Application of the SPECO Approach to the Hydrogen System

The SPECO approach to exergoeconomic analysis is applied to the case study in Section 9.2. Cost balances according to Eq. (9.25) are formulated for each component of the hydrogen system as follows:

For the electrolyzer, the exergy balance is given by:

$$c_1\dot{\Xi}_1 + c_{w,2}\dot{W}_2 + \dot{Z}_{Elect} = c_3\dot{\Xi}_3 + c_4\dot{\Xi}_4 + c_5\dot{\Xi}_5 \tag{9.32}$$

where c_1, $c_{w,2}$, c_3, c_4, and c_5 are the unit costs of the respective streams. $\dot{Z}_{Elect}$ is the investment cost rate in the electrolyzer.

Similarly, for the compressor, the cost balance is formulated as:

$$c_{w,6}\dot{W}_6 + \dot{Z}_{Com} = c_7\dot{\Xi}_7 - c_5\dot{\Xi}_5 \tag{9.33}$$

For the H_2 storage tank, the cost balance is given by:

$$c_7\dot{\Xi}_7 + \dot{Z}_{tank} = c_8\dot{\Xi}_8 + c_9\dot{\Xi}_9 \tag{9.34}$$

While for the fuel cell, the cost balance is given by:

$$c_9\dot{\Xi}_9 + c_{10}\dot{\Xi}_{10} + \dot{Z}_{FC} = c_{11}\dot{\Xi}_{11} + c_{12}\dot{\Xi}_{12} + c_{w,13}\dot{W}_{13} \tag{9.35}$$

Using the SPECO approach, the following unit costs are assigned zero, i.e.,

$$c_1 = 0, \tag{9.36}$$

$$c_{10} = 0, \tag{9.37}$$

where c_1 is the unit cost of water supplied to the electrolyzer, and it is assumed to be zero; it may not be zero in the case where the water is supplied by a pump or preheated. Also, in this case, the $\dot{\Xi}_1$ was zero, indicating the first term of Eq. (9.32) tends to zero. Also,

$$c_{w2} = c_{w6} = c_{w13} \tag{9.38}$$

where c_{w2}, c_{w6}, and c_{w13} are the unit costs of work (electricity energy) for the electrolyzer, compressor, and fuel cell, respectively.

From Eq. (9.32), $c_{w,2}$ is usually known; however, c_3, c_4, and c_5 are not. Applying the product rule, these unknowns are equated, i.e.,

$$c_3 = c_4 = c_5 \tag{9.39}$$

From Eq. (9.33), $c_{w,6}$ is usually known; however, the only unknown is c_7 and no rule is applied to the equation. From Eq. (9.34), there are two unknowns, c_8 and c_9, and the product rule is applied to the unknowns,

$$c_8 = c_9 \tag{9.40}$$

Finally, from Eq. (9.35), there are three unknowns, which are c_{11}, c_{12}, and $c_{w,13}$. Similarly, the product rule is applied,

$$c_{11} = c_{12} = c_{w,13} \tag{9.41}$$

Following the additional auxiliary equations, Table 9.3 summarizes the equations of the cost balances of the components and the additional auxiliary equations.

With the auxiliary equations, the unknown unit costs (c_3, c_4, c_5, c_7, c_8, c_9, c_{11}, c_{12}, and c_{13}) are simplified to four unknowns. The four cost balance equations are computed using a matrix formulation, as given in Eq. (9.42).

$$\begin{bmatrix} \left(\dot{\Xi}_3 + \dot{\Xi}_4 + \dot{\Xi}_5\right) & 0 & 0 & -\dot{W}_2 \\ -\dot{\Xi}_5 & \dot{\Xi}_7 & 0 & -\dot{W}_6 \\ 0 & -\dot{\Xi}_7 & \left(\dot{\Xi}_8 + \dot{\Xi}_9\right) & 0 \\ 0 & 0 & -\dot{\Xi}_9 & \left(\dot{\Xi}_{11} + \dot{\Xi}_{12} + \dot{W}_{13}\right) \end{bmatrix} \cdot \begin{bmatrix} c_3 \\ c_7 \\ c_8 \\ c_{13} \end{bmatrix} = \begin{bmatrix} \dot{Z}_{Elect} \\ \dot{Z}_{Com} \\ \dot{Z}_{tank} \\ \dot{Z}_{fuc} \end{bmatrix} \tag{9.42}$$

TABLE 9.3
The Components Auxiliary Equations Using the SPECO Approach

Component	Cost Balance	Auxiliary Equations
Electrolyzer	$c_1\dot{\Xi}_1 + c_{w,2}\dot{W}_2 + \dot{Z}_{Elect} = c_3\dot{\Xi}_3 + c_4\dot{\Xi}_4 + c_5\dot{\Xi}_5$	$c_1 = 0$, $c_{w2} = c_{w6} = c_{w13}$ $c_3 = c_4 = c_5$ (product-rule)
Compressor	$c_{w,6}\dot{W}_6 + \dot{Z}_{Com} = c_7\dot{\Xi}_7 - c_5\dot{\Xi}_5$	None
Storage tank	$c_7\dot{\Xi}_7 + \dot{Z}_{tank} = c_8\dot{\Xi}_8 + c_9\dot{\Xi}_9$	$c_8 = c_9$
Fuel cell	$c_9\dot{\Xi}_9 + c_{10}\dot{\Xi}_{10} + \dot{Z}_{FC} = c_{11}\dot{\Xi}_{11} + c_{12}\dot{\Xi}_{12} + c_{w,13}\dot{W}_{13}$	$c_{11} = c_{12} = c_{w,13}$ $c_{10} = 0$

9.4 EXERGOENVIRONMENTAL ASSESSMENT APPROACHES

One of the most effective approaches for quantitatively understanding the environmental costs associated with energy conversion systems is exergoenvironmental analysis [29]. This analysis introduces the compass of exergy, whereas conventional Life Cycle Assessment (LCA) assesses environmental impacts in a framework based on mass and energy [30, 31].

The exergoenvironmental approach demonstrates how thermodynamic inefficiencies give rise to real environmental pressure points by tracing the scarcity, quality, and irreversible dissipation of resources within a thermodynamic system. For H_2 systems in which each component has input exergy and output irreversibilities, the approach enables the integration of system design and environmental sustainability in a scientifically grounded and transparent manner. Generally, this approach involves three concrete steps: exergy evaluation, LCA, and exergoevironmental assessment.

In this section, the essential approaches, indicator formulations, and application trends of exergoenvironmental assessment in the context of the hydrogen conversion chain, as discussed in the previous sections, will be expanded. The exergoenvironmental framework assigns environmental impact rates to both exergy flow streams and exergy destruction. The approach introduces two central constructs: the environmental impact of exergy streams and that of exergy destruction.

The general environmental impact balance for a component is given by [4, 32]:

$$\sum_{j=1}^{n} \dot{B}_{j,in} + \dot{Y} = \sum_{j=1}^{m} \dot{B}_{j,out} \tag{9.43a}$$

$$\sum_{j=1}^{n} \left(b_j \dot{\Xi}_j\right)_{in} + \dot{Y} = \sum_{j=1}^{m} \left(b_j \dot{\Xi}_j\right)_{out} \tag{9.43b}$$

where $\dot{B}_{j,in}$ and $\dot{B}_{j,out}$ are the environmental impacts associated with input and output exergy streams j, respectively. b_j is the unit environmental impact. n and m are the number of streams of the input and output of the control volume, respectively. $\dot{Y}$ is the component-related environmental impact, given by [32]:

$$\dot{Y} = \dot{Y}^{CO} + \dot{Y}^{OM} + \dot{Y}^{DS} \tag{9.44}$$

where $\dot{Y}^{CO}$ is the environmental impact associated with the construction phases, $\dot{Y}^{OM}$ is the environmental impact associated with operation and maintenance, and $\dot{Y}^{DS}$ is the environmental impact associated with disposal.

The environmental impact balance of Eq. (9.43) may be given in terms of fuel and product concepts, i.e.,

$$\dot{B}_F + \dot{Y} = \dot{B}_P \tag{9.45a}$$

$$b_F\dot{\Xi}_F + \dot{Y} = b_P\dot{\Xi}_P \tag{9.45b}$$

where $\dot{B}_F$ and $\dot{B}_P$ are the environmental impacts associated with fuel and product exergies, respectively, while b_F and b_P are their respective unit environmental impacts.

The environmental impact of exergy destruction is given by:

$$\dot{B}_D = b_F\dot{\Xi}_P \tag{9.46}$$

Following the formulation of this analysis in Eqs. (9.43) to (9.46), the exergoenvironmental performance metrics include the overall environmental impact ($\dot{B}_{ov}$), relative difference of unit environmental impact (r_b), and exergoenvironmental factor (f_b). These are given in the following equations:

$$\dot{B}_{ov} = \dot{Y} + \dot{B}_D \tag{9.47}$$

$$r_b = \frac{b_P - b_F}{b_F} \tag{9.48}$$

$$f_b = \frac{\dot{Y}}{\dot{Y} + \dot{B}_D} = \frac{\dot{Y}}{\dot{B}_{ov}} \tag{9.49}$$

9.4.1 Applying Exergoenvironmental Analysis to the Hydrogen System

Exergoenvironmental analysis is applied to the case study of the H_2 conversion system of the previous section. Following Eq. (9.45), the environmental impact balances are formulated similarly to exrgoeconomic analysis as follows:

For the electrolyzer, the environmental impact balance is given by:

$$\dot{B}_1 + \dot{B}_2 + \dot{Y}_{Elect} = \dot{B}_3 + \dot{B}_4 + \dot{B}_5 \tag{9.50a}$$

$$b_1\dot{\Xi}_1 + b_{w,2}\dot{W}_2 + \dot{Y}_{Elect} = b_3\dot{\Xi}_3 + b_4\dot{\Xi}_4 + b_5\dot{\Xi}_5 \tag{9.50b}$$

TABLE 9.4
Environmental Impact Balances and Auxiliary Equations

Component	Environmental Impact Balance	Auxiliary Equations
Electrolyzer	$b_1\dot{\Xi}_1 + b_{w,2}\dot{W}_2 + \dot{Y}_{Elect} = b_3\dot{\Xi}_3 + b_4\dot{\Xi}_4 + b_5\dot{\Xi}_5$	$b_1 = 0$, $b_{w2} = b_{w6} = b_{w13}$ $b_3 = b_4 = b_5$ (Product-rule)
Compressor	$b_{w,6}\dot{W}_6 + \dot{Y}_{Com} = b_7\dot{\Xi}_7 - b_5\dot{\Xi}_5$	None
Storage tank	$b_7\dot{\Xi}_7 + \dot{Y}_{tank} = b_8\dot{\Xi}_8 + b_9\dot{\Xi}_9$	$b_8 = b_9$
Fuel cell	$b_9\dot{\Xi}_9 + b_{10}\dot{\Xi}_{10} + \dot{Y}_{FC} = b_{11}\dot{\Xi}_{11} + b_{12}\dot{\Xi}_{12} + b_{w,13}\dot{W}_{13}$	$b_{11} = b_{12} = b_{w,13}$ $b_{10} = 0$

For the compressor,

$$b_{w,6}\dot{W}_6 + \dot{Y}_{Com} = b_7\dot{\Xi}_7 - b_5\dot{\Xi}_5 \tag{9.51}$$

For the storage tank,

$$b_7\dot{\Xi}_7 + \dot{Y}_{tank} = b_8\dot{\Xi}_8 + b_9\dot{\Xi}_9 \tag{9.52}$$

For the fuel cell, the environmental balance is formulated as:

$$b_9\dot{\Xi}_9 + b_{10}\dot{\Xi}_{10} + \dot{Y}_{FC} = b_{11}\dot{\Xi}_{11} + b_{12}\dot{\Xi}_{12} + b_{w,13}\dot{W}_{13} \tag{9.53}$$

From Eqs. (9.50) to (9.53), there are nine unknown unit environmental impacts (i.e., b_3, b_4, b_5, b_7, b_8, b_9, b_{11}, b_8, and $b_{w,13}$), whereas there are four equations of the four components to be solved. Hence, the fuel and product rule is applied similarly to those of the exergoeconomic analysis to generate auxiliary equations. To solve the four equations simultaneously, the auxiliary equations are summarized in Table 9.4.

With Table 9.4, the equations are solved in Eq. (9.54),

$$\begin{bmatrix} \left(\dot{\Xi}_3 + \dot{\Xi}_4 + \dot{\Xi}_5\right) & 0 & 0 & -\dot{W}_2 \\ -\dot{\Xi}_5 & \dot{\Xi}_7 & 0 & -\dot{W}_6 \\ 0 & -\dot{\Xi}_7 & \left(\dot{\Xi}_8 + \dot{\Xi}_9\right) & 0 \\ 0 & 0 & -\dot{\Xi}_9 & \left(\dot{\Xi}_{11} + \dot{\Xi}_{12} + \dot{W}_{13}\right) \end{bmatrix} \cdot \begin{bmatrix} b_3 \\ b_7 \\ b_8 \\ b_{13} \end{bmatrix} = \begin{bmatrix} \dot{Y}_{Elect} \\ \dot{Y}_{Com} \\ \dot{Y}_{tank} \\ \dot{Y}_{fuc} \end{bmatrix} \tag{9.54}$$

9.5 HYDROGEN GENERATION APPLICATIONS AND UTILIZATIONS

One of the most valuable and versatile energy carriers for a climate-constrained world is H_2 [33]. Despite H_2 not being a primary energy source, it has a unique cross-cutting function in contemporary energy systems due to its capacity to transport, store, and transform energy across sectors [34].

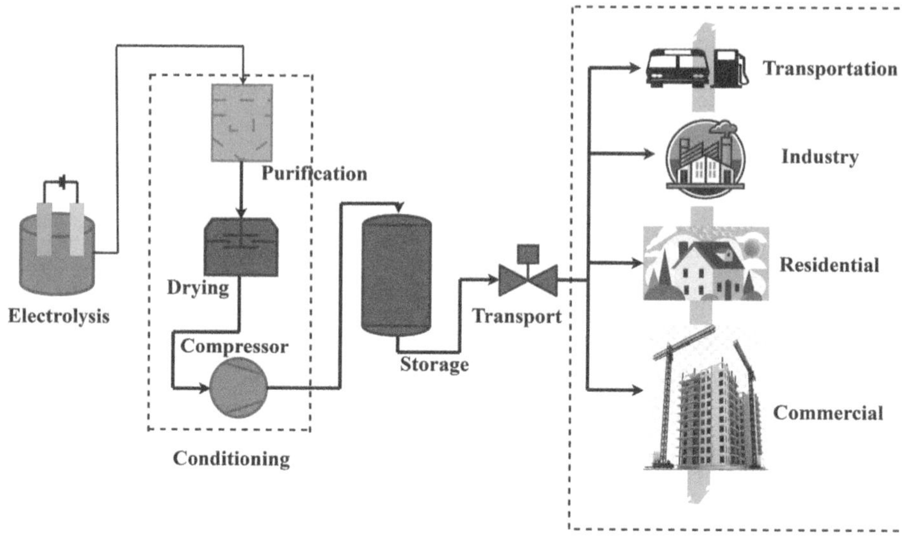

FIGURE 9.7 Electrolytic hydrogen value chain.

Hydrogen forms a bridge between renewable energy, emerging green technologies, and industrial decarbonization, serving roles ranging from clean industrial feedstock and long-duration energy storage to applications in mobility and chemical production [35].

Figure 9.7 depicts the H_2 value chain from production to utilization. Hydrogen is generated by three basic processes: thermochemical, electrochemical, and biological processes. Depending on the energy source and feedstock, each process generates H_2 with a different carbon footprint.

Electrolytic H_2 generation technologies include alkaline, Proton Exchange Membrane (PEM), and Solid Oxide Electrolysis (SOE). The PEM operates at about 50°C to 80°C and it has a dynamic response, which is suitable for renewable integration [36]. The alkaline electrolysis is more mature, cost-effective, and industrially adopted [37]. The SOE is high-temperature electrolysis, which has superior theoretical efficiency due to the integration of thermal energy [38].

The electrolytic processes, powered by low-carbon electrical energy, contribute to near-zero life-cycle emissions. Green H_2 deployment is dependent on the availability of renewable power, electrolyzer efficiency, water treatment, and stack longevity [39].

The thermochemical H_2 pathways include coal and biomass gasification, natural gas reforming, and water-splitting cycles, such as ceria-based cycles. Steam methane reforming produces blue H_2, which is particularly relevant during periods of transition. The biomass pathways have the potential to provide carbon-neutral H_2 when sourced sustainably [40–42].

The biological and photocatalytic routes are advancing through emerging research, including dark fermentation, photo-fermentation, and photocatalytic water splitting. Despite their promise, these technologies are still in the early stages with low conversion efficiencies.

Hydrogen is utilized in power systems, mobility and transportation, industrial processes, and residential and commercial sectors. In power systems, H_2 serves as a long-duration, high-capacity storage medium that complements the fluctuating output of renewable energy sources. Hydrogen energy can be utilized in power-to-hydrogen-to-power systems (P2H2P) [33].

In P2H2P systems, excess energy from renewable electricity electrolyzers generates H_2, which is stored and then reconverted in fuel cells, gas turbines, and combined-cycle power plants [43]. Hydrogen can be used for grid stability to strengthen grid flexibility by assisting frequency regulation, providing reserve capacity, supporting black-start capabilities, and coupling the energy sector [44]. It can also be used in hybrid renewable-H_2 systems, such as wind-H_2 farms, offshore H_2 generation utilizing floating platforms, and solar-driven electrolyzer clusters [45].

In mobility and transportation, H_2 offers a solution for decarbonizing mobility where battery limitations exist. It can be used in road, rail, maritime, and aviation transport. Figure 9.8 depicts the application of H_2 in the transportation sector.

Applications in industrial processes are poised to become the largest consumer of H_2. They find applications in steelmaking, ammonia and chemicals, refining, and high-temperature industrial heat [46–49]. One-third of the world's demand for fossil energy comes from industrial heat; H_2 offers a practical pathway to decarbonize high-grade heat.

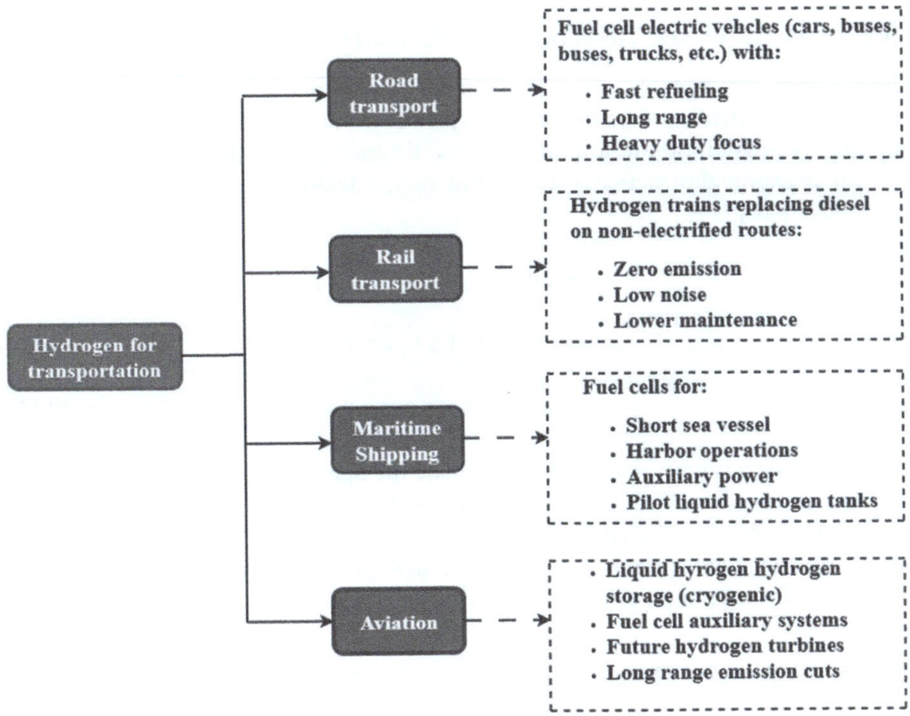

FIGURE 9.8 Hydrogen utilization across various transport modes.

Furthermore, in residential and commercial applications, H_2 supports power, heating, and combined heat and power (CHP) systems. Stationary fuel cell systems, such as the low-temperature fuel cells, are deployed for residential CHP applications, commercial microgrids, hospitals, and data centers.

9.6 OPTIMIZATION AND SUSTAINABILITY INDICES

The imperative to optimize H_2 technologies' performance grows as they move from laboratory to gigawatt-scale deployment [50]. The search for a balanced operating space where thermodynamic rigor, environmental responsibility, and economic viability converge defines optimization in H_2 and is not necessarily a mathematical exercise [51].

The three balanced operating space forms the foundation of sustainable H_2 engineering. In practical applications, engineers utilize a range of multidimensional indicators to assess H_2 generation, storage, and consumption systems, measuring cost-effectiveness, resource efficiency, and long-term environmental impacts [18].

Component degradation, thermodynamic losses, fluctuating renewable supply, and transient energy demands interact in a complex manner to operate H_2 systems. There are three broad optimization approaches, viz., deterministic thermodynamic optimization, stochastic optimization, and multi-objective optimization [6, 52]. These are discussed next.

9.6.1 Deterministic Thermodynamic Optimization

First principles equations with fixed boundary equations are the foundation of deterministic optimization. The procedure typically entails increasing efficiency, minimizing irreversibilities, or minimizing specific energy consumption [6].

For an exergy destruction objective function, a deterministic problem is generally formulated as [53]:

$$\text{minimize } f(x) = \dot{\Xi}_D(x) \tag{9.55a}$$

$$\text{subject to } g_i(x) \leq 0,\ i = 1, \cdots, k \tag{9.55b}$$

$$x_{\text{min}} \leq x \leq x_{\text{max}} \tag{9.55c}$$

where x is the design variable, which could be the operating temperature or pressure, current density, electrode thickness, compression ratio, or storage tank pressure. The general objective function is also illustrated in Figure 9.9. Commonly employed deterministic approaches include the Newton-Raphson, Lagrangian, and Sequential Quadratic Programming (SQP) methods.

9.6.2 Stochastic Optimization

Hydrogen systems face uncertainties, including variability in renewable electricity, material defects, fluctuating H_2 demand, and varying degradation rates. Such

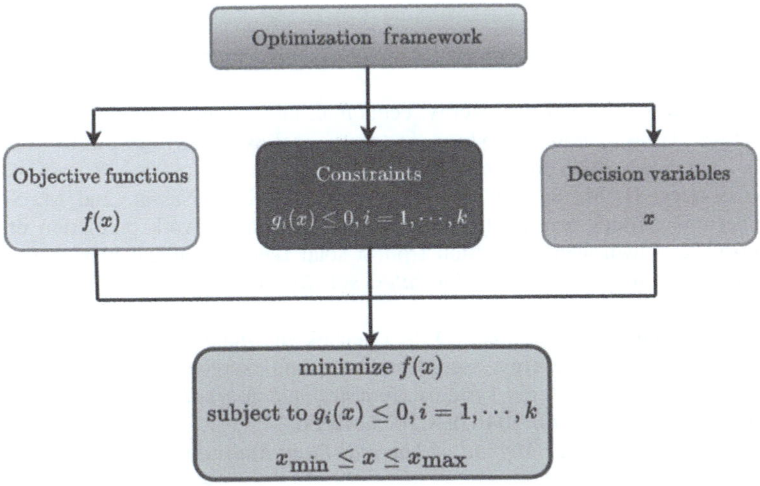

FIGURE 9.9 Optimization of hydrogen systems.

variables are accounted for by the Stochastic optimization, where probabilistic models are introduced. Some of these approaches are the Genetic Algorithm (GA), Simulated Annealing (SA), Particle Swarm Optimization (PSO), Bayesian optimization, and Monte Carlo-based uncertainty propagation [54]. To effectively optimize electrolyzer operating windows where overpotentials, stack lifetime, and current density must collectively be considered, the stochastic approach is particularly effective.

9.6.3 Multi-objective Optimization

There is always more than one objective in H_2 system problems. Designers focus on simultaneously maximizing efficiencies of the system components, minimizing their exergy destruction, minimizing the levelized cost of H_2, reducing emissions from the components, and extending their lifespan. These objectives are captured in the multi-objective optimization [55], i.e.,

$$\min\left[f_1(x),\ f_2(x),\ \cdots,\ f_n(x),\right] \tag{9.56}$$

where f_i represent the ith criterion, which may correspond to thermodynamic, economic, or environmental criteria.

Modern design may require integrating optimization of H_2 systems with sustainability indices. Such a framework involves defining system boundaries (such as the H_2 conversion case study in Section 9.2), performing exergy analysis to determine exergy destruction as a result of irreversibilities, applying an optimization technique, evaluating sustainability indices, and iterating until convergence on both sustainability and technical criteria. With this, the integrated framework guarantees that systems are resource-conscious, efficient, and future-proof.

REFERENCES

1. I. Dincer, and M. Rosen, *Exergy: Energy, Environment and Sustainable Development*, 3rd ed., Elsevier, 2020.
2. S. Esfandi *et al.*, 'Energy, exergy, economic, and exergoenvironmental analyses of a novel hybrid system to produce electricity, cooling, and syngas', *Energies*, vol. 13, no. 23, p. 6453, 2020. doi: 10.3390/en13236453.
3. A. El Jery, H. M. Salman, R. M. Al-Khafaji, M. F. Nassar, and M. Sillanpää, 'Thermodynamics investigation and artificial neural network prediction of energy, exergy, and hydrogen production from a solar thermochemical plant using a polymer membrane electrolyzer', *Molecules*, vol. 28, no. 6, p. 2649, 2023. doi: 10.3390/molecules28062649.
4. U. G. Azubuike, H. O. Njoku, O. V. Ekechukwu, and T. C. Jen, 'A comprehensive review on advanced exergy, exergoeconomic, and exergoenvironmental analyses for sustainable thermal energy systems', *International Journal of Exergy*, vol. 48, no. 1, pp. 16–53, 2025. doi: 10.1504/IJEX.2025.148556.
5. T.J. Kotas, *The Exergy Method of Thermal Plant Analysis*, Butterworth-Heinemann, 2013.
6. A. Bejan, G. Tsatsaronis, and M. J. Moran, *Thermal Design and Optimization*, Wiley, 1996.
7. A. Khosravi, R. N. N. Koury, L. Machado, and J. J. G. Pabon, 'Energy, exergy and economic analysis of a hybrid renewable energy with hydrogen storage system', *Energy*, vol. 148, pp. 1087–1102, 2018. doi: 10.1016/j.energy.2018.02.008.
8. O. V. Marchenko, and S. V. Solomin, 'The future energy: Hydrogen versus electricity', *International Journal of Hydrogen Energy*, vol. 40, no. 10, pp. 3801–3805, 2015. doi: 10.1016/j.ijhydene.2015.01.132.
9. O. V. Marchenko, and S. V. Solomin, 'Modeling of hydrogen and electrical energy storages in wind/PV energy system on the Lake Baikal coast', *International Journal of Hydrogen Energy*, vol. 42, no. 15, pp. 9361–9370, 2017. doi: 10.1016/j.ijhydene.2017.02.076.
10. S. Hamdy, *Cryogenic Energy Storage Systems: An Exergy-based Evaluation and Optimization*. Technische Universität Berlin, 2019. [Online]. Available: https://books.google.co.za/books?id=hdBCzQEACAAJ
11. M. Ozsaban, A. Midilli, and I. Dincer, 'Exergy analysis of a high pressure multistage hydrogen gas storage system', *International Journal of Hydrogen Energy*, vol. 36, no. 17, pp. 11440–11450, 2011. doi: 10.1016/j.ijhydene.2011.03.087.
12. A. H. Hassan, Z. Liao, K. Wang, M. M. Abdelsamie, C. Xu, and Y. Wang, 'Exergy and exergoeconomic analysis for the proton exchange membrane water electrolysis under various operating conditions and design parameters', *Energies*, vol. 15, no. 21, p. 8247, 2022. doi: 10.3390/en15218247.
13. T. Taner, 'Energy and exergy analyze of PEM fuel cell: A case study of modeling and simulations', *Energy*, vol. 143, pp. 284–294, 2018. doi: 10.1016/j.energy.2017.10.102.
14. R. O'Hayre, S. Cha, W. Colella, and F. B. Prinz, *Fuel Cell Fundamentals*, 1st ed., Wiley, 2016. doi: 10.1002/9781119191766.
15. Z. Yanxing, G. Maoqiong, Z. Yuan, D. Xueqiang, and S. Jun, 'Thermodynamics analysis of hydrogen storage based on compressed gaseous hydrogen, liquid hydrogen and cryo-compressed hydrogen', *International Journal of Hydrogen Energy*, vol. 44, no. 31, pp. 16833–16840, 2019. doi: 10.1016/j.ijhydene.2019.04.207.
16. H. Mejer, and S. E. Jørgensen, 'Exergy and ecological buffer capacity', in *State-of-the-Art in Ecological Modelling*, Elsevier, 1979, pp. 829–846. doi: 10.1016/B978-0-08-023443-4.50042-7.

17. G. Tsatsaronis, 'Thermoeconomic analysis and optimization of energy systems', *Progress in Energy and Combustion Science*, vol. 19, no. 3, pp. 227–257, 1993. doi: 10.1016/0360-1285(93)90016-8.
18. M. A. Rosen, and I. Dincer, 'Exergy–cost–energy–mass analysis of thermal systems and processes', *Energy Conversion and Management*, vol. 44, no. 10, pp. 1633–1651, 2003. doi: 10.1016/S0196-8904(02)00179-6.
19. R. A. Gaggioli, Ed., *Efficiency and Costing: Second Law Analysis of Processes*, in ACS symposium series, no. 235, American Chemical Society, 1983.
20. C. Frangopoulos, 'Thermoeconomic Functional Analysis: A Method for Optimal Design or Improvement of Complex Thermal Systems', PhD, Georgia Institute of Technology, 1983.
21. G. Tsatsaronis, and M. Winhold, 'Exergoeconomic analysis and evaluation of energy-conversion plants—II. Analysis of a coal-fired steam power plant', *Energy*, vol. 10, no. 1, pp. 81–94, 1985. doi: 10.1016/0360-5442(85)90021-0.
22. M. Von Spakovsky, 'A Practical Generalized Analysis Approach to the Optimal Thermoeconomic Design and Improvement of Real-World Thermal Systems', PhD, Georgia Institute of Technology, 1986.
23. M. J. Moran, *Availability Analysis: A Guide to Efficient Energy Use*, ASME Press, 1989.
24. G. Tsatsaronis, American Society of Mechanical Engineers Winter Meeting, & American Society of Mechanical Engineers Advanced Energy Systems Division. (1990). Computer-aided energy systems analysis: presented at the Winter Annual Meeting of the American Society of Mechanical Engineers, Dallas, Texas, November 25-30, 1990. ASME.
25. A. Valero, A. Torres, and L. Serra, 'A general theory of thermoeconomics: Part I: Structural analysis', presented at the International Symposium on Efficiency, Costs, Optimization and Simulation of Energy Systems, ECOS'92, Zaragoza, Spain, 1992, pp. 137–145.
26. G. Tsatsaronis, and J. Pisa, 'Exergoeconomic evaluation and optimization of energy systems—Application to the CGAM problem', *Energy*, vol. 19, no. 3, pp. 287–321, 1994. doi: 10.1016/0360-5442(94)90113-9.
27. A. Lazzaretto, and G. Tsatsaronis, 'SPECO: A systematic and general methodology for calculating efficiencies and costs in thermal systems', *Energy*, vol. 31, no. 8–9, pp. 1257–1289, 2006. doi: 10.1016/j.energy.2005.03.011.
28. M. Gorji-Bandpy, and V. Ebrahimian, 'Exergoeconomic analysis of gas turbine power plants', *International Energy Journal*, vol. 7, no. 1, pp. 35–41, 2006.
29. J. Buchgeister, 'Exergoenvironmental analysis – A new approach to support the design for environment of chemical processes?', *Chemical Engineering & Technology*, vol. 33, no. 4, pp. 593–602, 2010. doi: 10.1002/ceat.201000006.
30. M. Aghbashlo *et al*, 'Exergoenvironmental analysis of bioenergy systems: A comprehensive review', *Renewable and Sustainable Energy Reviews*, vol. 149, p. 111399, 2021. doi: 10.1016/j.rser.2021.111399.
31. E. J. C. Cavalcanti, 'Exergoeconomic and exergoenvironmental analyses of an integrated solar combined cycle system', *Renewable and Sustainable Energy Reviews*, vol. 67, pp. 507–519, 2017. doi: 10.1016/j.rser.2016.09.017.
32. L. Meyer, G. Tsatsaronis, J. Buchgeister, and L. Schebek, 'Exergoenvironmental analysis for evaluation of the environmental impact of energy conversion systems', *Energy*, vol. 34, no. 1, pp. 75–89, 2009. doi: 10.1016/j.energy.2008.07.018.
33. IEA, 'The Future of Hydrogen', Paris, 2019. [Online]. Available: https://www.iea.org/reports/the-future-of-hydrogen
34. I. Staffell *et al*, 'The role of hydrogen and fuel cells in the global energy system', *Energy & Environmental Science*, vol. 12, no. 2, pp. 463–491, 2019. doi: 10.1039/C8EE01157E.
35. M. El-Shafie, 'Hydrogen production by water electrolysis technologies: A review', *Results in Engineering*, vol. 20, p. 101426, 2023. doi: 10.1016/j.rineng.2023.101426.

36. M. Carmo, D. L. Fritz, J. Mergel, and D. Stolten, 'A comprehensive review on PEM water electrolysis', *International Journal of Hydrogen Energy*, vol. 38, no. 12, pp. 4901–4934, 2013. doi: 10.1016/j.ijhydene.2013.01.151.
37. K. Zeng, and D. Zhang, 'Recent progress in alkaline water electrolysis for hydrogen production and applications', *Progress in Energy and Combustion Science*, vol. 36, no. 3, pp. 307–326, 2010. doi: 10.1016/j.pecs.2009.11.002.
38. S. K. Mazloomi, and N. Sulaiman, 'Influencing factors of water electrolysis electrical efficiency', *Renewable and Sustainable Energy Reviews*, vol. 16, no. 6, pp. 4257–4263, 2012. doi: 10.1016/j.rser.2012.03.052.
39. R. Bhandari, C. A. Trudewind, and P. Zapp, 'Life cycle assessment of hydrogen production via electrolysis – A review', *Journal of Cleaner Production*, vol. 85, pp. 151–163, 2014. doi: 10.1016/j.jclepro.2013.07.048.
40. I. Dincer, and C. Acar, 'Review and evaluation of hydrogen production methods for better sustainability', *International Journal of Hydrogen Energy*, vol. 40, no. 34, pp. 11094–11111, 2015. doi: 10.1016/j.ijhydene.2014.12.035.
41. Y. T. Shah, *Chemical Energy from Natural and Synthetic Gas*, CRC Press, 2017. doi: 10.1201/9781315302355.
42. C. Acar, and I. Dincer, 'Review and evaluation of hydrogen production options for better environment', *Journal of Cleaner Production*, vol. 218, pp. 835–849, 2019. doi: 10.1016/j.jclepro.2019.02.046.
43. B. P. Koirala, H. Cai, J. De Koning, P. Heer, and K. Orehounig, 'Flexibility assessment of power-hydrogen-power (P2H2P) system in multi-energy districts', *Journal of Physics: Conference Series*, vol. 2600, no. 7, p. 072007, 2023. doi: 10.1088/1742-6596/2600/7/072007.
44. M. Ahang, P. C. D. Granado, and A. Tomasgard, 'Investments in green hydrogen as a flexibility source for the European power system by 2050: Does it pay off?', *Applied Energy*, vol. 378, p. 124656, 2025. doi: 10.1016/j.apenergy.2024.124656.
45. International Renewable Energy Agency. *Green Hydrogen: A Guide to Policy Making*. International Renewable Energy Agency, 2020.
46. V. Vogl, M. Åhman, and L. J. Nilsson, 'Assessment of hydrogen direct reduction for fossil-free steelmaking', *Journal of Cleaner Production*, vol. 203, pp. 736–745, 2018. doi: 10.1016/j.jclepro.2018.08.279.
47. M. Appl, *Ammonia: Principles and Industrial Practice*, Wiley-VCH, 1999. doi: 10.1002/9783527613885.
48. J. G. Speight, *The Refinery of the Future*, 2nd ed. Gulf Professional Publishing, 2020.
49. I. Moradpoor, T. Koivunen, S. Syri, and J. Hirvonen, 'The benefits of integrating industrial hydrogen production with district heating in cold climates with different building renovation levels', *Energy*, vol. 303, p. 131953, 2024. doi: 10.1016/j.energy.2024.131953.
50. IEA, 'Global Hydrogen Review 2022', 2022. [Online]. Available: https://www.iea.org/reports/global-hydrogen-review-2022
51. M. Dincer, and M. Agelin-Chaab, 'Sustainability analysis of electrolysis based green hydrogen production pathways: A life cycle perspective', *International Journal of Hydrogen Energy*, vol. 138, pp. 617–625, 2025. doi: 10.1016/j.ijhydene.2025.05.150.
52. T. Tezer, 'Multi-objective optimization of hybrid renewable energy systems with green hydrogen integration and hybrid storage strategies', *International Journal of Hydrogen Energy*, vol. 142, pp. 1249–1271, 2025. doi: 10.1016/j.ijhydene.2025.03.006.
53. M. Cavazzuti, 'Deterministic optimization', in *Optimization Methods*, Springer, 2013, pp. 77–102. doi: 10.1007/978-3-642-31187-1_4.
54. P. Nikolaidis, and A. Poullikkas, 'A comparative overview of hydrogen production processes', *Renewable and Sustainable Energy Reviews*, vol. 67, pp. 597–611, 2017. doi: 10.1016/j.rser.2016.09.044.

55. M. Ming, R. Wang, Y. Zha, and T. Zhang, 'Multi-objective optimization of hybrid renewable energy system using an enhanced multi-objective evolutionary algorithm', *Energies*, vol. 10, no. 5, p. 674, 2017. doi: 10.3390/en10050674.

APPENDIX

```
          ///// Scilab code for computation of the system in Fig. (9.2)  //////
                                 ////////       Constants        //////
T_0 = 298.15;                      // reference temperature (K)
P_0 = 101325;                     // reference temperature (Pa)
W_2 = 100e3;                     // Electrical power to the electrolyzer (W)
m_5 = 5.82e-4;                   // mass flow rate hydrogen (assumption) (kg/s)
T_4 = 300;                         // Electrolyzer stack temperature (K)
T_8 = T_4;                        // Temperature of the storage tank (assumption) K
P_4 = 3e6; P_5 = P_4;  // Electrolyzer operating pressure (Pa)
P_7 = 20e6;                      // Compressor outlet pressure (Pa)
R_H2 = 4124;                  // Specific gas constant for H2 (Jkg-1K-1)
R_O2 = 259.8;                // Specific gas constant for H2 (Jkg-1K-1)
Q_Tan = 0.5e3;               // heat loss in the tank (use Q = UA(T0 - T) for
accurate calculation) W
P_9 = (2/10)*1e6        // pressure of fuel cell inlet (Pa)
 M_5 = 2.016e-3;         // Molar mass of hydrogen (kg/mol)
M_3 = 32e-3;                 // Molar mass of oxygen of oxygen (kg/mol)
e_5_ch = 120e6;           // Specific chemical exergy of hydrogen (J/kg).
e_3_ch = 0.124e6;       // Specific chemical exergy of oxygen (J/kg)
delta_H = 285.83e3;  // Enthalpy change in the electrolyzer (J/mol)
k = 1.41;                           // Heat capacity ratio for hydrogen
eta_eff = 0.75;                // Isentropic efficiency of the compressor
M_H2O = 0.018015; // Molar mass of H2O (kg mol-1)
c_p = 4181;                      // heat capacity of water at T=353 (Jkg-1K-1)
 T_11 = 353;                     // Fuel cell temperature assumed (K)

                            //// mass of oxygen flow rate ////
  n_H2 = m_5/M_5;          // molar flow rate of hydrogen (mol/s)
  n_O2 = n_H2/2;              // molar flow of oxygen (stoichiometric) (mol/s)
  m_3 = n_O2 * M_3;    // mass flow rate of oxygen (kg/s)

      ///// Chemical and physical exergy of water at ambient conditions (stream
1)  ////
 E_1 = 0;      // Chemical and physical exergy of water at ambient conditions =
zero (J/s)
                          // Not zero if the stream conditions exceed those of ambient
(in case of preheating, etc)

    //// Total exergy rate of the hydrogen stream (stream 5) ////
e_5_ph = R_H2*T_0*log(P_5/P_0);      // Specific exergy of hydrogen stream
(stream 5) (J/kg)
E_5 = m_5*(e_5_ch + e_5_ph);            // Total Exergy rate of hydrogen stream
(J/s)

  //// Total exergy rate of the oxygen stream (stream 3) ////
e_3_ph = R_O2*T_0*log(P_5/P_0);      // Specific exergy of stream 3 (J/kg)
E_3 = m_3*(e_3_ch + e_3_ph);            // Total Exergy rate of oxygen (J/s)

  //// Electrolyzer heat rate  ////
Q_Ele = W_2 - (n_H2*delta_H);    // Heat rate of electrolyzer (J/s)

  //// Work done by the compressor ////
w_is = ((R_H2*T_4)/(k - 1))*(((P_7/P_5)^((k -1)/k)) - 1); // isentropic work (J/
kg)
W_ac = m_5*(w_is/eta_eff); W_6 = W_ac;                         // Actual work of
the compressor (J/s)
```

```
  //// The total exergy of the compressor output (stream 7)    ////
E_7_p = m_5*(R_H2*T_0*log(P_7/P_5));
                                                              // Physical exergy
(ideal gas pressure term approximation) J/s
                                                              // This is approx.
computation to capture the pressure term
E_7 = E_5 + E_7_p;                                  // Total exergy of the stream 7 (J/s)

   //// Throttling exergy for the fuel cell inlet conditions ///
E_th = E_5 + m_5*(R_H2*T_0*log(P_9/P_0));
E_9 = E_th;                                          //Throttle outlet exergy (J/s)

      ////     Components exergy computations ////
                //// Electrolyzer   ////
  Ex_Ele_F =  E_1 + W_2;                                                 // Fuel
exergy of the electrolyzer (J/s)
  Ex_Ele_P = E_3 + (Q_Ele*(1 - (T_0/T_4))) + E_5;  // Product exergy of the
electrolyzer (J/s)
  Ex_d_Ele = Ex_Ele_F - Ex_Ele_P;                                    // Exergy
destruction in the electrolyzer (J/s)
  Ex_eff_Ele = (Ex_Ele_P/Ex_Ele_F)*100;                // Exergy efficiency in
the electrolyzer (%)

                /////    Compressor    ////
  Ex_Com_F =  W_5;                                                   // Fuel exergy
of the compressor (J/s)
  Ex_Com_P = E_7 - E_5;                                            // Fuel exergy of
the compressor (J/s)
  Ex_d_Com = Ex_Com_F - Ex_Com_P;                  // Exergy destruction in the
compressor (J/s)
  Ex_eff_Com = (Ex_Com_P / Ex_Com_F)*100;  // Exergy efficiency in the
compressor (%)

               ////    Hydrogen storage tank      ////
   Ex_Tan_F =  E_7;                                                     // Fuel exergy
of the stroage tank (J/s)
   Ex_Tan_P = (Q_Tan*(1 - (T_0/T_8))) + E_th;     // Fuel exergy of the stroage
tank (J/s)
   Ex_d_Tan = Ex_Tan_F - Ex_Tan_P;                        // Exergy destruction in
the stroage tank (J/s)
   Ex_eff_Tan = (Ex_Tan_P/Ex_Tan_F)*100;          // Exergy efficiency in the
stroage tank (%)

              ////    Fuel cell      ////
  E_10 = 0;                                          // Exergy of ambient air stream
entering the fuel cell (J/s)
                                                        // Both physical and chemical
exergies are zero, but if O2 is fed to
                                                        // the fuel cell, the chemical
exergy is not equal to zero.
  W_13 = (50/100)*E_9;          // Fuel cell electrical output (given fuel-cell
electrical efficiency
                                                       // 50% wrt H2 chemical exergy).
This is an assumption. (J/s)
  Q_fuc = 25e3;                        // Waste or useful heat rate from the
fuel cell (Assumption) (J/s)

   /// Exergy of water exiting the fuel cell ///
  n_H2O = n_H2;                          // mola flow rate of water (mol/s)
                                                       // from stiochiometric, 1 mole
of H2O = 1 mole of H2
  m_fc_w = n_H2O * M_H2O;      // Mass flow rate of water exiting fuel cell
(kg/s)
  E_12 = m_fc_w*(c_p*(T_11 - T_0 - (T_0*(log(T_11/T_0)))));
```

```
                                                      // Water ex. exiting fuel
cell. (J/s)
  E_11 = (Q_fuc*(1 - (T_0/T_11)));                //heat release from fuel cell.
(J/s)
    Ex_fuc_F =  E_9 + E_10;                            // Fuel exergy of the
fuel cell (J/s)
    Ex_fuc_P = E_11 + E_12 + W_13;              // Fuel exergy of the fuel cell
(J/s)
    Ex_d_fuc = Ex_fuc_F - Ex_fuc_P;              // Exergy destruction in the
fuell cell (J/s)
    Ex_eff_fuc = (Ex_fuc_P/Ex_fuc_F)*100;    // Exergy efficiency of the fuel
(%)

             //// System efficiency /////
  Ex_sys1 = (W_13 / (E_1 + W_2 + W_6 + E_10))*100;            // System exergy
efficiency with
// venting of oxygen and heat from
// the fuel cell is not useful (%)
  Ex_sys2 = ((W_13 + E_11 + E_3) / (E_1 + W_2 + W_6 + E_10))*100;
// System exergy efficiency when
// oxygen and heat are useful (%)

          //// Results    ///////
Comp_Fuel_Ex = [Ex_Ele_F, Ex_Com_F, Ex_Tan_F, Ex_fuc_F];
// Component fuel exergy rates
Comp_Product_Ex = [Ex_Ele_P, Ex_Com_P, Ex_Tan_P, Ex_fuc_P];
// Component product exergy rates
Comp_Ex_d = [Ex_d_Ele, Ex_d_Com, Ex_d_Tan, Ex_d_fuc];
// Component exergy destruction
Comp_Ex_eff = [Ex_eff_Ele, Ex_eff_Com, Ex_eff_Tan, Ex_eff_fuc];
// Component exergy efficiencies
Sys_Ex = [Ex_sys1, Ex_sys2];
// System exergy efficiencies
```

10 Research and Future Directions in Application of Green Hydrogen and CFD

10.1 PREAMBLE

Within the researchers' and practitioners' communities concerning energy alternatives and environmental impact mitigation, few domain interplays are as prominent as the combination of green hydrogen and Computational Fluid Dynamics (CFD). The integration of green hydrogen and CFD represents the intersection of innovative clean energy and the modeling technology advancements necessary to achieve a sustainable, low-carbon future.

Generated by the electrolysis of water with renewable energy sources (wind, solar, and hydro), green hydrogen represents a carbon-free energy vector applicable to every form of transportation, power generation, and a variety of industrial processes. CFD allows researchers and engineers to model and optimize the complex multi-physics systems governing hydrogen production, storage, transport, and end use.

The combination of green hydrogen and CFD has become increasingly important in addressing the technical, economic, and infrastructural challenges associated with the widespread use of hydrogen. The combination of green hydrogen and CFD offers the opportunity to develop a better understanding of the fundamental processes, as well as predictive models for hydrogen systems and their components, enabling the safe and efficient use of these systems and thereby facilitating the advancement of a wide range of hydrogen technologies in response to global challenges associated with reducing carbon emissions [1, 2].

Green hydrogen helps facilitate energy production without the environmental impacts of fossil fuels. The production of hydrogen from the electrolysis of water is done using green energy, which does not produce greenhouse gases. This differs from the production of grey or blue hydrogen, which is produced from the natural gas reforming process, a process which does use carbon capture. Renewables-based hydrogen gas can be considered a carrier and fuel for multiple applications like aviation, heavy trucking, steelmaking, and chemicals manufacturing (in electrolysis, electrification is not a viable option). The value chain of hydrogen production and use must be profitable.

The design of the electrolyzer (membrane, heat and mass transfer, integration with other renewable systems) involves complex variables that must be evaluated and optimized. CFD is a powerful modeling and analysis tool for optimizing complex systems from micro-catalyzers to reactors and entire plants. These systems represent

DOI: 10.1201/9781003727194-10

physical and political systems that may not be easily evaluated [3, 4]. Such activities are invaluable to the improvement of electrolyzers, the most prominent pieces of equipment in electrolytic hydrogen production, in which the CFD model in production engineering gets most of its work.

10.1.1 CFD Applications in Hydrogen Production

These three types of electrolyzers, alkaline, Proton Exchange Membrane (PEM), and Solid Oxide Electrolyzers (SOE), have unique and different thermodynamic, fluid, and electrochemical operations. CFD modeling makes it possible to simulate the processes using the governing equations of mass, momentum, and energy conservation with electrochemical kinetics. For example, in PEM electrolyzers, CFD has applications in modeling the gas bubble formation, flow profile, and heat removal provisions in the anode and cathode chambers. These model simulations may be used to specify and detect the current and temperature density profiles which are the main cause of local overheating and efficiency collapse [5, 6].

CFD modeling helps researchers in redesigning flow channels and gas diffusion layers with the aim of designing systems that reduce the energy barrier in order to obtain optimal hydrogen production. This principle also applies to alkaline electrolyzers, where CFD assists in optimizing the rate of circulation of the electrolyte, the distance between electrodes, the mechanisms of bubble removal, and the arrangement of the electrodes to improve the overall performance of the system and ensure stability during operation.

Beyond electrolyzers, CFD also finds applications in the integration of hydrogen systems with renewable energy sources. Due to the intermittent nature of solar and wind energy, CFD-based models help in designing efficient coupling systems and thermal management solutions that stabilize hydrogen production under fluctuating energy inputs. The dynamic interaction between renewable energy supply and electrolysis demand can be optimized using transient CFD simulations combined with control algorithms. Moreover, CFD can model heat recovery and utilization processes in hybrid systems where waste heat from other industrial processes is used to enhance the efficiency of SOE. Such integration supports the development of smart hydrogen production plants that maximize renewable energy utilization and overall exergy efficiency [4, 7].

10.1.2 CFD in Hydrogen Storage and Transportation

Hydrogen storage represents another area where CFD is indispensable. Storing hydrogen safely and efficiently remains one of the key challenges for its large-scale adoption. The three primary storage methods—compressed gas, liquid hydrogen, and solid-state storage—each involve complex thermodynamic and transport phenomena. CFD simulations provide detailed insights into pressure distribution, temperature gradients, and phase transitions occurring within storage tanks. For compressed hydrogen storage, CFD can model leak detection, heat dissipation during rapid filling, and structural integrity under different loading conditions.

In cryogenic liquid hydrogen storage, CFD assists in predicting boil-off gas generation, stratification, and sloshing behaviour, enabling the design of insulated tanks that minimize energy losses [8, 9]. Solid-state hydrogen storage, particularly in metal hydrides and adsorbent materials, benefits from CFD analysis of heat and mass transfer within porous media, facilitating the optimization of desorption kinetics and reactor geometry. These computational insights are vital for ensuring the safety and performance of hydrogen storage infrastructure, especially for mobile and stationary applications.

Hydrogen distribution and transportation present new difficulties that CFD can successfully handle. Delivery systems, refueling stations, and hydrogen pipes must be built to resist high pressures without leaking or becoming embrittled. By capturing phenomena including turbulence, compressibility effects, and material interactions, CFD offers a predictive understanding of hydrogen flow behaviour under various operational situations. Critical spots vulnerable to hydrogen-induced cracking can be found using sophisticated CFD models in conjunction with fatigue and fracture mechanics investigations. To guarantee quick and safe vehicle fueling procedures, CFD also helps optimize the aerodynamic and thermal design of hydrogen refueling nozzles [1, 10]. Computational models will be essential to ensure that transportation systems in future hydrogen-based infrastructure are not only effective but also compliant with safety and environmental regulations.

10.1.3 CFD in Hydrogen Combustion and Safety

CFD research is also very beneficial for the combustion and use of hydrogen in power and transportation applications. Compared with traditional fuels, hydrogen's combustion behaviour is more complex due to its high diffusivity, low ignition energy, and broad flammability range. Researchers can capture the mechanisms of heat release, flame propagation, and pollutant generation under various operating circumstances using CFD models.

CFD-based combustion models are used in gas turbine applications to design burners that guarantee high thermal efficiency and low NOx operation. CFD makes it possible to accurately forecast the properties of hydrogen combustion in real-world systems by combining comprehensive chemical kinetics with turbulence models like Large Eddy Simulation (LES). These models accelerate the transition to zero-carbon energy systems by supporting the development of boilers, industrial furnaces, and engines that are compatible with hydrogen [2, 6].

Furthermore, by lowering computing costs without sacrificing predictive accuracy, hybrid CFD and Machine Learning (ML) models are becoming potent tools to speed up combustion simulations. In all hydrogen applications, safety is still the priority, and CFD is essential for risk assessment and accident avoidance. Because of its high diffusivity and low molecular weight, hydrogen disperses quickly in the event of a leak, but in confined spaces, this can cause flames or explosions. CFD-based dispersion models help with safety design and emergency response planning by simulating leak scenarios, ventilation efficiency, and explosive propagation. Refineries, gas stations, and home hydrogen installations all require these kinds of tests. The safe distances, venting arrangements, and structural reinforcements needed to reduce risks are all determined with the aid of CFD simulations. The development of worldwide safety standards and regulatory frameworks that will direct the future hydrogen economy is facilitated by these predictive capabilities [11, 12].

10.1.4 Emerging Technologies

Research-wise, multi-scale and multiphysics modeling is one of the most promising avenues for CFD applications for green hydrogen. Conventional CFD techniques are frequently constrained by simple assumptions and computational expenses. But the area is being revolutionized by the combination of data-driven modeling, Artificial Intelligence (AI), and High-Performance Computing (HPC). From full-scale system dynamics to atomic-level electrochemical reactions, these methods enable the simultaneous simulation of phenomena occurring across scales.

CFD with ML support can minimize simulation time while precisely determining the ideal design parameters. For instance, using previously calculated CFD data, deep neural networks can be trained to forecast flow and heat patterns within electrolyzers or storage systems, allowing for real-time optimization during operation [3, 4]. An important step toward intelligent hydrogen infrastructure is the creation of adaptive systems that can self-learn and improve performance on their own thanks to the combination of CFD and AI.

Combining CFD with Lifecycle Assessment (LCA) and Techno-Economic Analysis (TEA) is another new area of study. It is crucial to comprehend the economic and environmental effects of hydrogen technologies as they approach commercialization. To assess system sustainability and cost-effectiveness, CFD outputs, such as energy efficiency and exergy destruction profiles, can be fed into LCA and TEA models. Researchers can find design enhancements that reduce lifecycle costs and carbon footprint while simultaneously improving technical performance thanks to this integrated methodology. In the developing hydrogen economy, such comprehensive modeling frameworks will play a crucial role in directing investment plans, industrial design standards, and governmental decisions [1, 5].

CFD advances our knowledge of how fluid flow and material characteristics interact in hydrogen systems. For instance, CFD simulations can capture how microstructural alterations affect electrical conductivity and gas diffusion in SOE. These discoveries direct the creation of sophisticated ceramic materials with improved ionic conductivity and durability. Similar to this, CFD helps determine the best compositions and morphologies for catalysts and membranes used in fuel cells for hydrogen production. Such interdisciplinary approaches, which integrate CFD with materials science, chemistry, and nanotechnology to produce performance and lifespan advancements, will be crucial to the future of green hydrogen research [13].

Another area where CFD-based research is anticipated to grow is energy system integration. Green hydrogen will serve as a link between the fuel, heat, and electricity sectors in a future decarbonized energy environment. The linkage of hydrogen production units with power grids, thermal plants, and storage systems can be simulated using CFD, allowing for optimal operation under changing supply and demand situations. CFD optimization of reactors and gas mixing dynamics greatly helps the idea of Power-to-Gas (P2G) systems, which transform excess renewable electricity into hydrogen or synthetic methane. Additionally, CFD is essential to the design of hydrogen-based microgrids because it guarantees effective mass and heat transfer across subsystems like heat exchangers, fuel cells, and electrolyzers [6, 14]. As countries move toward integrated energy systems, CFD will remain at the forefront of modeling these complex, interconnected infrastructures.

The promise of green hydrogen as a clean fuel for heavy-duty vehicles, shipping, and aviation is becoming more and more apparent in the transportation sector. Designing effective hydrogen engines, turbines, and propulsion systems requires the use of CFD simulations. For instance, CFD models are used in aviation to forecast the properties of hydrogen combustion at high altitudes and low pressures, as well as the aerodynamic impacts of cryogenic fuel storage on aircraft design. CFD helps analyze exhaust plume dispersion, hull temperature control, and hydrogen fuel cell integration in maritime applications. Innovation in clean transportation systems is accelerated by CFD's capacity to replicate real-world operating conditions, guaranteeing performance optimization, safety, and dependability [15].

Large-scale infrastructure planning is also required for the shift to a hydrogen economy, and CFD helps with the design of pipelines, storage terminals, and hydrogen refueling stations. By simulating hydrogen flow through pipeline networks, computational modeling can find possible bottlenecks and optimize compressor and valve configurations. Since some future networks may carry blends of natural gas and hydrogen, it can also examine how various gases interact in blended systems.

Planners may create data-driven plans for national hydrogen infrastructure that optimize efficiency and safety by combining CFD with Geographic Information Systems (GIS) and optimization algorithms [16]. Future CFD and green hydrogen research must focus on sustainability and policy alignment in addition to technical developments. The importance of hydrogen in facilitating deep decarbonization is becoming more evident as countries commit to net-zero ambitions. Policymakers can quantify the potential emissions reduction attainable through different hydrogen deployment scenarios with the help of CFD-based modeling. CFD models can determine the best combination of hydrogen production, storage, and consumption methods that are in line with local resource availability and climate objectives by modeling energy flows across sectors. These insights are essential for creating hydrogen roadmaps, investment plans, and legal frameworks that strike a balance between economic expansion and environmental sustainability [8, 17].

10.1.5 Challenges and Future Research Directions

Nevertheless, there are still several obstacles to the widespread use of CFD in hydrogen research. CFD simulations of hydrogen systems sometimes entail considerable complexity because of multi-phase flow, electrochemical reactions, and nonlinear coupling effects, despite notable advancements in processing capacity. Detailed boundary conditions and experimental validation are necessary for accurate modeling, although they are not always easily accessible. Moreover, turbulence, heat transmission, and material degradation uncertainties are introduced by the CFD models' scalability from the laboratory to the industrial scale. Standardized validation procedures, open-access databases, and ongoing investments in experimental facilities are necessary to meet these problems [2, 9]. Collaborative efforts between government, business, and academic institutions are crucial to advancing the field and guaranteeing the stability and dependability of CFD tools for hydrogen research.

Global climate policy, the growth of renewable energy, and the convergence of digital technologies will all influence the future of green hydrogen and CFD. CFD is predicted to undergo a revolution thanks to quantum computing and exascale simulation capabilities, which will significantly reduce computation time while improving accuracy. This will enable predictive maintenance and optimization at previously unheard-of scales by enabling researchers to simulate entire hydrogen reactors or transportation networks in real time. By enabling continuous monitoring and predictive insights, digital twins, virtual replicas of physical hydrogen systems driven by CFD and real-time sensor data, are poised to revolutionize operations management [4, 10]. By predicting system performance, identifying irregularities, and recommending remedial measures, digital twins can guarantee dependability and effectiveness throughout the hydrogen value chain.

The advancement of CFD and green hydrogen in the future will also depend heavily on education and capacity growth. The next generation of engineers and researchers will be equipped to handle new issues through training programs that integrate data analytics, renewable energy systems, and computational modeling.

The creation of publicly accessible hydrogen datasets and open-source CFD tools will democratize access to these technologies, fostering innovation in both developed and developing countries. By promoting the interchange of modeling techniques, experimental data, and best practices, international collaboration will further improve research outcomes [18].

A completely optimized, sustainable energy ecosystem where computational intelligence powers every step, from production and storage to distribution and utilization, is the long-term goal for the convergence of green hydrogen and CFD research. Adaptive, effective, and robust energy networks that can react quickly to changes in the economy and environment would be made possible by such a system. Humanity is on the verge of an energy revolution that not only lowers carbon emissions but also promotes technological innovation, economic expansion, and global energy security by fusing the predictive power of CFD with the sustainability of green hydrogen [19].

In conclusion, one of the most promising areas in the pursuit of a sustainable future lies in the nexus of CFD and green hydrogen. CFD offers the analytical and predictive skills required to fully realize the promise of hydrogen, even if it offers a route to decarbonization across industries. While new technologies like AI-assisted simulations, digital twins, and multi-scale modeling are opening the door for previously unheard-of breakthroughs, ongoing research is expanding our knowledge of electrolysis, storage, transport, and combustion processes. Through interdisciplinary cooperation and ongoing innovation, the remaining issues, which range from system integration and safety to cost reduction and scalability, can be successfully resolved.

The future of energy will be determined by how cleverly we design, simulate, and maximize their use rather than just the sources we select. In this sense, the combination of green hydrogen and CFD is a key component of the clean energy revolution, providing both scientific rigour and the innovative potential to revolutionize the world's energy landscape for future generations.

10.2 EVOLVING HYDROGEN TECHNOLOGIES

One of the most significant technological frontiers of the twenty-first century is the development of hydrogen into a green and sustainable form, which has become a key component of the worldwide clean energy transition. Green hydrogen provides an unmatched chance to decarbonize difficult-to-abate industries including steel production, chemical manufacture, aviation, and heavy transportation. Green hydrogen is created from renewable electricity by water electrolysis. Its usage as an energy carrier and long-term storage medium promises to bridge the gap between large-scale industrial demand and renewable energy output. These developments have been sped up by the increasing use of CF) in hydrogen research, which enables predictive modeling and design optimization that reduce development cycles and boost system efficiency [3, 20]. CFD continues to influence the development of hydrogen technologies by offering vital insights into fluid flow, heat transfer, electrochemical processes, and safety mechanisms. The dynamic research landscape that is driving the global energy sector toward a carbon-neutral future is defined by the interaction between developing hydrogen technologies and CFD-driven simulation.

From the early days of hydrogen production based on fossil fuels to processes based on renewable energy, the development of hydrogen technologies has taken an amazing path. Due to their maturity and affordability, conventional hydrogen generation techniques like coal gasification and Steam Methane Reforming (SMR) continue to be the most popular, but they emit large amounts of carbon dioxide. Electrolysis powered by renewable electricity will take the place of these carbon-intensive technologies in the shift to green hydrogen. This process is carried out by electrolyzers, which have continuously evolved from basic laboratory equipment to extremely effective industrial machines. Each of the three main types, SOE, PEM electrolyzers, and alkaline electrolyzers, represents a different stage in the development of hydrogen production technology. By simulating the movement of ions, gases, and heat via electrodes, membranes, and electrolyte layers, CFD has been essential in optimizing these systems. Engineers have been able to increase overall performance and durability by minimizing deterioration, improving current distribution, and reducing energy losses thanks to these models [7, 9].

CFD research has given important insights into gas evolution and water management in PEM electrolyzers, which are essential for guaranteeing stable operation and high efficiency. Bubble dynamics simulations in flow channels and porous transport layers can be used to pinpoint regions where mass transfer is limited and pressure drops, which can result in an unequal distribution of current. CFD allows designers to accomplish consistent reactant supply and effective gas removal by modifying the flow field shape. Furthermore, a thorough understanding of how temperature gradients affect ion conductivity and membrane durability is provided by CFD models that integrate electrochemical reactions with fluid flow and heat transfer. Engineers can produce optimized designs that reduce overpotential losses and increase operational life thanks to these insights [1, 12]. The adjustment of electrode spacing, electrolyte circulation, and bubble detachment mechanisms in alkaline electrolyzers has also been influenced by CFD research. Researchers can forecast how hydrogen and oxygen bubbles develop and escape, affecting system resistance and energy efficiency, by using multi-phase flow modeling.

Because of its high working temperature, which enables the co-electrolysis of water and carbon dioxide to produce syngas, a mixture of hydrogen and carbon monoxide that may be used for the generation of synthetic fuel, the SOE represents a more advanced step in the evolution of hydrogen. In order to analyze the intricate thermo-electrochemical events that occur within these systems, CFD models have become essential tools. In order to maintain stability, high-temperature operation presents linked difficulties of mass diffusion, heat transport, and material stress. Engineers can forecast performance under dynamic operating settings by integrating these phenomena into full multiphysics frameworks through CFD simulations. The use of CFD and ML hybrid models to optimize materials, reduce thermal gradients, and scale the technology for industrial applications is where SOE research is headed [4, 10].

Beyond production, a crucial area of research is the development of hydrogen storage devices. Compressed gas, cryogenic liquid, and solid-state storage are the three main types of hydrogen storage, each requiring special design considerations. Analyzing the intricate thermodynamics and flow dynamics connected to these storage techniques requires the use of CFD. CFD aids in the prediction of temperature rise and pressure distribution for compressed hydrogen during quick filling procedures, guaranteeing structural integrity and safety. CFD simulations that show how various geometries and insulating materials affect heat dissipation and hydrogen density are useful for storage tank design [3, 8]. CFD helps comprehend boil-off gas production, stratification, and convective heat transmission within tanks for cryogenic liquid hydrogen, which is held at temperatures below –253°C. Engineers can create sophisticated insulation systems that lower evaporation losses and increase storage efficiency by modeling these processes.

Another quickly developing area where CFD is essential is solid-state hydrogen storage, which entails the absorption or adsorption of hydrogen into materials like metal hydrides, carbon nanotubes, or Metal–Organic Frameworks (MOFs). Heat and mass transfer are intricately connected phenomena in these systems' absorption and desorption processes. CFD models take into consideration varying permeability, reaction kinetics, and temperature management while simulating the behavior of hydrogen diffusion in porous media. Reactors with optimum shape, heat exchangers, and operating cycles can be designed thanks to these models. Future research on hydrogen storage is probably going to concentrate on hybrid systems, which integrate various storage techniques to balance cost, safety, and energy density. In order to integrate these many technologies into effective, scalable storage systems, CFD-based optimization will continue to be essential [11, 18].

Hydrogen transport and distribution have also made great strides thanks to CFD-enabled fluid mechanics and structural analysis studies. High pressure must be tolerated by pipelines used for the transportation of hydrogen while limiting leaks and avoiding material deterioration like hydrogen embrittlement. Engineers can investigate flow characteristics, pressure losses, and transient behavior under various operating settings using CFD simulations. Researchers can forecast how hydrogen molecules interact with steel and composite materials at the atomic level, preventing structural breakdowns, by combining CFD data with material models. In order to optimize nozzle shape, pressure control, and heat management during rapid refueling, CFD has also been employed in the design of hydrogen refueling stations. Hydrogen car refueling is made safe and effective by the capacity to replicate temperature gradients, flow turbulence, and gas compressibility [14, 21].

Hydrogen consumption and combustion have evolved simultaneously. Because of its fast flame speed, broad flammability range, and low ignition energy, hydrogen requires specific engine and burner designs. By offering comprehensive models of flame propagation, turbulence–chemistry interactions, and pollutant generation, CFD has completely changed our understanding of hydrogen combustion. CFD models aid in the development of low-NO_x burners for gas turbines that can minimize emissions while ensuring stable combustion. Combustion dynamics can be accurately predicted under actual operating conditions by combining comprehensive chemical kinetics with sophisticated turbulence models like LES. Additionally, CFD helped build boilers and internal combustion engines that run on hydrogen, opening the door for their use in industrial and power production applications. By combining CFD with experimental data, innovative ideas can be effectively validated and improved, hastening their commercial adoption [2, 5].

As hydrogen technologies advance, safety continues to be a top priority. Hydrogen is prone to leaks due to its high diffusivity and low molecular weight, and its broad flammability range raises the possibility of an explosion. When examining leak dispersion, ventilation efficiency, and explosive propagation, CFD is essential. Engineers can design safer storage and processing facilities by predicting hydrogen concentration fields, ignition zones, and overpressure effects through the simulation of accidental release scenarios. By offering quantitative risk assessments for a range of hydrogen applications, CFD also contributes to the creation of safety laws and design standards. In order to allow predictive safety management systems, it is anticipated that future developments in CFD-based safety modeling would include real-time data from sensors and digital twins [3, 9].

Energy system integration is part of the growing synergy between CFD and hydrogen technology. Green hydrogen provides a way to store excess electricity and balance grid fluctuations as renewable energy supply grows increasingly localized. Designing integrated systems that connect electrolyzers to renewable energy sources like solar and wind farms is made easier with the help of CFD. CFD can model power supply transient variations, component thermal behavior, and efficiency under different loads through dynamic simulations. CFD makes it possible to optimize energy conversion paths in hybrid systems that integrate fuel cells, heat recovery units, and hydrogen generation [6]. "Power-to-X" systems, which use hydrogen as a bridge to produce fuels, chemicals, and even ammonia, are based on this integration. Researchers can find ideal operating conditions that increase efficiency and minimize environmental effects by using CFD to simulate these interactions.

The next stage in CFD-assisted hydrogen research is represented by the development of digital twin technology. A digital twin is a real-time virtual duplicate of a physical system that combines AI, CFD models, and sensor data to track performance, anticipate failures, and improve operations. Digital twins can continually update CFD-based models in hydrogen reactors using real-time data from sensors that measure flow rate, temperature, and pressure. This makes it possible to make decisions in real time for energy optimization, safety management, and preventative maintenance. By combining AI and ML with CFD, predicted accuracy is increased and computing time is decreased, resulting in intelligent hydrogen systems that can self-optimize. As part of the Industry 4.0 revolution, digital twins will probably become commonplace in hydrogen production, storage, and transportation systems [4, 5].

The development of hydrogen technologies is largely dependent on material science, and CFD has been crucial in bridging the gap between material behavior and system performance. For example, hydrogen diffusion through metallic and polymeric membranes can be modeled using CFD simulations in conjunction with molecular dynamics, aiding in the development of materials resistant to penetration or embrittlement. CFD models forecast how material characteristics affect temperature gradients, water transport, and current distribution in membranes and electrodes in fuel cells. These discoveries direct the creation of novel materials with enhanced mechanical strength, conductivity, and durability [20, 22]. CFD-driven multiphysics simulations that integrate fluid flow, electrochemistry, and mechanical stress analysis at various scales will become more and more important in the future of materials research in hydrogen systems.

CFD is also changing how hydrogen technologies are assessed from an economic and environmental standpoint. To give thorough evaluations of system performance and sustainability, CFD data are combined with LCA and TEA. For instance, LCA models can directly incorporate CFD simulations of waste heat recovery and electrolyzer efficiency to measure the carbon footprint and energy payback period of hydrogen systems. In a similar vein, TEA models that use data from CFD can forecast cost savings that can be attained through design optimization. These multidisciplinary methods will be essential for directing investment and policy choices in the developing hydrogen economy [23].

Future CFD and hydrogen technology research is anticipated to concentrate on a number of important areas. First, more precise prediction and management of hydrogen systems will be possible with the creation of multi-scale modeling frameworks that connect micro-scale electrochemical processes to macro-scale system dynamics. Second, real-time optimization and adaptive control of hydrogen plants will be made possible by further advancements in the integration of CFD with AI and ML [4, 24]. Third, by permitting simulations of hitherto unheard-of complexity and enabling the simultaneous modeling of entire hydrogen networks, exascale computing and quantum simulation technologies will transform CFD. Fourth, model accuracy and dependability will be improved by combining CFD with experimental data using data assimilation techniques, closing the performance gap between simulation and real-world applications.

Additionally, it is anticipated that hydrogen technologies will advance into new areas, such as offshore hydrogen generation, where CFD models can maximize the performance of compressors and electrolyzers operating in dynamic oceanic circumstances. Similar to this, CFD will be used in the study of hydrogen blending with natural gas for pipeline transportation in order to simulate intricate fluid interactions and guarantee infrastructure compatibility. CFD will continue to be essential for developing propulsion systems that employ liquid or gaseous hydrogen as a fuel in aircraft and marine applications because it offers comprehensive insights into combustion stability, thermal management, and structural integrity [18, 21].

Another crucial area where CFD may offer helpful direction is the environmental impact of hydrogen technologies. CFD can measure the overall sustainability of hydrogen-based systems by simulating emission profiles, energy flows, and thermal efficiency in great detail. This will help the world's efforts to reach net-zero emissions by the year 2050. Policymakers will be able to evaluate the viability and effects of adopting hydrogen on regional and global energy systems with the use of CFD integration into environmental modeling frameworks [5, 25].

Maintaining innovation in CFD and hydrogen technologies requires both personnel development and education. A new generation of engineers skilled in both computational modeling and renewable energy technologies is required due to the increasing complexity of multidisciplinary hydrogen systems. In order to prepare students for careers in this quickly developing industry, universities and research organizations are progressively integrating CFD-based hydrogen modeling into their courses. The creation of collaborative research platforms and open-access CFD tools will democratize access to sophisticated modeling skills, promoting innovation in both developed and emerging economies [26].

Future plans call for the seamless integration of CFD-driven digital intelligence to power the production, storage, distribution, and use of green hydrogen. CFD will be a key component of design and innovation in the ongoing development of hydrogen technologies, from electrolyzers and fuel cells to storage and transportation systems. Hydrogen systems will become safer, more economically viable, and more efficient as research advances, bringing the world closer to a carbon-neutral energy landscape. In addition to redefining how we generate and consume energy, the combination of developing hydrogen technologies and CFD will lay the groundwork for a resilient, sustainable, and just global energy future.

10.3 CFD PROGRESS IN REACTIVE AND MULTIPHASE FLOWS

In the field of green hydrogen production, storage, and consumption, CFD has become an essential tool for researching reactive and multiphase flows. Many important processes in hydrogen technologies, such as electrolysis, combustion, and fuel cell operation, are defined by the behavior of fluids involving chemical reactions and various interacting phases. The mathematical modeling, numerical approaches, and computational techniques that underpin CFD have advanced remarkably over the past few decades, enabling researchers to simulate complex phenomena like turbulence, phase change, bubble dynamics, and electrochemical reactions with previously unheard-of accuracy. A greater comprehension of the physical mechanisms governing hydrogen systems has been made possible by the ongoing improvement of CFD, spurring advancements in scalability, safety, and efficiency. CFD integration in reactive and multiphase flow research has emerged as a key component of green hydrogen development, bridging the gap between theory, experimentation, and industrial implementation as global energy priorities change toward decarbonization [20, 27].

Hydrogen-related technologies revolve around reactive fluxes, which include catalytic reactions in reforming and synthesis processes, combustion in engines and turbines, and electrochemical reactions in electrolyzers and fuel cells. By resolving the governing equations of mass, momentum, and energy conservation in conjunction with chemical reaction kinetics, CFD offers a potent framework for examining these processes. For example, CFD simulations have made it possible to thoroughly examine flame propagation, ignition delay, and pollutant generation under different conditions in hydrogen combustion. Compared to traditional hydrocarbons, hydrogen possesses distinct combustion properties, including a broad flammability range, a rapid flame speed, and a low ignition energy. For renewable energy applications, these characteristics offer both possibilities and difficulties. Researchers have used CFD modeling to optimize engine and burner designs to maximize the clean combustion potential of hydrogen while reducing problems like flashback, thermoacoustic instability, and NO_x emissions [12, 26]. The use of sophisticated turbulence–chemistry interaction models, such as LES and Flamelet Generated Manifolds (FGM), has significantly increased the predictive accuracy of hydrogen combustion simulations, enabling engineers to create next-generation combustion systems with reduced environmental impact and increased efficiency.

Reactive flow modeling concentrates on electrochemical reactions combined with multiphase transport phenomena in electrolyzers and fuel cells. Ions, electrons, and gases are moved across porous electrodes and membranes in these devices, which frequently present problems with heat production and water management. These interactions may be seen in great detail because of CFD simulations, which offer insights that are almost impossible to get through experimentation. For example, CFD models show how gas bubbles develop and move inside flow channels in PEM electrolyzers, illustrating how bubble aggregation can impede reactant flow and lower efficiency [7, 28]. By modeling thermal gradients, reactant transport, and current density distributions, CFD helps control the high-temperature environment in SOE. Similar to this, CFD modeling of reactive multiphase flows in fuel cells has proved essential for improving membrane hydration and water transport, both of which are necessary for sustaining longevity and performance. CFD assists in determining design choices that improve gas diffusion, lower energy losses, and extend system longevity by combining electrochemical kinetics with multiphase flow and heat transfer equations.

Throughout the hydrogen value chain, multiphase flow dynamics are equally important. Many interacting phases with different physical properties are involved in processes including gas–liquid flow in electrolyzers, liquid–solid interactions in hydrogen storage materials, and gas–solid flows in reactors. Using techniques like the Volume of Fluid (VOF) method, Euler-Lagrange modeling, and population balance equations, CFD offers the computational framework to resolve these intricate interactions. For example, CFD simulations monitor the growth and coalescence of gas bubbles on electrode surfaces during hydrogen synthesis via water electrolysis. The local current density, pressure distribution, and overall efficiency are all greatly impacted by the detachment and rise of bubbles. Researchers can minimize mass transfer constraints and increase hydrogen output by using CFD to modify electrode surface morphology and flow field geometry for quicker bubble removal. Multiphase CFD models capture the impact of temperature gradients, diffusion resistance, and reaction kinetics on hydrogen yield in high-temperature reactors by accounting for the interaction of gaseous reactants with solid catalysts [5, 29].

Another area where CFD-based multiphase flow research has significantly advanced is hydrogen storage, whether via chemical or physical techniques. CFD simulations aid in the prediction of phase change behavior and heat transmission during liquid hydrogen filling and evaporation in cryogenic storage. Accurate modeling of multiphase flow phenomena is necessary because of the intricate interactions between convection, conduction, and boiling. The design of insulation systems that reduce energy losses can be guided by CFD models, which can forecast temperature stratification, pressure rise, and boil-off gas generation. CFD models combine mass and heat transport with reaction kinetics to simulate hydrogen absorption and desorption in solid-state storage systems that use metal hydrides or adsorbent materials. These processes, which involve exothermic reactions that affect temperature distribution and gas diffusion across porous media, are intrinsically multiphase. The practicality of solid-state hydrogen storage solutions is increased by CFD-based optimization of reactor shape, heat exchanger placement, and material selection, which guarantees quicker cycles of charging and discharging [10, 21].

Safety research in hydrogen systems has also changed as a result of advances in CFD modeling of reactive and multiphase flows. Because of its high diffusivity and low molecular weight, hydrogen can quickly disperse in the event of a leak, but it can also create combustible or explosive mixes in some circumstances. With great spatial and temporal resolution, CFD models enable engineers to forecast leak dispersion, igniting behavior, and explosion propagation. For example, hydrogen discharge in confined places like storage facilities or fuel cell enclosures has been simulated using CFD models, offering crucial insights into ventilation design and ignition prevention. Accurate prediction of hydrogen plume behavior under different temperature and pressure circumstances is ensured by the ability to describe buoyancy-driven flow and turbulence. CFD supports the creation of safety standards and regulations in explosion modeling by capturing the change from deflagration to detonation and quantifying the impacts of overpressure [14, 30]. These uses demonstrate how CFD helps with both performance optimization and the dependable and safe implementation of hydrogen technologies.

Despite significant advancements, the nonlinear coupling of transport processes, chemical kinetics, and phase interactions makes modeling reactive and multiphase flows in hydrogen systems computationally challenging. Bubble nucleation and surface responses are examples of micro-scale processes that are difficult for traditional CFD methods to accurately resolve. However, the boundaries of simulation capabilities are being expanded by developments in algorithm development and HPC. To handle huge, complicated models with millions of grid cells, modern CFD solvers use GPU acceleration, parallel computing, and adaptive mesh refinement. Furthermore, hybrid techniques that combine CFD with Lattice Boltzmann methods (LBM) or Molecular Dynamics (MD) are becoming increasingly effective tools for capturing micro-scale phenomena in macro-scale simulations. For example, combining CFD and MD can improve model fidelity across scales by offering atomistic insights into membrane transport features or hydrogen adsorption on catalyst surfaces [31, 32].

CFD research in reactive and multiphase flows is also changing due to ML and AI. As reduced-order surrogates, data-driven models trained on CFD data may forecast flow and reaction behavior in real time. These models can be integrated into digital twins and control systems since they significantly cut calculation time without sacrificing accuracy. AI-enhanced CFD models may continually learn from operational data in hydrogen reactors to improve temperature, flow, and chemical composition forecasts. The basis for smart hydrogen systems is laid by these adaptive modeling capabilities that provide the foundation for intelligent hydrogen systems, which facilitates defect identification, process optimization, and predictive maintenance. CFD, AI, and the Internet of Things (IoT) will probably come together in the future to create resilient, self-optimizing energy infrastructures [9, 33].

Multi-scale and multiphysics coupling is another area of interest in CFD research for hydrogen applications. Numerous hydrogen-related processes take place over a wide variety of temporal and geographical scales, from meter-scale reactor flows to electrochemical reactions at the nanoscale. Hybrid modeling frameworks that integrate several computational methods are necessary to bridge these scales. To capture both micro-scale mass transfer and macro-scale flow dynamics, for instance, continuum-level CFD models can be combined with pore-scale simulations utilizing Lattice Boltzmann or Direct Numerical Simulation (DNS). Similar to this, combining CFD with Finite Element Analysis (FEA) allows for the simultaneous modeling of structural deformation and fluid flow, which is very helpful for examining thermal stresses in SOE or high-pressure vessels [12, 13]. Designing next-generation hydrogen systems that function dependably in dynamic environments will require multiphysics models that incorporate fluid dynamics, heat transfer, electromagnetism, and chemical kinetics.

In reactive and multiphase flows, experimental validation is still a crucial part of CFD research. Even if simulations are becoming more accurate, CFD predictions still need to be verified by comparison with experimental data. Particle Image Velocimetry (PIV), Laser-Induced Fluorescence (LIF), and X-ray tomography are examples of advanced diagnostic instruments that offer precise measurements of temperature, concentration, and velocity fields. These datasets are used to calibrate turbulence or response sub-models and improve CFD models. The reliability of CFD predictions in hydrogen research has increased due to the increasing usage of open-access experimental databases and standardized validation procedures. This validation environment is being further strengthened by cooperative efforts between academia, research institutions, and industry, guaranteeing that CFD will continue to be a reliable tool for hydrogen technology design and analysis [17, 31].

The importance of CFD in comprehending reactive and multiphase flows will only increase as green hydrogen's position in global energy systems continues to grow. Predictive models for sophisticated electrolyzers, fuel cells, and combustion systems are being developed; heat management and mass transfer in multiphase reactors are being optimized; and hydrogen storage and transportation are being made safer and more sustainable. The cost-effectiveness and environmental advantages of new hydrogen technologies will also be quantified through the integration of CFD with techno-economic and LCA models. In order to account for variations in material qualities, operating conditions, and model parameters, future CFD frameworks will probably include Uncertainty Quantification (UQ) approaches. Large-scale hydrogen system deployment will be made easier as a result of the improved robustness of design choices and risk assessments [16, 21].

The twin objectives of accuracy and efficiency will ultimately drive the development of CFD in reactive and multiphase flow simulation. It will be feasible to simulate entire hydrogen plants at fine resolution by combining all unit operations into a single virtual environment thanks to developments in exascale computing, quantum simulation, and hybrid data-driven modeling. The development of hydrogen technology will become a digitally driven process thanks to such comprehensive models, which will allow for real-time optimization, safety prediction, and environmental monitoring. Engineers will be able to interactively perceive complex reactive and multiphase phenomena by combining CFD with augmented reality and virtual prototyping, which will accelerate training and creativity [32, 34].

In the end, CFD's advancement in reactive and multiphase flow research is not just a technical accomplishment but also a key factor in the green hydrogen revolution.

CFD enables researchers and engineers to create safer, more effective, and cleaner technologies by offering profound insights into the chemistry and physics of hydrogen systems. Green hydrogen's future will rely on ongoing advancements in CFD modeling, driven by interdisciplinary cooperation, high-performance computation, and a steadfast dedication to sustainability. Together, CFD and hydrogen technology will contribute to the development of a low-carbon, resilient energy landscape that satisfies the needs of both the current and future generations.

10.4 RESEARCH GAPS AND STRATEGIC FRAMEWORKS

Despite tremendous advancements, there are still large research gaps that restrict scalability, performance improvement, and cost competitiveness in the fields of CFD and green hydrogen. Addressing these shortcomings through strategic frameworks that incorporate advanced simulation, experimental validation, and policy coordination is essential to the shift to a global hydrogen economy. A special chance to decipher the intricacies of multi-scale, multiphysics, and multi-sectoral interactions in production, storage, transit, and use is provided by the interaction between CFD modeling and green hydrogen development. Nevertheless, the conversion of CFD insights into useful, deployable technologies is still lacking. Therefore, developing precise frameworks that connect computational models with experimental design, commercial implementation, and national energy objectives is essential to the future success of hydrogen innovation [30].

Due to constraints on modeling accuracy, scalability, data integration, and computing costs, there are now research gaps in the fields of green hydrogen and CFD. Although CFD has been useful in investigating hydrogen production systems like water electrolysis, the models frequently rely on oversimplifying assumptions that undermine realism. Ion conduction, gas evolution, and heat dissipation are some of the transport mechanisms involved in the electrochemical reactions that take place concurrently at various sizes in PEM and SOE. Current CFD models often ignore local turbulence, bubble nucleation, and concentration gradients by decoupling these interactions or assuming homogeneous flow fields. These oversimplifications reduce model applicability to industrial-scale systems and impair predicted accuracy [5, 19]. The difficulty is in creating fully coupled, multiphysics CFD frameworks that are computationally practical while capturing the intricate connection between fluid flow, phase shift, and electrochemical kinetics.

Apart from modeling constraints, a significant deficiency is the absence of consistent experimental data to verify CFD models. Small-scale laboratory experiments that don't accurately reflect industrial circumstances are used to validate the majority of CFD models of hydrogen systems. Uncertainties in boundary conditions, material qualities, and reaction kinetics lead to differences between numerical predictions and experimental outcomes. The calibration and benchmarking of CFD models are hampered by the lack of high-fidelity datasets with temporal and spatial precision. Proprietary restrictions in industrial hydrogen initiatives, which restrict open data exchange, make this data scarcity even worse. Coordinated international efforts to create open-access datasets with comprehensive measurements of flow, temperature, and reaction parameters in hydrogen systems are desperately needed to close this gap

[28,29]. Reproducibility throughout the research community would be facilitated by these repositories, which would offer uniform benchmarks for model validation.

The integration of CFD with lifecycle and techno-economic assessments is a significant research gap. Thermodynamics, fluid dynamics, and economic considerations all interplay with hydrogen technologies, making them complicated systems. The majority of current CFD research ignores system-level effects on cost, efficiency, and emissions in favor of targeted optimization, such as enhancing flow distribution or lowering pressure drop. Evaluating design choices holistically is hampered by the lack of integrated models that connect CFD outputs to lifecycle performance measures. Optimizing an electrolyzer for optimum hydrogen yield, for instance, can result in higher capital costs or material degradation, which would counteract the advantages. Hybrid modeling frameworks that integrate CFD simulations with LCA and TEA are necessary to close this gap. Data-driven decision-making that strikes a balance between technological, financial, and environmental goals can be made possible by such interdisciplinary approaches [8, 12].

Another area with unresolved issues is multiphase flow behavior and thermal management. Phase transitions between liquid, gas, and solid states are common in hydrogen systems, including fuel cell water management, cryogenic storage, and bubble evolution in electrolysis. Current CFD models frequently use empirical correlations, which fall short of capturing these phenomena's transient and chaotic nature. For instance, mass transfer and current distribution are greatly impacted by bubble coalescence and detachment during electrolysis, although most models handle these as averaged effects. Macroscopic simulations become uncertain due to the lack of a thorough mechanistic understanding of multiphase interactions at the microscale. Furthermore, little is known about heat transport in hydrogen systems, especially when there is a changing load and an intermittent renewable power source [27]. Designing effective and durable hydrogen technologies will need the development of sophisticated CFD models that incorporate phase-change thermodynamics, non-equilibrium kinetics, and dynamic operating circumstances.

CFD simulation of material deterioration and surface interactions in hydrogen environments is another crucial gap. System durability may be significantly impacted by phenomena including embrittlement, corrosion, and catalyst poisoning that result from the interaction of hydrogen with structural materials. The chemical and mechanical connection at the fluid–solid interface is ignored by traditional CFD methods, which regard materials as inert barriers. However, these interactions are crucial in determining longevity and safety in practical systems. Therefore, reactive boundary models that account for hydrogen absorption into materials, surface reaction processes, and stress-induced microstructural evolution are essential for future hydrogen CFD research [2, 23]. To connect atomic and continuum scales and develop a thorough understanding of material–fluid interactions, CFD must be integrated with molecular dynamics and Density Functional Theory (DFT) simulations.

Another area of research that is highly unknown is the function of turbulence modeling in hydrogen systems. The low density and high diffusivity of hydrogen flows lead to intricate turbulence–chemistry interactions that defy traditional turbulence models. DNS is still computationally costly for large-scale systems, and Reynolds-averaged Navier-Stokes (RANS) models frequently fail to resolve transient characteristics in reactive flows. By capturing large-scale turbulence structures, LES provides a compromise, although it still needs a lot of processing power. For hydrogen-specific conditions, adaptive hybrid models combining RANS, LES, and data-driven turbulence closures are required. To guarantee dependability across many flow regimes, from high-velocity combustion in turbines to microchannels in electrolyzers, these models need to be verified against experimental data [24, 25].

The restricted integration of CFD and AI for real-time system optimization is another significant gap. Although CFD offers comprehensive physical insights, its application in operational management and predictive maintenance is limited due to its computational complexity. By producing reduced-order models that approximate complex fluid dynamics with little computational effort, AI and ML can enhance CFD. However, AI-CFD coupling research for hydrogen applications is still in its early stages. Instead of directly integrating AI with CFD solvers, the majority of current research uses AI for data analysis or optimization. Creating intelligent CFD frameworks with AI algorithms that dynamically modify mesh density, boundary conditions, or turbulence models in response to changing system behavior is the next area of research. Real-time monitoring, control, and optimization of hydrogen systems would be made possible by such adaptive modeling, especially in digital twin contexts [20, 31].

Another significant obstacle is the uncertainty surrounding the scalability of CFD models. The nonlinear scaling effects that appear in industrial systems are frequently not captured by laboratory-scale simulations. Geometric and operational complexity causes parameters like flow distribution, heat dissipation, and reaction kinetics to behave differently at bigger dimensions. As a result, systems that are optimized at small scales could perform poorly at larger scales. The creation of scale-bridging CFD frameworks that include micro-, meso-, and macro-scale simulations is necessary to address this. Coarse-graining techniques should be used in multi-scale modeling systems so that detailed microscale simulation data can guide macroscale models without incurring excessive processing costs. Reliable extrapolation from laboratory to industrial systems is made possible by this hierarchical approach, which can guarantee that CFD predictions hold true across scales [21, 26].

The absence of thorough CFD modeling for integrated hydrogen energy systems represents another research gap. Beyond particular technologies, hydrogen serves as a bridge across the fuel, heat, and electrical sectors. However, rather than examining system-wide interactions, the majority of CFD studies concentrate on isolated components like electrolyzers, storage tanks, or combustors. For example, there is still much to learn about integrating CFD models of electrolyzers with fuel cell operation, waste heat recovery systems, and renewable energy unpredictability. Understanding how brief variations in one subsystem impact the performance of others is limited by this compartmentalized approach. The creation of integrated CFD-based energy system simulators would make it possible to optimize the production, distribution, and use of hydrogen holistically, increasing grid stability and efficiency [15].

Important research gaps are also revealed by safety modeling. Uncertainties still exist in modeling ignition likelihood, deflagration-to-detonation transitions, and thermal radiation, despite the fact that CFD has made substantial progress in predicting hydrogen dispersion and explosion dynamics. The empirical presumptions used in current safety models might not hold true in a variety of geometrical and environmental circumstances. Specifically, nothing is known about how confinement, barrier designs, and ventilation patterns affect the intensity of explosions. Reliable predictive safety methods require advanced CFD studies that incorporate UQ and Probabilistic Risk Assessment (PRA). These models should be incorporated into hydrogen infrastructure design frameworks to guarantee adherence to changing safety regulations and standards [22].

Strategic frameworks for the seamless integration of research, technology development, and policy must be devised in order to close these gaps and propel future advancements. The five interrelated pillars of scientific innovation, computational progress, experimental validation, industrial partnership, and policy alignment should form the foundation of a comprehensive strategy framework for developing green hydrogen and CFD research. Together, these pillars serve as the cornerstone of a coordinated worldwide strategy for hydrogen innovation.

The first pillar, scientific innovation, is centered on using multi-scale, multiphysics modeling to advance basic understanding. This entails creating next-generation CFD models that are capable of capturing heat transfer, fluid dynamics, chemical kinetics, and electrochemical reactions all at once. Discovery will be accelerated by the use of hybrid modeling approaches that integrate data-driven learning with first-principles physics. The creation of open-source CFD platforms specifically designed for hydrogen systems would further democratize access to sophisticated modeling tools and promote international cooperation between academia and business [1].

HPC, cloud-based simulation environments, and quantum computing are all integrated in the second pillar, computational development. In order to enable high-resolution simulations of complete hydrogen plants, including production, storage, and usage, future CFD frameworks need to take advantage of exascale computational capabilities. By more effectively resolving nonlinear transport equations, quantum-inspired algorithms may further reduce computational complexity [9]. Concurrent developments in visualization technologies, such as augmented reality interfaces and immersive 3D modeling, will improve the interpretation of CFD results and make it easier for engineers and policymakers to convey knowledge.

Establishing standardized facilities and data sources for benchmarking CFD models is necessary for the third pillar, experimental validation. Large-scale hydrogen testbeds with sophisticated diagnostic equipment that can record precise flow, temperature, and composition data should be funded by governments and research groups. Metadata and metrics for quantifying uncertainty should be included with this data and made publicly accessible via open-access platforms. In order to improve model accuracy and credibility, cooperative efforts between academic institutions and business sectors can guarantee ongoing feedback between simulations and experiments.

The fourth pillar, industrial partnership, acknowledges that incorporating CFD into the product design cycle is essential to the commercialization of hydrogen technology. Using simulations to direct material selection, component geometry, and process control, industries should make CFD-driven design optimization a normal procedure. With the help of CFD and real-time sensor data, digital twins can offer predictive insights on maintenance requirements and performance degradation

[17, 34]. In order to ensure that engineers have the computational and analytical abilities required for next-generation hydrogen systems, strategic relationships between academia and industry should place a high priority on knowledge transfer and workforce development.

The creation of supportive conditions for research and deployment is the emphasis of the fifth pillar, policy alignment. CFD-based evidence should be incorporated into regulatory frameworks that control the production, transit, and storage of hydrogen. Projects that integrate CFD research with real-world applications in grid balancing, sustainable manufacturing, and renewable energy integration should be given priority by funding sources. International cooperation can promote cross-border technology transfer, standardize standards, and exchange best practices through institutions like Mission Innovation and the International Energy Agency (IEA). The creation of international centers of excellence devoted to hydrogen CFD research would strengthen knowledge and quicken advancement [26].

The strategy framework must prioritize resilience, inclusion, and sustainability in addition to these pillars. Throughout the hydrogen lifespan, CFD research should minimize emissions, energy consumption, and material waste in addition to optimizing performance. Technological innovation will be in line with more general sustainability objectives thanks to interdisciplinary cooperation between engineers, chemists, economists, and environmental scientists. Incorporating developing countries into hydrogen research projects is crucial because it promotes fair access to clean energy technologies and helps achieve global decarbonization goals [2, 33].

In conclusion, whereas CFD has changed the understanding and development of hydrogen technologies, considerable research gaps remain across modeling accuracy, data availability, scalability, and system integration. Coordinated strategic frameworks that bring together computational science, experimental research, industrial application, and policy support are necessary to close these gaps. The synergy between cutting-edge CFD modeling and practical implementation, propelled by cooperative innovation and digital transformation, will determine the future of green hydrogen. By putting these strategic frameworks into practice, the scientific community can fully realize hydrogen's potential as a pillar of global energy sustainability, paving the way for an efficient, reasonably priced, and universally accessible clean energy future.

10.5 CONCLUSION

One of the most promising areas in the worldwide clean energy transition is the developing junction of CFD with green hydrogen. While CFD delivers the rigorous analytical framework required to comprehend, develop, and optimize the intricate systems that create, store, transport, and use green hydrogen, green hydrogen offers a flexible, carbon-free energy carrier capable of decarbonizing sectors that are challenging to electrify directly. When combined, they create a potent synergy: CFD gives the knowledge to make deep decarbonization technically feasible, safe, and economically competitive, while hydrogen provides the means.

CFD has already started to change how technologies are conceptualized and developed throughout the hydrogen value chain. CFD allows for the detailed multiphysics modeling of alkaline, PEM, and SOE in production, shedding light on phenomena that directly impact durability and efficiency, such as gas bubble behavior, flow distribution, heat management, and current density uniformity. CFD assists designers in addressing thermodynamics and safety issues in pipelines, refueling infrastructure, and compressed, liquefied, and solid-state systems in storage and transportation. The development of low-emission turbines, engines, and power systems specifically designed for hydrogen and hydrogen-based fuels is facilitated by CFD-guided combustion and fuel cell modeling. High-fidelity simulations accelerate development timelines, minimize trial-and-error, and clearly identify performance bottlenecks that may be removed.

However, current skills are insufficient to fully realize the potential of hydrogen, and this is where future research paths become crucial. Heat transfer, fluid flow, electrochemistry, and material behavior are all intricately linked in green hydrogen systems, which are by nature multi-scale and multiphase. Predictive accuracy is still hampered by many current models' simplistic assumptions, small experimental datasets, or scale-specific methods, particularly at an industrial scale. To ensure that CFD results can accurately inform billion-dollar infrastructure and policy decisions, the next generation of research must strive toward fully coupled, multiphysics and multi-scale frameworks, improved experimental validation, and comprehensive UQ.

Future directions are indicated by emerging digital technology. What is computationally possible is already changing thanks to HPC, AI-assisted turbulence and response modeling, and reduced-order surrogates educated on intricate simulations. CFD and real-time sensor data will be combined in digital twins of electrolyzers, storage farms, refueling stations, and integrated hydrogen hubs to enable dynamic risk management, predictive maintenance, and continuous optimization. As these techniques advance, CFD will become a key operational intelligence layer integrated into live hydrogen systems, rather than merely a design and research tool.

To guide this development, strategic frameworks will be crucial. How well technological advancements are applied in the real world will depend on interdisciplinary cooperation among CFD specialists, electrochemists, materials scientists, control engineers, economists, and legislators. Learning across borders and institutions can be accelerated via common modeling platforms, open data, and standardized validation scenarios. The integration of lifecycle evaluation and TEA with CFD outputs will guarantee that computer-optimized designs also achieve optimal performance in terms of cost, sustainability, and regulatory compliance. In parallel, training a workforce capable of working at the nexus of advanced simulation and sustainable energy systems would require education and capacity building.

A robust, effective, and truly low-carbon energy environment is the ultimate aim that unites research and future directions in the use of green hydrogen and CFD. The value of hydrogen depends on how wisely we create, transport, and use it; it is not a guarantee of sustainability on its own. CFD allows us to rebuild these processes from the ground up, not only to better understand them. The combined application of green hydrogen and CFD can go from a promising notion to a fundamental pillar of the global energy system, helping to assure a cleaner, safer, and more equal future, as models get more accurate, data becomes more plentiful, and digital infrastructure becomes more widespread.

REFERENCES

1. M. Calabrese, M. Portarapillo, A. Di Nardo, V. Venezia, M. Turco, G. Luciani, and A. Di Benedetto, 'Hydrogen safety challenges: A comprehensive review on production, storage, transport, utilization, and CFD-based consequence and risk assessment', *Energies*, vol. 17, no. 6, p. 1350, 2024.
2. X. Yang, T. Xi, Y. Qin, H. Zhang, and Y. Wang, 'Computational fluid dynamics–discrete phase method simulations in process engineering: A review of recent progress', *Applied Sciences*, vol. 14, no. 9, p. 3856, 2024.
3. F. N. Al-Otaibi. (2023). Computational Fluid Dynamics (CFD) Modelling for Process Intensification of Fluidized Dry Methane Reforming (Doctoral dissertation, Khalifa University of Science).
4. Y. Yang, D. Jing, and L. Zhao, 'Computational fluid dynamics modeling of reactive multiphase flow for suspended photocatalytic water splitting of hydrogen production system', *Applied Thermal Engineering*, vol. 173, p. 115220, 2020.
5. M. Y. Arshad, A. S. Ahmad, J. Mularski, A. Modzelewska, M. Jackowski, H. Pawlak-Kruczek, and L. Niedzwiecki, 'Pioneering the future: A trailblazing review of the fusion of computational fluid dynamics and machine learning revolutionizing plasma catalysis and non-thermal plasma reactor design', *Catalysts*, vol. 14, no. 1, p. 40, 2024.
6. S. Kaheel, K. A. Ibrahim, G. Fallatah, V. Lakshminarayanan, P. Luk, and Z. Luo, 'Advancing hydrogen: A closer look at implementation factors, current status and future potential', *Energies*, vol. 16, no. 24, p. 7975, 2023.
7. K. Adeli, M. Nachtane, A. Faik, D. Saifaoui, and A. Boulezhar, 'How green hydrogen and ammonia are revolutionizing the future of energy production: A comprehensive review of the latest developments and future prospects', *Applied Sciences*, vol. 13, no. 15, p. 8711, 2023.
8. N. Padoin, T. Matiazzo, H. G. Riella, and C. Soares, 'A perspective on the past, the present, and the future of computational fluid dynamics (CFD) in flow chemistry', *Journal of Flow Chemistry*, vol. 14, no. 1, pp. 239–256, 2024.
9. A. Y. Shash, N. M. Abdeltawab, D. M. Hassan, M. Darweesh, and Y. G. Hegazy, 'Computational methods, artificial intelligence, modeling, and simulation applications in green hydrogen production through water electrolysis: A review', *Hydrogen*, vol. 6, no. 2, p. 21, 2025.
10. S. J. Li, L. T. Zhu, X. B. Zhang, and Z. H. Luo, 'Recent advances in CFD simulations of multiphase flow processes with phase change', *Industrial & Engineering Chemistry Research*, vol. 62, no. 28, pp. 10729–10786, 2023.
11. H. Ashraf, M. Mae, and R. Matsuhashi, 'Advancements in green hydrogen production: A comprehensive review of system integration, power grid applications, and cost optimization', *International Journal of Energy Research*, vol. 2025, no. 1, p. 5321698, 2025.
12. L. T. Zhu, X. Z. Chen, B. Ouyang, W. C. Yan, H. Lei, Z. Chen, and Z. H. Luo, 'Review of machine learning for hydrodynamics, transport, and reactions in multiphase flows and reactors', *Industrial & Engineering Chemistry Research*, vol. 61, no. 28, pp. 9901–9949, 2022.
13. S. Sreejith, A. K. Panigrahy, J. Ajayan, N. V. Uma Reddy, and R. S. Dev, 'A comprehensive review of green hydrogen technologies for clean energy futures', *Journal of the Korean Physical Society*, vol. 88, no. 1, pp. 1–25, 2025.

14. V. Thakur, P. Ahir, P. Kumar, S. Sharma, A. Thakur, and S. Kumar, 'Innovations and advancements in green hydrogen production: A review', *Recent Innovations in Chemical Engineering*, vol. 18, no. 3, pp. 165–193, 2025.
15. L. Cheng, Z. Guo, and G. Xia, 'A review on research and technology development of green hydrogen energy systems with thermal management and heat recovery', *Heat Transfer Engineering*, vol. 45, no. 4–5, pp. 300–322, 2024.
16. S. O. Akpasi, I. M. Smarte Anekwe, E. K. Tetteh, U. O. Amune, S. I. Mustapha, and S. L. Kiambi, 'Hydrogen as a clean energy carrier: Advancements, challenges, and its role in a sustainable energy future', *Clean Energy*, vol. 9, no. 1, pp. 52–88, 2025.
17. M. T. Ahad, M. M. H. Bhuiyan, A. N. Sakib, A. Becerril Corral, and Z. Siddique, 'An overview of challenges for the future of hydrogen', *Materials*, vol. 16, no. 20, p. 6680, 2023.
18. R. Poudineh. (2024). *Can UK Green Hydrogen Contract for Difference (CfD) Match the Cost-Saving Success of Renewable Electricity?* (No. 39). OIES Paper: ET.
19. S. S. H. Dehshiri, and B. Firoozabadi, 'Wind energy integrated green hydrogen system as sustainable solution to decarbonize Iranian industrial cities', *Energy*, vol. 306, p. 132450, 2024.
20. A. Iranzo, 'CFD applications in energy engineering research and simulation: An introduction to published reviews', *Processes*, vol. 7, no. 12, p. 883, 2019.
21. F. Z. Wang, I. L. Animasaun, T. Muhammad, and S. S. Okoya, 'Recent advancements in fluid dynamics: Drag reduction, lift generation, computational fluid dynamics, turbulence modelling, and multiphase flow', *Arabian Journal for Science and Engineering*, vol. 49, no. 8, pp. 10237–10249, 2024.
22. A. I. Atteya, D. Ali, M. Hossain, and N. Sellami, A comprehensive review on the potential of green hydrogen in empowering the low-carbon economy: Development status, ongoing trends and key challenges. *Green Energy and Environmental Technology*, no. 4, 2023. doi:10.5772/geet.23.
23. T. Sebbagh, M. E. Şahin, and C. Beldjaatit, 'Green hydrogen revolution for a sustainable energy future', *Clean Technologies and Environmental Policy*, vol. 26, no. 12, pp. 4017–4040, 2024.
24. S. H. Osman, N. S. M. Yatim, O. S. J. Elham, N. Shaari, and Z. Zakaria, 'Three decades of hydrogen energy research: A bibliometric analysis on the evolution of green hydrogen technologies', *Sustainable Energy & Fuels*, vol. 9, no. 12, pp. 3182–3202. 2025.
25. M. Jaradat, S. Almashaileh, C. Bendea, A. Juaidi, G. Bendea, and T. Bungau, 'Green hydrogen in focus: A review of production technologies, policy impact, and market developments', *Energies*, vol. 17, no. 16, p. 3992, 2024.
26. A. Flah, H. Kraiem, M. Jayachandran, C. Z. El-Bayeh, B. S. Goud, V. Blazek, and M. M. Irfan, 'Advancing sustainable energy transition through green hydrogen valleys', *IEEE Access*, vol. 13, pp. 31442–31471, 2025, doi: 10.1109/ACCESS.2025.3542554.
27. J. Zhang, and J. Li, 'Revolution in renewables: Integration of green hydrogen for a sustainable future', *Energies*, vol. 17, no. 16, p. 4148, 2024.
28. S. Hasan, and R. Fuentes, 'Designing effective hydrogen markets: Policy recommendations from electricity and gas market reform', *Energy Policy*, vol. 207, p. 114816, 2025.
29. L. Gazzaneo, F. Longo, G. Mirabelli, M. Pellegrino, and V. Solina, 'Towards green and smart ports: A review of digital twin and hydrogen applications in maritime management', *Applied System Innovation*, vol. 8, no. 6, p. 165, 2025.
30. H. M. Mustafa, R. M. Kamal, and R. M. Mohamed, 'Enhancing and optimization of green hydrogen mixtures in natural gas to optimize transportation, power, and economy', *Egyptian Journal of Chemistry*, vol. 67, no. 11, pp. 491–501, 2024.
31. Z. Feng, I. Eiubovi, Y. Shao, Z. Fan, and R. Tan, 'Review of digital twin technology applications in hydrogen energy', *Chain*, vol. 1, no. 1, pp. 54–74, 2024.

32. V. Madadi Avargani, M. Habibzadeh, H. Abdlla Maarof, S. Zendehboudi, and X. Duan, 'Harnessing renewable energy for hydrogen production: Advances, challenges, and opportunities', *Industrial & Engineering Chemistry Research*, vol. 64, no. 25, pp. 12368–12418, 2025.
33. R. Angelico, F. Giametta, B. Bianchi, and P. Catalano, 'Green hydrogen for energy transition: A critical perspective', *Energies*, vol. 18, no. 2, p. 404, 2025.
34. A. Franco, 'Green hydrogen and the energy transition: Hopes, challenges, and realistic opportunities', *Hydrogen*, vol. 6, no. 2, p. 28, 2025.

Index

Note: Page number in **bold** and *italics* represent tables and figures, respectively.

For Product Safety Concerns and Information please contact our EU representative GPSR@taylorandfrancis.com Taylor & Francis Verlag GmbH, Kaufingerstraße 24, 80331 München, Germany